In accordance with the latest Syllabus prescribed by the Council for the Indian Certificate of Secondary Education Examination, New Delhi.

MOST LIKELY

ICSE QUESTION BANK

CHEMISTRY

CLASS X

By

A PANEL OF AUTHORS

OSWAL PUBLISHERS

1/12, Sahitya Kunj Colony, M.G. Road, Agra - 282002

EDITION
YEAR 2021

DISCLAIMER

With the ambition of providing standard academic resources, we have exercised extreme care in publishing the content. In case of any discrepancies in the matter, we request readers to excuse the unintentional lapse and not hold us liable for the same. Suggestions are always welcome.

ISBN : 978-81-95133-37-6

PUBLISHED BY

OSWAL PUBLISHERS

1/12, Sahitya Kunj Colony, M.G. Road, Agra - 282002

(0562) 2527771-4, +91 7534077222

Contact@oswalpublishers.com

www.oswalpublishers.com

The cover of this book has been designed using resources from Freepik.com

Printed At Upkar Printing Unit, Agra

PREFACE

It is a matter of immense pride for us to present our 'ICSE MOST LIKELY QUESTION BANK' series, especially prepared for students appearing for Board examinations in the oncoming year.

This book is a perfect capsule for building self-confidence during exam. preparation. Based on chunking strategy, the 'Categorywise–Chapterwise' format with its exhaustive set of questions allow the students to cover every category in a chapterwise manner.

Covering easy categories first boosts student's self-esteem and the ascending score braces them to take up challenging categories without fear. This prepares the student to see the exam paper as achievable at all times.

With its simple language and style, it is a one-stop solution for smart study. We are confident that the book will enable the candidates to develop a better understanding of the curriculum and help them organize their learning process. This book shall definitely prove to be a fruitful tool for the students and encourage them towards scholastic excellence.

Constructive suggestions for further improvement of the book are always welcome.

Note: Questions marked with '*' are frequently asked in previous years board examinations.

—**Publisher**

LET'S GO ORGANIC

Growing microgreens can be a good, easy and economical way for a student to start practicing plantation at home.

Microgreens are highly nutritious plants that are harvested after the sprout-stage and just before the maturity phase.

Microgreens may include mustard, coriander, basil seeds, dried peas, beans, broccoli, radish, spinach, beetroot, cauliflower and others.

It is more convenient to start growing microgreens that require low maintenance like mustard, coriander, peas, basil and others.

Start with finding a flat tray or a shoe-box or any tetra-pack carton.

Make holes in the box for easy drainage of water.

Spread two inches of soil in the box. You can get ready-made soil from the garden or nursery. Sprinkle your desired seeds over the soil.

You can spread a mixture of coco-peat and fertile topsoil or a stack of moist tissue papers over it. Remember to keep the tissue paper moist till the seeds start germinating.

Cover the box with newspaper and keep it in a place where there is no direct sunlight.

Spray water twice a day atleast.

On third and fourth day, once the leaves start to appear, start exposing the germinating microgreens to indirect sunlight.

You can try to expose them to direct sunlight for not more than 4 hours.

By ninth or tenth day, your microgreens will be ready, harvest them!

Microgreens can be eaten with sandwiches, wraps, salads, smoothies, juices or as garnishers on pizza, soup and curries.

CONTENTS

Note: Questions marked with '*' are frequently asked in previous years board examinations.

SUGGESTIONS FOR STUDENTS

1. Utilize the 15 minutes' reading time given in the question paper thoroughly and choose the correct questions to attempt.
2. Read the chapters thoroughly to get a very clear concept of the subject matter. Ensure all aspects of the syllabus are covered and avoid selective study of topics.
3. Practice writing balanced chemical equations. Chemical reactions of compounds must not be learnt in isolation. A comparative study brings forth a pattern and with a little understanding, writing chemical equations becomes easy. In a chemical equation, write an acid with its strength (dilute/concentration).
4. Remember to distinguish between a pair of compounds using a single reagent, stating that one gives the characteristic reaction while the other does not. Learn all reactions along with the specific conditions.
5. Expose yourselves to variety of questions by solving past years' Question Papers.
6. Laws and definitions must be learnt well and reproduced with the correct terms. Highlight key words.
7. Practice answering observation-based questions. Answer reasoning questions in points by focusing on the main reason.
8. While answering questions on periodic table do not try identifying the element unless specified in the question. Writing down the electronic configuration with the atomic numbers known, helps in answering questions on trends in properties across the periodic table
9. Learn symbols of elements and their valences. Practice drawing of dot diagrams. Ionic bond formation must include depicting the ions formed with appropriate charges, pattern and with a little understanding, writing chemical equations becomes easy.
10. Learn trends in properties across a period and along a group, with the help of charts or schematic representations.
11. Acquire knowledge of solubility of salts (in NaOH, NH_4OH), especially in deciding the methods of preparation of salts.
12. Develop the habit of solving numerical problems stepwise and try to solve as many types of numerical problems, as possible.
13. Read the questions carefully and answer what is required, e.g., do not write products for observation-based questions or the formula for a question where 'name' is asked.
14. Practice drawing of Structural formulae of organic compounds with all the bonds properly visible, keeping the tetra valency of carbon in mind and the IUPAC rules for naming. Show functional group in expanded form and not in the condensed form.
15. Create charts with a comparative study on Hydrocarbons – alkanes, alkenes and alkynes including the equations and conditions and for various industrial processes.
16. Be thorough with observations in practical chemistry and use colours to highlight observations along with tests to differentiate between substances. When a question demands observations, it is not necessary to identify the substance. Observations could include specific smell or specific colour of precipitate or loss of colour and so on.
17. Focus on observations such as specific smell or specific colour of precipitate or change in colour between reactants and products. Follow confirmatory test for colourless gases. Study the typical reactions of acids as word equations such as Metal + Acid $\rightarrow$ Salt + H_2O, OR Carbonate + Acid $\rightarrow$ Salt + H_2O + CO_2
18. Remember the rules for selective discharge of ions and learn to apply them.
19. Various roles of different substances and electrodes used must be understood in metallurgy compounds.
20. Study alloys in a tabulated form highlighting the main component and the other metals.

Word of Advice

Chapter 1. Periodic Properties and Variations of Properties

Topics Found difficult By Students

- Usage of the sign < and > in showing trends
- Electronegativity and Electron affinity
- Arranging elements as per the trends in properties across a period and down a group.
- Identification of terms such as electronegativity etc. on the basis of descriptions given.
- Scientific terms for process/properties.
- Reasons behind trends in properties across or down the periodic table.
- Electron affinity and Ionization potential.
- Application of trends in periodic Table.
- Confusion between period/group – and determination of period / group from the atomic number.

1. Some students confused Helium with Hydrogen with highest ionisation potential.
2. Some students wrote electron affinity instead of electronegativity. Some students simply written the abbreviation form of electronegativity.
3. A few students chose Sulphur over Chlorine, as highest electronegative element.
4. A few students mentioned that Aluminium is brittle which is incorrect.
5. Some students identified the elements instead of using the given letters of alphabets. In a few cases, the position of metals and non-metals was not clear. Despite the clear instructions not to repeat elements, some students repeated certain elements.
6. Some students wrote that inert gases are stable without mentioning electronic configuration / 8 electrons in the valence shell. Some students wrote atomic number 8 for a rare gas.
7. Many students could not explain the reason for increase in ionisation potential across a period, from left to right.
8. Many students did not write the correct reason to explain alkali metals are good reducing agents.
9. Some students wrote only 'ionisation' or selected the incorrect term 'electron affinity indicating that they were confused between ionisation energy and electron affinity.
10. Properties of electrovalent compounds on the basis of electrostatic force between ions, were not perhaps understood well.

Chapter 2. Chemical Bonding

Topics Found difficult By Students

- Properties of Ionic and covalent compounds
- Writing balanced chemical equations for the given chemical reactions
- Electron dot diagram
- Definition of lone pair of electrons.
- Formulae of compounds when electronic configurations were given.
- Bonding – single covalent bonds and triple bond.
- Structure of hydronium ion.
- Writing of formula, given the atomic number of an element.

1. Some students had no idea that CCl_4 is a covalent compound.
2. A few students got confused between covalent and coordinate bonding.
3. Some chose methane instead of ammonia as polar covalent compound.
4. A number of students did not draw arrows to show transfer of electrons. Some drew only one arrow instead of two to show transfer of two valence electrons of divalent ions, such as calcium. In several cases, electrons of calcium were not represented differently from oxygen, or the charge on calcium ion and oxide ion was not written. Some drew it as a covalent molecule.

5. Several students were confused regarding the number of electrons shared, in the covalent compound.
6. Some students forgot to draw the lone pair of electrons on water.
7. Some students could not relate the term trivalent metal.
8. Electrons of sodium were not differently represented from chlorine; transfer of electron and ion formation was not shown. A few students drew the orbital diagram instead of the electron dot diagram.
9. Some students forgot the lone pair of electrons of N atom.
10. Coordinate bond was not shown by some students. Overall positive charge was not shown on the ion in few cases. Some drew the structure of ammonia instead of ammonium.
11. Intermolecular force of attraction instead of strong electrostatic force of attraction between ions was stated to be responsible for the high melting point in ionic compounds.
12. Many students could not define a lone pair of electrons completely. They missed out the key words in the definition like pair of electrons or not shared/unshared or covalent molecule.
13. Observations in the electron dot diagram of hydronium ion drawn by some students were - the lone pair or the charge was missing or same kind of dots were used for electrons of both O and H or the coordinate bond was not indicated by an arrow.
14. Many students selected the incorrect example of water or ammonia for a compound that does not have a lone pair of electrons.
15. Some students gave the option of being always soluble in water while others mentioned conducts electricity when it is in molten state, as common characteristics of covalent compound.
16. Some students wrote the symbol of ions without the charge or giving wrong charge.
17. Some of the errors made by students were, 'coordinate bond' instead of 'covalent bond' in the case of water, indicating that students got confused with hydronium ion.
18. Students did not write the electronic configuration of the element and made errors while writing the chemical formulae of the compounds.
19. For the structure of ammonium ion, the two particles *i.e.*, NH_3 and H^+, used in the formation of NH_4^+ were not shown. Co-ordinate bond was not shown or the charge on the ammonium ion was missing

Chapter 3. Acids, Bases and Salts

Topics Found difficult By Students

- Identification of gases and compounds
- Methods of preparation of salts
- Identifying the salts from the given observations
- Connecting the type of solution to its pH value based on certain observations.
- Gases involved in reactions.
- Neutralisation and titration.

1. Some students stated that the hydrogen gas burns with a blue flame instead of stating that a burning splint is put off with a pop sound.
2. The words water of crystallisation was missing in most of the answers. Instead students wrote moisture is removed. A few even used an incorrect term 'Amorphous'
3. A few students got confused between efflorescence and deliquescence. Spelling errors were also noticed in some answers. Some referred to it as effervescence.
4. Some students mixed the examples of neutral, acidic and basic salts.
5. Instead of writing that $CaCl_2$ turns into colourless solution, students wrote that it turns hydrated. Some students stated that it turns powdery having got confused with efflorescent substances.
6. A few students did not understand the relation between acidity and alkalinity with the pH of the solution.
7. Several students gave the answer for amphoteric oxide reduced by carbon as Al instead of Pb. They seemed to have got confused between Al and Pb as both form amphoteric oxides.
8. Students failed to note that a Neutral solution has pH = 7, while alkaline solutions have lower H^+ ion concentration and pH goes beyond 7

Chapter 4. Analytical Chemistry

Topics Found difficult By Students
- Distinguishing between substances on the basis of test/reagent given.
- Specific tests to identify/distinguish between substances.
- Scientific reasons for certain observations.
- Difference between precipitate and solution.
- Identifying substances based on analytical chemistry.
- Solubility of precipitates in excess NaOH and excess NH_4OH.

1. Some students did not write the solubility of the blue precipitate of copper hydroxide in excess of ammonium hydroxide solution; instead wrote 'deep or inky blue solution' while some wrote "deep blue precipitate is formed".
2. Some students gave only the name of the product and some did not give the confirmatory test for carbon dioxide.
3. The word 'residue' was missing in the observations.
4. Some students were not able to distinguish between soluble and insoluble salts. A few even chose ferric chloride instead of ferrous chloride.
5. Many students wrote the wrong formula for $NaAlO_2$.
6. Some wrote that Nitric oxide turns lead acetate paper silvery black.
7. Some students used copper sulphate as the reactant or used sodium sulphate instead of sulphuric acid, for the preparation of copper sulphate.
8. A large number of students made errors in writing the formula of ammonium sulphate or ferrous sulphate.
9. On heating lead nitrate strongly in a test tube, products formed were listed by many students without any noticeable observations.
10. Some students wrote yellow precipitate of PbO is obtained instead of yellow residue.
11. Some students interchanged the colours of the two ions, ferrous and ferric.
12. Several students got confused between neutralisation and precipitation.
13. Some students chose the salt copper nitrate and lead nitrate instead of calcium nitrate, as a reactant which does not react with ammonium hydroxide.
14. Some students wrote incorrect formula of potassium zincate and many wrote one of the products as $Zn(OH)_2$ or ZnO or K_2O.
15. For solubility of the precipitate $Pb(OH)_2$ in excess NaOH, some wrote insoluble in excess instead of soluble in excess or White solution instead of colourless solution.
16. A few students identified the product $BaSO_4$ but missed out the word precipitate or the colour white.
17. Many students ignored the solution state of both $Pb(NO_3)_2$ and $Zn(NO_3)_2$ and gave dry heating test for each, which was contrary to the requirement of the question. Some students used the reagent solution NaOH in place of NH_4OH for distinguishing one from the other.
18. A few students used concentrated H_2SO_4 to distinguish between NaCl and $NaNO_3$ overlooking the fact that both were in solution form and not in solid state.
19. Some students used two steps to obtain $ZnCO_3$ from $ZnSO_4$: first reaction of $ZnSO_4$ with HCl and second reaction of $ZnCl_2$ with Na_2CO_3.
20. Some of the common errors were: reddish brown precipitate instead of dirty green precipitate. A few students mentioned "dirty green ferrous hydroxide "but failed to mention the word precipitate or insoluble compound.
21. Most of the students ignored the colour change or the change from the crystalline state to amorphous state and instead mentioned the phenomenon of dehydration or the formation of anhydrous copper sulphate.
22. Many students could not use the word precipitate in the correct place or failed to record the formation of ink blue solution with NH_4OH and instead recorded the formation of deep blue precipitate with excess NH_4OH, without mentioning the blue precipitate of $Cu(OH)_2$

23. Many students related $Pb(NO_3)_2$ formation to precipitation instead of neutralisation and the answer to $ZnCO_3$ formation thus went wrong as it involved precipitation. Some other students referred to neutralisation in the case of $NaNO_3$ from NaOH instead of titration and hence the choice for formation of $Pb(NO_3)_2$ went wrong.

24. Students either forgot to state the colour of the ppt. Some students had doubts about the solubility of the precipitate.

Chapter 5. Mole Concept & Stoichiometry

Topics Found difficult By Students

- Conversion of molecular formula to empirical formula.
- Numerical problems based on mole concept and application of Gay Lussac's and Avogadro's Laws.
- Relation between Gram molecular mass and molar volume
- Concept of g atom and g molecule and the relation with mass and volume.
- Calculations based on chemical equations.
- Calculation of moles and molecules, application of mole concept

1. A few made errors in the calculation of molecular weights. Many students got confused between molecular weight and molecular formula.

2. Many calculated the number of moles instead of molecules. A few did not use Avogadro's number. A large number of students made incorrect substitution and/or calculated mass instead of moles.

3. Most of the students did not state the law, asked in question. They only named it.

4. A few calculated the molecular weight of Na_3AlF_6 incorrectly while others did not substitute the values properly. Instead of multiplying 3 × 23 for sodium, some students calculated the percentage of sodium with only 23.

5. Many students took the mass of one mole of oxygen gas as 16 instead of 32 and hence, could not calculate number of moles correctly.

6. In a number of cases, calculations for weight of 10^{22} atoms of carbon were incorrect indicating inadequate understanding of the concept.

7. Some students used Gay Lussac's law instead of mole concept or calculated the molecular mass incorrectly.

8. Many students determined the empirical formula correctly but failed in calculating the value of *n* and hence, erred in the molecular formula. Some students calculated the number of atoms by dividing atomic weight by the individual percentage. Therefore, empirical formula went incorrect.

9. Many students ignored the stoichiometric coefficients

10. Many students did not calculate the volume of unused O_2.

11. Many others got confused between the number of gram atoms and number of atoms and ended up calculating number of atoms.

12. Some students made errors in calculating the relative molecular mass of $CuSO_4.5H_2O$ wherein 5 molecules of water were calculated as 5 × 2 + 16 instead of 5(2 + 16). Another error made was in the calculation of percentage of water of crystallisation wherein some students forgot to multiply the mass of water by 5.

13. Students did not see the relation between gram molecular mass and molar volume of 22.4l at STP (for any gas) and substitute the wrong values in the equations.

Chapter 6. Electrolysis

Topics Found difficult By Students

- Oxidation / Reduction reaction applied to electrolysis
- Reactions and observation at cathode and anode
- Oxidation and reduction in terms of loss and gain of electrons
- Ionic equations, especially involving the discharge of ions/formation of ions.
- Gases involved in reactions.

- ➢ Selective discharge of ions in electrolysis.
- ➢ Oxidising and reducing agents.
- ➢ Enumeration of results of electrolysis.
- ➢ Ionic equations involving oxidation / reduction.
- ➢ Applications of electrolysis.

1. A few students wrote Nickel as the inert electrode.
2. Most students mentioned that graphite is cheaper and easily available or inert or that platinum is expensive, instead of stating that graphite remains unaffected by bromine vapour, for the electrolysis of lead bromide.
3. A few students missed associating the word 'pure' with electrorefining of impure metal.
4. Some students were not able to distinguish between oxidation and reduction reactions.
5. Some wrote atoms in place of ions as conducting particles in electrolyte.
6. Most students wrote cations migrate towards cathode and anions towards anode or anions are negatively charged and cations are positively charged. Some were confused and wrote that cations gain electrons while anions lose electrons.
7. Many students could not give the correct observation at the cathode/anode. Some students did not associate the correct electrode with the observation given in question. Some wrote the cathode and anode reactions, instead of the observations, while some wrote observations in the reverse way.
8. Several students were not able to identify the spectator ions - all possible ions were mentioned.
9. A few made the error of writing CuO_2 for copper (II) oxide (CuO) and 2H instead of H_2.
10. Some students wrote incorrect responses for electrolysis of copper sulphate, like Cathode diminishes in size, Blue colour of copper sulphate fades. In a few cases, the change was noted without mention of the electrode.
11. Some students listed examples instead of stating the presence of ions or molecules depending on the type of electrolyte.
12. A few students either wrote oxidation or reduction instead of redox reaction.
13. Many students confused between dissociation and ionisation.
14. Colour of bromine was written as orange/red by some students; some forgot to add the word vapour or gas with the colour of bromine; A few students wrote that grey metal is deposited at the anode, when molten lead bromide is electrolyzed using graphite electrodes.
15. Instead of white precipitate many students wrote reddish brown fumes of NO_2. Some wrote the equation without stating the colour of the precipitate. A few students expressed the answer by leaving out the effect of heating on the solubility of $PbCl_2$.
16. Most students wrote about the conductivity of dilute HCl or acetic acid but could not compare the conductivity of both the acids
17. While writing the name of the gas produced at the anode during the electrolysis of acidified water many students overlooked the term anode and wrote hydrogen and oxygen.
18. Many students incorrectly named the substances as silver argentocyanide instead of sodium argentocyanide or named silver nitrate as the electrolyte.
19. Several students misinterpreted the question and wrote equations instead of the observations at the electrodes. Some interchanged the electrodes.
20. Students made the wrong choice of oxidising agents. They perhaps did not see the association between metals being electron donors and thus their behaviour as reducing agents.
21. Knowledge of basic particles in all the three: weak, strong and non-electrolytes was lacking in some students.
22. The common error committed by students was that copper cannot conduct electricity in molten or aqueous state due to absence of ions.

Chapter 7. Metallurgy

Topics Found difficult By Students
- Components of alloys and their application.
- Froth flotation and roasting.
- Thermite welding, ores and main components of alloys.

1. Students wrote that the ore mixes with oil and settles down, in Froth Flotation method.
2. A few students confused Calcination with Roasting.
3. In writing the chemical formula of bauxite, many students either wrote incorrect number of water molecules or did not write them at all.
4. Some students wrote the name of the gas evolved on electrolytic reduction of alumina as CO_2 instead of O_2. A few students wrote the symbol of nascent oxygen, which was not acceptable.
5. Most students listed all the components of the alloys instead of giving the main components as asked in the question. Many students were not sure of the uses of alloys. Students made random choices as the main metals in the alloys were not known or learnt.
6. Most students wrote the name of the common ore of aluminum from which the metal is extracted as alumina or cryolite instead of Bauxite. Some students gave chemical formula of Bauxite as $Al_2O_3.5H_2O$ wherein the number of water molecules was incorrect.
7. A few students named the process by which impure ore of aluminum gets purified by using concentrated solution of an alkali as Hall's process or Hoope's Process in place of Baeyer's process.
8. Some students confuse the terms galvanising and electroplating,
9. Zinc sulphide or Galena was written by many students instead of Zinc blende.
10. Incomplete answers for powdered coke are sprinkled over the electrolytic mixture were given, such as to prevent burning of electrodes / prevent radiation/reacts with oxygen.
11. Students gave incorrect equations such as $2Al^{3+} + 6e^- \rightarrow Al$ Or $Al^{3+} - 3e^- \rightarrow Al$
12. Some students named calcium as they might have missed out on the word non–metal present in fluorspar.

Chapter 8. (a) Study of Compounds : Hydrogen Chloride

Topics Found difficult By Students
- Catalysts used in various industrial processes.
- Scientific terms for process/properties.
- Gases involved in reactions.
- Conditions of Industrial processes.

1. Some students wrote, "Because it comes in contact with ammonia gas" for the statement: HCl gas fumes in moist air. A few even wrote that it is due to the reaction between HCl and NH_3. Several students got confused and wrote "dense white fumes are formed".
2. Many students gave inadequate explanation for unfeasibility of direct absorption of HCl gas, without mentioning the effect back suction. A few students wrote thistle funnel instead of inverted funnel arrangement.
3. Some students wrote that the HCl gas was collected by downward displacement of air.
4. Some students identified the drying agent for HCl gas as Conc. H_2SO_4 instead of CaO and P_2O_5.
5. A few students mentioned incorrect catalyst such as Fe/ V_2O_5/CaO, to oxidise ammonia.
6. Many students got confused between magnesium sulphite, sulphide and sulphate. Some others erred by writing one of the products as H_2SO_3 instead of $H_2O + SO_2$.
7. Some students did not mention the necessary condition of temperature *i.e.,* below 200 °C to ensure Na_2SO_4 is not formed which has a tendency to get fused with the glass and ended up writing the equation which was not required as per the question asked : $2NaCl + H_2SO_4 \rightarrow Na_2SO_4 + 2HCl$
8. Many students could not write the correct colour of the flame on burning ammonia.
9. HCl does turn blue litmus solution red; the formation of fountain due to extreme solubility in water was ignored by many students.

Chapter 8. (b) Study of Compounds : Ammonia and Nitric Acid

Topics Found difficult By Students
- Reaction of ammonia with excess chlorine and vice versa.

1. Some students wrote the products of ammonia with excess Chlorine as NCl_3 and HCl.
2. In the reaction of ammonia with heated copper oxide, the gas evolved was incorrectly written as N, NO or NO_2.
3. Industrial method for the preparation of ammonia was incorrectly written as Ostwald's process/Baeyer's process. The catalyst listed was platinum/nickel, etc.
4. Some wrote methods of preparation of NH_3 gas using NaOH or using $(NH_4)_2SO_4$ instead of NH_4Cl. Many students chose NaOH as the alkali in the laboratory preparation of ammonia instead of $Ca(OH)_2$.
5. A few got confused with the formula of ammonia as NH_4.
6. Students ignored the word excess NH_3 and wrote equation as $2NH_3 + 3Cl_2 \rightarrow N_2 + 6HCl$.
7. Instead of illustrating the reducing nature of NH_3, students indicated basic nature of NH_3 with the equation $NH_3 + HCl \rightarrow NH_4Cl$ or Alkali formation with ammonia as $NH_3 + H_2O \rightarrow NH_4OH$. Some students wrote the correct equation but failed to balance it.
8. Many students did not mention the word 'dissolves' or dissolution of the gas. Instead, they simply wrote, due to NO_2 for atmospheric exposure of conc. Nitric acid.
9. A few students stated the reason for 'in the preparation of Nitric acid, the complete apparatus is made up of glass as rubber will melt due to heat or cork will break.
10. Many students listed HCl or H_2SO_4 as dilute acid acting as oxidising agent instead of nitric acid.

Chapter 8. (c) Study of Compounds : Sulphuric Acid

1. The word concentrated was missing in some cases, wherever necessary. Students only mentioned the substance as sulphuric acid.
2. A few students got confused between the drying agent of ammonia and HCl gas and selected CaO instead of conc.H_2SO_4.
3. Some students wrote the formula of the Na_2SO_4 as '$NaSO_4$'.
4. Several students represented Zinc sulphide (ZnS) incorrectly as $ZnSO_3$ or $ZnSO_4$.
5. Haber's Process was written instead of Contact Process for Industrial method for the preparation of sulphuric acid.
6. Many students did not specify whether sulphuric acid used for the preparation of hydrogen chloride gas in the laboratory should be concentrated or dilute. In response to the preference of this acid over other acids, some wrote that it does not react with HCl gas.
7. Many students wrote drying instead of dehydration.
8. Several students stated acidic property of the acid instead of oxidising property.
9. Many students wrote the formula for sugar as $C_6H_{12}O_6$ (glucose) which is incorrect.
10. Many students forgot the two-step conversion of SO_3 to H_2SO_4 through the formation of oleum and diluting it in the second step; instead, they treated SO_3 directly with water to get sulphuric acid.
11. A few students wrote concentrated Nitric acid, produce carbon from cane sugar, instead of concentrated sulphuric acid.
12. Students mentioned formation of sugar charcoal or carbon without stating its appearance, when sulphuric acid is added to sugar. Some of them described the observation as dehydration of sugar which was incorrect.

Chapter 9. Organic Chemistry

Topics Found difficult By Students

- Drawing of isomers
- Difference between the terms dehydration and dehydrogenation.
- Difference in structure of an aldehyde and alcohol.
- IUPAC nomenclature and structures of organic compounds.
- Specific terms in organic chemistry IUPAC and trivial naming of organic compounds.

1. Some students wrote the condensed formula instead of the structural formula. Many students did not understand the difference between straight and branched chain. Some showed carbon skeleton but missed showing the complete structural formula.

2. Some students wrote the IUPAC name as methanal instead of ethanal.

3. Some students were confused between 'ene' and 'yne'.

4. A few students mentioned esterification reaction as 'ester' or 'neutralisation reaction'.

5. Most students drew the first isomer correctly but could not draw the other two isomers of pentane. Some students drew only two isomers. Branched structures were not drawn correctly. Some missed out H atoms or the count of carbon atoms went incorrect. Many students did not understand the difference between straight and branched chain.

6. Several students were confused between dehydration and dehydrogenation.

7. Some students wrote the product as $C_2H_2Cl_2$ instead of $C_2H_4Cl_2$ in the chemical equation.

8. Many students were confused in the functional groups: alcohol, aldehyde and ketone.

9. Some students mentioned the presence of a linear planar structure or three covalent bonds rather than presence of triple covalent bond in alkynes. Some gave incorrect answers such as, unsaturated compound or, undergoes addition reaction and so on.

10. Some students wrote the abbreviation as EF for empirical formula.

11. Instead of stating the general term 'Acid' many students gave the specific name of an acid which were not acceptable.

12. In writing a balanced chemical equation for the preparation of Ethene from bromoethane, the formula for bromoethane was written as C_2H_6Br in place of C_2H_5Br by some students. Several students used chloroethane instead of bromoethane.

13. In writing a balanced chemical equation for the preparation of Ethyne using calcium carbide, some students wrote the product as CaO instead of $Ca(OH)_2$.

14. The formula of sodium acetate was written incorrectly by some students.

15. Some students incorrectly drew the structural formula when the name of the organic compound was asked.

16. Students could not write the correct product of chlorination of ethyne. Some students wrote incorrect numbering of the C atoms having the chlorine atoms.

17. Several students wrote H_2 is evolved in the hydrogenation reaction of ethene, which was an incorrect answer.

18. Many students did not associate the fact that alkanes show substitution reactions.

19. Some students wrote self-linking instead of Catenation.

20. Some wrote 1,2-dimethylethene instead of 2-butene or but-2-ene. Some simply wrote butane or butan-2-ene.

21. Some students made mistakes related to the homologous series by giving the general name instead of specific name; the characteristic bond was referred to as unsaturated/saturated instead of triple covalent bond and single covalent bond respectively. The IUPAC name of the first member of the alkynes was given as methyne.

22. Some students confused between carbonic acid and carboxylic group.

23. Many students wrote the equation or named the products or wrote the term esterification instead of writing the relevant observation of fruity smell.
24. Diethyl ether was confused with acetone or ketone group (oxygen linking the 2 alkyl groups).
25. Students made errors by linking the C in the carboxylic group with H.
26. Some students failed to associate the word 'glacial' with acetic acid.
27. Some students wrote 'Isotopes' or 'Homologous series' instead of 'Isomerism'.
28. In several cases the chemical reaction was written using aqueous KOH instead of with alcoholic KOH.
29. Students got confused between hydration and hydrogenation and failed to understand that hydrogenation is addition of hydrogen and hydration is addition of water.
30. Incorrect IUPAC nomenclature written by some students. For e.g. 2 propene or prop -2-ene instead of propene, but -2-yne or butyne instead of but-2-yne, ethanol or 2- ethanol or ethanoic acid instead of ethanal.
31. Some wrote alkenes instead of alkanes, for substitution reactions.

Miscellaneous

1. A common error in State the observations was that students either wrote the equation or named the product formed instead of stating the observations. In some cases, incomplete observations were written. Several students wrote the correct observation without mentioning the reagent.
2. Students did not mention the word 'gas for the compound evolved in the observations.
3. Some students did not balance the equations correctly.
4. Key words were missing in answers.
5. Many students overlooked the word 'not' mentioned in the question or missed/incorrectly read the options, and wrote incorrect answers.
6. Several students did not mention the condition or wrote incorrect condition required for an equation. Some used greater than sign instead of less than sign or vice versa for temperature required for the that reaction.
7. Practical chemistry not given adequate attention and hence solubility of salts were not clear to students. Students made errors in identifying the salt mentioned in questions related to practical chemistry.
8. Some students got confused between the symbols '>' and '<' symbols incorrectly to arrange the given items according to the instructions given in the question, which was not required. Some students simply arranged the items in reverse order. Some students made unnecessary use of formula instead of choosing the names given.

Chapter at a Glance

Set 1

Chapter 1. Periodic Properties and Variations of Properties

➤ **Periodic Table :** It is a table that classifies all the known elements in accordance with their properties in such a way that elements with similar properties are grouped together in the same vertical column and dissimilar elements are separated to one another.

The 115 known elements are arranged in the Periodic Table in order of their increasing atomic number.

The vertical columns are called 'groups'.

The horizontal rows are called 'periods'.

Approaches To Periodic Classification of Elements

Theory	Year	Statements
Dobereiner's Triads	1817	The atomic weight of the middle element in a triad is arithmetic mean of the two.
Newland's Law of Octaves	1864	If the chemical elements are arranged according to increasing atomic weight, those with similar physical and chemical properties occur after each interval of seven elements.
Mendeleev's Periodic law	1869	The properties of the elements are periodic functions of their atomic weights.
Modern periodic law of Mossley	1913	The physical and chemical properties of the elements are periodic functions of their atomic numbers.

Characteristics of Group 1A and VII A

Properties	Group 1A	Group VIIA
Name	Alkali metals	Halogens
Atomic and ionic radii	Increases down the group	Increases down the group
Ionisation energy	Decrease down the group	Decrease down the group
Electropositive nature	Strongly electropositive Increases down the group Cs is most electropositive	Weak electropositive elements Increases down the group
Electronegativity nature	Weak electronegative elements Decrease down the group	Strongly electronegative elements Decrease down the group F is the most electronegative
Oxidation state	+1	Variable oxidation state- -1, +1, +3, +5, +7

➤ **Periodicity :**

Properties	Definition	Trend along groups	Trend along period
Valence electron	Electrons present in outermost shell of an atom	Remains same	Increases from 1-8
Number of shells/ orbit	Electrons revolve around nucleus in definite circular path or shells	Remains same	Increases from 1-8
Atomic radii/size	The distance between the centre of the nucleus to the outermost shell of an atom	Increases	Decreases
Metallic character	Tendency to lose valence electrons and form positive ion	Increases	Decreases

Non-metallic character	Tendency to gain valence electrons and form negative ion	Decreases	Increases
Ionisation energy	The energy required to completely remove an electron from a gaseous atom or ion.	Decreases	Increases
Electron Affinity	The energy change that occurs when an electron is added to a gaseous atom.	Decreases	Increases
Electronegativity	The tendency of an atom to attract electrons to itself when combined in a compound	Increases	Increases
Ionic Radii	The distance between the nucleus and the electron in the outermost shell of an ion.	Increases	Decreases
Acidic Character	It is the measure of an element to produce hydrogen gas when react with certain salts.	Increases	Decreases
Oxidising Property	The tendency of an element to gain electrons	decrease	Increase
Reducing property	The tendency of an element to lose electrons	Increases	Decreases
Melting point	The temperature at which the solid and liquid forms of a pure substance can exist in equilibrium.	Decrease, then Increase	Increase, then decrease
Boiling Point	The temperature at which its vapour pressure is equal to the pressure of the gas above it.	Decrease, then Increase	Increase, then decrease

➤ **Diagonal relationship :** These elements are called bridge elements. The relationship between two elements that are diagonally present, is called Diagonal relationship.

	IA	IIA	IIIA	IVA
2nd Period	Li	Be	B	C
3rd Period	Na	Mg	Al	Si

(Li → Mg, Be → Al, B → Si)

➤ **Atomic number (Z) :** The number of protons in the nucleus of the atom or the number of electrons revolving around the nucleus in isolated electrically neuctal atom.

➤ **Mass number (A) :** It is the total number of protons and neutrons present in the nucleus of the atom.

➤ **Relation between atomic number (Z) and mass number (A) :**

$$A = p + e = Z + n$$

Symbol for the Element : ${}^{A}_{Z}X$

➤ **Metalloids :** Elements which can neither be classified as metals nor as non-metals. They are semiconducting elements that is, they are insulators at normal temperature and conduct electricity when heated.

Chapter 2. Chemical Bonding

➤ **Atom :** The smallest particle of an element which does not exist in the free state and has a tendency to combine with the atom(s) of the other metal to form a molecule.

➤ **Molecule :** The smallest particle of matter which can exist in free state. It is formed when two or more atoms combine chemically.

➤ **Sharing of Electrons :** A set of eight electrons in the outermost shell confers stability. Atoms of other elements interact to attain 8 electrons in their outermost shells by electron transfer or by sharing of electrons.

The maximum number of electrons that can be accommodated in K shell is 2, for L shell is 8, for M shell is 18 and for N shell is 32.

The 2, 8, 18, 32 are terms as Magic Numbers.

In the outermost shell of any atom there could be a maximum of eight electrons to attain the noble gas configuration, this rule is known as Octet rule.

- **Ions :** When a chemical compound, in fused state or aqueous solution, breaks up into electrically charged atoms or group of atoms, then they are collectively called ions.

  Formation of Ions : Cations (Positively charged) and Anions (Negatively charged).

- **Chemical Bond :** The attractive force which holds the two atoms together in a molecule is known as chemical bond.

- **Types of Bond :** The important types of bonds are :

  (i) Electrovalent bond

  (ii) Covalent bond

  (iii) Co-ordinate bond

  Some other special types of bonds are :

  (i) Metallic bond

  (ii) Vander Waal's bond

- **Electrovalent or Ionic Bond [Heteropolar Bond] :** The chemical bond that is formed between two atoms by transfer of one or more electrons. e.g. NaCl, $MgCl_2$, CaO

  They have high melting and boiling point due to strong electrostatic forces of attraction between their ions.

- **Formation of Ionic Bond :** In case of [NaCl] sodium chloride :

$$Na^{\times} + \cdot\ddot{\underset{..}{Cl}}: \longrightarrow Na^{+} + Cl^{-} \text{ or } NaCl$$

Sodium atom : A = 23; Z = 11 Na : 2, 8, 1 $Na - 1e^{-} \rightarrow Na^{+}$ (2, 8)	Chlorine atom : A = 35; Z = 17 Cl : 2, 8, 7 $Cl + 1e^{-} \rightarrow Cl^{-}$ (2, 8, 8) $Na^{\circ} + Cl : \rightarrow (Na)^{+} + (:Cl:)^{-} \rightarrow NaCl$

- **Covalent Bond :** The chemical bond that is formed between two combining atoms by mutual sharing of one or more electrons. e.g.: H_2, NH_3, H_2O etc.

  The melting and boiling points of such compounds are lower than those of electrovalent or ionic compounds.

- **Formation of Covalent Bond :** Hydrogen [H_2]

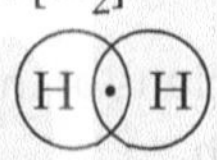

or H – H

Hydrogen atom : A = 1 ; Z = 1

One pair of shared electron is shown by a single line, and is called a single covalent bond.

- **Non-polar covalent Bond :** A covalent bond in which shared pairs of electrons are equally distributed between two atoms is called a non-polar bond. e.g. CH_4, SiH_4.
- **Polar Covalent Bond :** A covalent bond in which shared pair of electrons are unequally distributed between the atoms is called a polar covalent bond. e.g. HCl, HBr, HF.
- **Vander Waal's Force :** Very weak forces existing between the molecules of non-polar covalent compounds.
- **Intermolecular Forces :** Weak electrostatic forces between the molecules of polar covalent compounds.
- **Lone Pair of Electrons :** The unshared pair or pairs of electrons around an atom.

  e.g. In ammonia molecule N atom has one lone pair of electron and in water molecule O atom has two lone pairs of electrons.

- **Co-ordinate Bond :** A special type of covalent bond which is formed by one side sharing of one electron pair between the two atoms. The atom which donates the electron pair is called donor while the other atom which accepts the electron pair is called acceptor. Due to slightly polar character it is also called semi-polar or dative bond.

e.g.

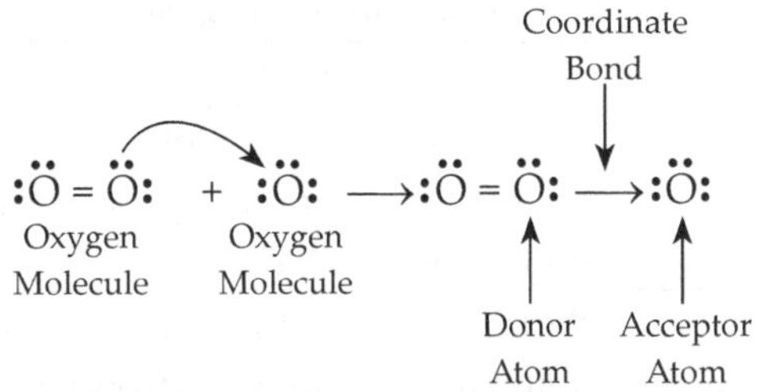

Chapter 3. Acids, Bases and Salts

➢ **Acid :** A compound which on dissolving in water furnishes a proton [Hydrogen ion (H^+)] as the only positively charged ion or it is a compound which dissolves in water to form hydronium ion, *i.e.*, HCl, H_2SO_4 etc.

For example :

$$HCl \rightleftharpoons H^+ + Cl^-$$

$$H + H_2O \rightleftharpoons H_3O^+$$

Hydrogen ion — Hydronium ion

➢ **Classificaiton of acids on the basis of their origin :**

Organic acid : An acid derived from an animal or vegetable origin is called organic acid.

Examples : Acetic acid (CH_3COOH), Oxalic Acid $(COOH)_2$

Inorganic (Mineral) acids : The acids derived from minerals are called inorganic or mineral acid.

Examples : HCl, H_2SO_4, HNO_3

➢ **General Properties of Acids :**

(i) Sour to taste

(ii) Turns blue litmus red

(iii) Form salt and water when reacting with alkalies

(iv) Liberation of hydrogen in reactions with active metals :

(v) Liberation of CO_2 from carbonates and hydrogen carbonates :

(vi) Neutralization of a base :

Classification of Acid

Property	Types	Definition	Examples
Strength	Strong Acid	Dissociates completely in aqueous solution, producing high conc. of H^+ ions	HCl, HNO_3, H_2SO_4
	Weak Acid	Partial decomposition in aqueous solution, producing low conc. of H+ ions	HCN, CH_3COOH, H_2CO_3
Basicity	Monobasic Acid	Produce one H^+ ion per molecule, upon dissociation.	HCl
	Dibasic Acid	Produce two H^+ ions per molecule, upon dissociation.	H_2SO_4
	Tribasic Acid	Produce two H^+ ions per molecule, upon dissociation.	H_3PO_4
Concentration	Concentrated Acid	Contain very little or no water.	99.9% H_2SO_4
	Dilute Acid	Contain large amount of water.	20% H_2SO_4

Base : A base is a compound which combines with the hydronium ion (H_3O^+) of an acid to form salt and water only, or it is the oxide or hydroxide of a metal which will neutralize the acid.

For example : Na_2O, NaOH, $Cu(OH)_2$, etc.

Classification of Base			
Property	**Types**	**Definition**	**Examples**
Strength	Strong Base	Dissociates completely in aqueous solution, producing high conc. of OH^- ions	KOH, NaOH, $Ba(OH)_2$
	Weak Base	Partial decomposition in aqueous solution, producing low conc. of OH- ions	$Mg(OH)_2$, NH_4OH, $Cu(OH)_2$
Basicity	Monoacidic Base	Produce one OH^- ion per molecule, upon dissociation.	KOH
	Diacidic Base	Produce two OH^- ions per molecule, upon dissociation.	$Ca(OH)_2$
	Triacidic Base	Produce two OH^- ions per molecule, upon dissociation.	$Al(OH)_3$
Concentration	Concentrated Base	Contain very little or no water.	99.9% NaOH
	Dilute Base	Contain large amount of water.	20% NaOH

➢ **New Concept of Acids and Bases**

(a) **Acids :** Compounds that are electron deficient, *i.e.,* acids donate protons. e.g., BF_3, $AlCl_3$, all (+ve) ions etc.

(b) **Bases :** Compounds which contain lone pair of electrons i.e., base accepts protons, e.g., NH_3, all (–ve) ions etc.

➢ **Alkalies :** A compound, which on dissolving in water furnishes OH^- ions as only negative ions, e.g., NaOH, KOH etc.

➢ **pH of solution :** The negative logarithm to the base 10 of the hydrogen ion concentration expressed in gram ions per litre or mole per litre. pH scale is a measure of degree of acidity or alkalinity of a solution.

$$pH = \log_{10} (H^+)$$

pH scale ranges from 0 to 14.

➢ **Neutralisation :** A chemical reaction in which the hydronium ions (H_3O^+) derived from an acid and the hydroxyl ions (OH^-) derived from the base combine to form undissociated water molecules.

For example : HCl (Acid) + NaOH (Base) $\longrightarrow$ NaCl (Salt) + H_2O (water)

➢ **Salts :** Is an ionic compound which when dissolved in water, yields a positive ion other than hydrogen ion (H^+) and a negative ion other than hydroxyl ion (OH^-).

For example : $\underset{\text{Acid}}{HCl} + \underset{\text{Base}}{NaOH} \longrightarrow \underset{\text{Salt}}{NaCl}$

Classification of Salts		
Types	**Definition**	**Examples**
Normal Salt	Does not contain any ionisable or replacable hydrogen atoms.	NaCl, AgCl, $FeCl_3$
Acidic Salt	Contains ionisable or replacable hydrogen ions.	$NaHCO_3$, $KHSO_3$, $Ca(HCO_3)_2$
Basic Salt	Contains ionisable or replacable hydroxyl ions.	$CuCO_3$, $PbCO_3$, $Pb(OH)_2$

Chapter at a Glance

Properties of Salts		
Property	**Definition**	**Examples**
Hydrolysis of salts	Reaction in which one of the ions from a salt reacts with water, forming either an acidic or basic solution. • Strong acid + Strong base → Neutral solution • Strong acid + Weak base → Acidic solution • Weak acid + Strong base → Basic solution	$F^-(aq) + H_2O(l) \rightleftharpoons HF(aq) + OH^-(aq)$ $NH^+_4(aq) + H_2O(l) \rightleftharpoons H_3O^+(aq) + NH_3(aq)$
Anhydrous Salt	Does not contain any water of crystallisation	NaCl, $NaNO_3$
Hydrated salts	Contain definite number of water molecules, loosely attached.	$Na_2CO_3.\ 10H_2O$, $CuSO_4.5H_2O$
Water of crystallisation	Definite number of water molecules chemically combined with crystals	$CuSO_4.5H_2O$ (blue vitriol), $MgSO_4.7H_2O$ (Epsom salt)
Efflorescence	Upon exposure to atmosphere, salts lose their Water of crystallisation and form powder.	$Na_2CO_3.10H_2O$ (washing soda) $MgSO_4.7H_2O$ (Epsom salt)
Deliquescence	Upon exposure to atmosphere, salts absorb moisture from air and lose crystalline form and form saturated solution.	$FeCl_3$, $MgCl_2.\ 6H_2O$
Hygroscopy	Upon exposure to atmosphere, salts absorb moisture from air, without dissolving in it.	$CaCl_2$, CaO

➢ **Amphoteric Nature of Some Metals or their Oxides :**
When metals (such as Al or Zn) or their oxides, react with acids as well as alkalies to form salt and water or hydrogen as the products, the metals or their oxides are said to be amphoteric in nature.

Chapter 4. Analytical Chemistry

➢ Water soluble bases are called alkalies. e.g. NaOH, KOH, etc. Water insoluble bases are called non-alkalies. e.g. CaO, ZnO, CuO, etc. Salts of metals like Ca, Zn, Pb, Mg, Al, Cu, Fe, etc., react with alkalies like NaOH, NH_4OH to form precipitate of the metal hydroxides. Some of these metal hydroxides react with excess of alkalies to form water soluble complex compounds like sodium zincate, sodium plumbate, potassium meta-aluminate, etc.

Effects of hydroxides on solution of salts can be better understood by the table given below :

Aqueous solution of salt	Cation present	Salt with colour	Effect of sodium hydroxide solution	Reaction	Effect of ammonium hydroxide solution	Reaction
Mg	Mg^{2+}	$MgCl_2$ (white)	A white ppt, of $Mg(OH)_2$, insoluble in excess of NaOH is formed.	$MgCl_2 + 2NaOH \rightarrow 2NaCl + Mg(OH)_2$	No. ppt.	No reaction
Fe	Fe^{2+}	$FeSO_4$ (green)	A dirty green ppt. of $Fe(OH)_2$, insoluble in excess of NaOH, is formed.	$FeSO_4 + 2NaOH \rightarrow Na_2SO_4 + Fe(OH)_2$	A dirty green ppt. of $Fe(OH)_2$, insoluble in excess of NH_4OH, is formed.	$FeSO_4 + 2NH_4OH \rightarrow (NH_4)_2SO_4 + Fe(OH)_2$

	Fe^{3+}	$FeCl_3$ (Brown)	A red brown ppt. of $Fe(OH)_3$, insoluble in excess of NaOH, is formed.	$FeCl_3 + 3NaOH \rightarrow 3NaCl + Fe(OH)_3$	A red brown ppt. of $Fe(OH)_3$ insoluble in excess of NH_4OH, is formed.	$FeCl_3 + 3NH_4OH \rightarrow 3NH_4Cl + Fe(OH)_3 \rightarrow$ Contd.
Cu	Cu^{2+}	$CuSO_4$ (Blue)	Light blue ppt. of $Cu(OH)_2$, insoluble in excess of NaOH is formed.	$CuSO_4 + 2NaOH \rightarrow Na_2SO_4 + Cu(OH)_2$	Light blue ppt. of $Cu(OH)_2$, which dissolves in excess of ammonium hydroxide to give a dark blue solution of tetra-ammine copper (II) sulphate. $[Cu(NH_3)]_4SO_4$, is formed.	$CuSO_4 + 2NH_4OH \rightarrow (NH_4)_2SO_4 + Cu(OH)_2 \rightarrow Cu(OH)_2 + (NH_4)_2SO_4 + 2NH_4OH \rightarrow Cu(NH_3)_4SO_4$
Zn	Zn^{2+}	$ZnSO_4$ (white)	A white ppt. of $Zn(OH)_2$, which dissolves in excess of NaOH to give a colourless solution of sodium zincate, Na_2ZnO_2, is formed.	$ZnSO_4 + 2NaOH \rightarrow Na_2SO_4 + Zn(OH)_2 \rightarrow Zn(OH)_2 + 2NaOH \rightarrow Na_2\ ZnO_2$	A white ppt. of $Zn(OH)_2$, which dissolves in excess of NH_4OH to give a colourless solution, is formed.	$ZnSO_4 + 2NH_4OH \rightarrow (NH_4)SO_4 + Zn(OH)_2 \rightarrow Zn(OH)_2 + (NH_4)_2SO_4 + 2NH_4OH \rightarrow Zn(NH_3)_4SO_4$
Pb	Pb^{2+}	Pb $(NO_3)_2$ (white)	A white ppt. of $Pb(OH)_2$, which dissolves in excess of NaOH to give a colourless solution of sodium plumbate, Na_2PbO_2, is formed.	$Pb(NO_3)_2 + 2NaOH \rightarrow 2NaNO_3 + Pb(OH)_2 \rightarrow Pb(OH)_2 + NaOH$ (Excess) $\rightarrow Na_2PbO_2$	A white ppt. of $Pb(OH)_2$, insoluble in excess of NH_4OH is formed.	$Pb(NO_3)_2 + 2NH_4OH \rightarrow (NH_4)SO_4 + Pb(OH)_2$
Al	Al^{3+}		A white ppt of $Al(OH)_3$ soluble in excess of NaOH to give colourless solution of $NaAlO_2$ is formed.	$Al_2O_3 + 2NaOH \rightarrow 2NaAlO_2 + H_2O$	A white gelatinous ppt. of $Al(OH)_3$ insoluble in excess of NH_4OH is formed.	$AlCl_3 + 3NH_4OH \rightarrow Al(OH)_3 + 3NH_4Cl$

➢ **Action of Alkali on Metals :** These metals react with caustic alkalies like NaOH, KOH on heating, resulting in the liberation of H_2 gas.

$$Zn(s) + 2NaOH(aq.) \longrightarrow \underset{\text{Sodium zincate}}{Na_2ZnO_2(aq.)} + H_2(g)\uparrow$$

$$2Al + 2NaOH + 2H_2O \longrightarrow \underset{\text{Sodium meta aluminate}}{2NaAlO_2} + 2H_2\uparrow$$

Metals like iron, magnesium, copper do not react with alkalies.

➢ **Action of Alkali on Metal Oxides :**

The amphoteric metal oxides react with strong caustic alkali like NaOH, KOH to form complex salt and water.

$$ZnO + 2NaOH \longrightarrow \underset{\text{Sodium zincate}}{Na_2ZnO_2} + H_2O$$

$$PbO + 2NaOH \longrightarrow \underset{\text{Sodium plumbate}}{Na_2PbO_2} + H_2O$$

Chapter 5. Mole Concept & Stoichiometry

➢ Avogadro's law : Under the same conditions of temperature and pressure, equal volumes of all the gases contain the same number of molecules.

Example : $$\underset{\text{2 vols.}}{2SO_2(g)} + \underset{\text{1 vol.}}{O_2(g)} \longrightarrow \underset{\text{2 vols.}}{2SO_3(g)}$$

Chapter at a Glance

➢ **Gay-Lussac's law of Gaseous volumes :** At the same condition of temperature and pressure, the volume of gases taking part in a chemical reaction either as a reactants or products bear a simple whole number ratio to one another.

Example : $H_2(g) + Cl2(g) \longrightarrow 2HCl(g)$

1 vol. 1 vol. 2 vols.

Ratio of reacting gases and products is 1 : 1 : 2 which is a simple ratio.

➢ **Some Useful Relations :**

(i) One mole of atom $= 6{\cdot}023 \times 10^{23}$ atoms
= gram atomic mass of an element
= 1 gm atom of the element

(ii) One mole of molecule $= 6{\cdot}023 \times 10^{23}$ molecules
= gram molecular mass of the compound
= 1 gm molecule of the compound

(iii) One mole of a gas = 22·4 litres at S.T.P.

(iv) $$\text{Moles of an element} = \frac{\text{Mass of the element}}{\text{Atomic mass}}$$

(v) $$\text{Moles of a compound} = \frac{\text{Mass of the compound}}{\text{Molecular mass}}$$

(vi) $$\text{Mass of one atom} = \frac{\text{Atomic mass}}{6.023 \times 10^{23}}$$

(vii) $$\text{Mass of one molecule} = \frac{\text{Molecular mass}}{6.023 \times 10^{23}}$$

(viii) Number of molecules $=$ moles $\times 6{\cdot}023 \times 10^{23}$

(ix) Total number of atoms $=$ Number of atoms in one molecule $\times$ moles $\times 6{\cdot}023 \times 10^{23}$

(x) $$\text{Vapour density} = \frac{\text{Mass of certain volume of the gas}}{\text{Mass of an equal volume of hydrogen}}$$

(xi) Molecular weight $= 2 \times$ VD

(xii) $$1 \text{ amu} = \frac{1}{12} \text{ of } C - 12$$

(xiii) $$\text{Percentage composition} = \frac{\text{Weight of an element in a molecule of a compound}}{\text{Gram molecular weight of compound}} \times 100$$

(xiv) Molecular formula = Empirical formula $\times n$

➢ The chemical equation $2H_2(g) + O_2(g) \rightarrow 2H_2O(l)$. gives us the following information :

(i) Hydrogen and oxygen combine in 2 : 1 molecular ratio and 2 : 1 volume ratio.

(ii) 2 moles H_2 combine with 1 mole O_2 to form 2 moles H_2O.

(iii) 4g H_2 combine with 32g O_2 to form 36g H_2O.

(iv) 4kg H_2 combine with 32kg O_2 to form 36kg H_2O.

(v) 4tons of H_2 combine with 32tons of O_2 to form 36tons of H_2O.

➢ A balanced chemical equation gives an idea of the ratio-proportions of the amount of reactants consumed and that of the products formed. Therefore, if we know the amount of any one substance of a reaction, we can compute the amount of other substances with the help of the balanced chemical equation. We have following types of calculations.

(i) Mole—mole calculations

(ii) Mass—mass calculations

(iii) Mass—volume calculations

(iv) Volume—volume calculations

Chapter 6. Electrolysis

- **Electrolysis :** The process due to which, a chemical compound in aqueous or fused state conducts direct electric current and at the same time undergoes chemical decomposition, due to the discharge of charged ions, is called electrolysis.
- **Electrodes :** The graphite or metal plates or rods, through which current enters or leaves an electrolyte, are called electrodes. They are of two types : Cathode and Anode.
- Any alloy or any metal which does not react chemically with the electrolyte or with the products of electrolysis can be used to make electrodes. Commonly used substances are graphite (carbon), platinum, steel, copper etc.

Cathode	Anode
Connected to negative terminal of battery	Connected to positive terminal of battery
Has excess of electron	Deficiency of electrons
Current from electrolyte leaves through cathode	Current enters the electrolyte from anode
Cations migrate towards cathode	Anions migrate towards anode
Cations accepts electrons from cathode	Anions donate electrons to anode

- Current flows in a direction opposite to the flow of electrons.
- **In a battery :** Electrons enter from +ve terminal and leave from –ve terminal.
- **In an electrolytic cell :** Electrons enter from –ve terminal and leave from +ve terminal.
- **Electrical Conductors :** All metals e.g., Cu, Ag, Al etc. All alloys e.g., steel and graphite (a non-metal)
- **Electrical non-conductors or insulators :** Non-metals except graphite e.g., S, P etc.
  All covalent compounds in pure state e.g., AgCl.
  All ionic compounds in solid state e.g., solid NaCl.
  Substances like plastic, rubber, polythene etc.
- **Electrolytic Conductors (Electrolytes) :** All ionic compounds in molten state or in aqueous solution e.g., molten NaCl, aqueous NaCl, Polar covalent compounds in aqueous solution e.g. aqueous HCl.
- **Electrolytic non-conductors (Non-Electrolytes) :** All ionic compounds in solid state.
  All non-polar and polar compounds in pure state. Compounds like rubber, plastic etc.
- **Ionisation :** The process of formations of positively and negatively charged ions from molecules which are not initially in the ionic state is called ionisation. e.g. Polar covalent compound, HCl

$$HCl \rightleftharpoons H^+ + Cl^-$$

- **Dissociation :** The process of separation of ions which are already present in an ionic compound is called electric dissociation. e.g. electrovalent compounds, NaCl

$$NaCl \rightleftharpoons Na^+ + Cl^-$$

- **Reduction :** A process of gaining of electrons by a cation to form neutral atom (or gaining of electrons by an electronegative atom to form an anion) is called reduction. Reduction takes place at cathode.
- **Oxidation :** A process of loss of electrons by an atom to form a cation (or loss of electrons by an anion to form atom) is called oxidation. Oxidation occurs at anode.
- **Applications of Electrolysis :**
  1. Electroplating
  2. Electro-refining
  3. Electrometallurgy

Conductors	Electrolytes
Allows current to pass through them in solid state	Allows current to pass through them in fused state or in aqueous solution
Do not undergo any chemical reaction during electrolysis	Undergo chemical reaction during electrolysis
Flow of electricity is due to free electrons	Flow of electricity is due to discharge of ions

➢ **Electrochemical Series :** In this series, the metals are arranged in decreasing order of their tendency to lose electrons. The most active metal (K) which can lose electron most readily is kept at the top while the least reactive metal (Au) is placed at the bottom of the series.

Cations : K^+, Na^+, Ca^{2+}, Mg^{2+}, Al^{3+}, Zn^{2+}, Fe^{2+}, Pb^{2+}, H^+, Cu^{2+}, Hg^{2+}, Ag^+ (To cathode)

Anions : SO_4^{2-}, NO_3^-, Cl^-, Br^-, I^-, OH^- (To anode)

Chapter 7. Metallurgy

➢ **Element :** A pure substance that consists of only one type of atoms, e.g., Ag, N, O, etc.

➢ **Metal :** A metal is an electropositive element which is hard, sonerous, lustrous, malleable, ductile with tensile strength and also good conductor of heat and electricity. Metals form basic or amphoteric oxides.

➢ **Non-metal :** A non-metal is an electronegative element which occurs as solids, liquids and gases. They are generally poor conductors of heat and electricity. Solids are usually soft and brittle. Non-metals form acidic or neutral oxides.

➢ **Metalloid :** Elements which exhibit the properties of both metals and non-metals are called metalloids. They are also known as semi–metals, e.g., Si, Ge, As.

➢ **Mineral :** The metallic compounds occuring in nature are called minerals.

➢ **Ore :** A naturally occuring mineral from which a metal can be conveniently extracted, is called an ore.

➢ **Matrix or gangue :** The unwanted impurities which are associated with ore are called gangue or matrix. e.g. Stone, clay etc.

➢ **Metallurgy :** It is process of extracting metal in a pure state on a large scale from its ore by physical and chemical process.

It is carried out in the following steps :

➢ **Alloy :** It is a homogenous mixture of two or more metals (or metal and non-metal) combined in a definite proportion in their molten state.

➢ **Amalgam :** It is an alloy of mercury with one or more other metals. Most metals form amalgams which may be liquid or solid.

➢ **Physical properties of metals and non-metals :**

Metals	Non-metals
1. Generally hard solid and vapourize only at high temperature.	Generally soft; and usually gases or liquids.
2. Have metallic lusture and can be polished.	Lack lustre.
3. Melting and boiling points are high (except alkali metals and Hg)	Low melting and boiling points.
4. High density except alkali metals like Na, K etc.	Low density.
5. Generally malleable and ductile and have a very high tensile strength.	Non-malleable and non-ductile.
6. Alloys and amalgams.	Do not form alloys.
7. Hard, but not brittle.	Brittle in nature.
8. Generally good conductors of heat and electricity.	Bad conductors of heat and electricity.
9. Do not generally dissolve in liquids except by chemical reaction.	Generally soluble in water or other solvents and do not react with them chemically.

➢ Chemical properties of metals and non-metals :

Metals	Non-metals
1. Have 1, 2, 3 valence electrons.	Usually have 4, 5, 6, 7 valence electrons.
2. Electropositive in nature.	Electronegative in nature.
3. Good reducing agents.	Good oxidising agents.
4. Form basic Oxides.	Generally form acidic oxides.

5. Do not usually react with H_2. Metals placed above hydrogen in the activity series, react with dilute HCl and dilute H_2SO_4 to liberate H_2 gas.	Do not react with dil. acids to liberate hydrogen.
6. Readily with Cl_2 to form chlorides.	Usually form chlorides, which are volatile covalent liquid or gases.
7. Form electrovalent or ionic compounds. which conduct electricity in fused state or in solution, due to the free electrons or ions.	Form covalent compounds which do not conduct electricity because there are no free electrons or ions to carry the current.

➤ **Activity Series :**

The activity series of metals is a list of metals arranged in the order of their decreasing Chemical Activity.

➤ Various steps involved in the extraction of pure metals from their ores.

I	Crushing and grinding of the ore-pulverisation of the ore	Electromagnetic separation method
		Hydraulic washing method
II	Concentration or dressing of the ore. Removal of impurities from the ore	Froth flotation method
		Leaching process :
III	Conversion of the concentrated ore into metallic oxide	Calcination
		Roasting
IV	Reduction of the metallic oxide into free metal	Reduction by carbon, carbon monoxide
		Reduction by electrolysis
		Reduction by Al, Aluminothermic process
V	Purification or refining of impure (crude) metals	Liquation process
		Fractional distillation process
		Oxidation process
		Electrolytic refining

Chapter 8. (a) Study of Compounds : Hydrogen Chloride

➤ **Hydrogen chloride (HCl) :**

Formula : HCl Nature : Covalent. Molecular Mass : 36·5 [*i.e.,* HCl = 1 + 35·5 = 36·5]

Occurrence : (i) Present in gastric juices. (ii) Present in volcanic gases.

➤ **Laboratory preparation of HCl gas :**

$$NaCl + H_2SO_4 \xrightarrow{Heat} \underset{\text{Sod. hydrogen sulphate}}{NaHSO_4} + HCl\ (g)$$

$$NaCl + NaHSO_4 \xrightarrow{Heat} \underset{\text{Sod. sulphate}}{Na_2SO_4} + HCl\ (g)$$

➤ **Physical Properties of HCl :**

(i) It is a colourless pungent smelling gas with acidic taste and which fumes strongly in air.

(ii) It is heavier than air (V. D. = 18·25) and is extremely soluble in water, at room temperature 1 volume of water dissolves 450 volumes of the gas.

(iii) Hydrogen gas can be liquified into a colourless liquid at 40 atm. pressure and 10°C temperature.
Boiling point of liquid HCl is –83°C.
Freezing point of liquid HCl is –113°C.

➤ **Chemical Properties of HCl :**

(i) It is neither combustible nor supporter of combustion.

(ii) Due to its acidic nature, it turns moist litmus paper red.

(iii) It rapidly dissolve in water, to form hydronium ions.

$$HCl + H_2O \rightleftharpoons H_3O^+ + Cl^-$$

Chapter at a Glance

(iv) When hydrogen chloride gas is passed through silver nitrate solution, it forms white precipitate of silver chloride.

$$AgNO_3 + HCl \rightarrow \underset{\text{White ppt.}}{AgCl\downarrow} + HNO_3$$

(v) Reaction with NH_3 : Dry HCl gas combines with ammonia (NH_3) gas to give dense white fumes of ammonium chloride.

$$\underset{\text{Ammonia}}{NH_3(g)} + \underset{\text{Hydrogen chloride}}{HCl(g)} \rightarrow \underset{\text{Ammonium chloride (White fumes)}}{NH_4Cl}$$

(vi) Reaction of HCl gas with ammonia can be used as a test for HCl.

➤ **Uses of hydrogen chloride :**

(i) It is used in manufacturing of glucose from starch.

(ii) It is used for the extraction of glue from animal tissue and bones.

(iii) It is used in the manufacture of dyes, drugs, paints and is widely used in photography.

Chapter 8. (b) Ammonia and Nitric Acid

➤ **Ammonia :**

Molecular formula ⟶ NH_3

Vapour Density ⟶ 8·5

Molecular Mass ⟶ 17

Nature ⟶ Non-poisonous, basic in nature.

➤ **Occurrence :** It occurs in free state and is present in atmospheric air, natural water (traces), sewage water (higher proportion) decaying plant and animal tissues, urine and excreta of animals.

In combined form ammonia is found as ammonium chloride (NH_4Cl) (Nausadar) and ammonium sulphate [$(NH_4)_2SO_4$].

Sources : Ammonia is used commonly in three forms :

(i) Gaseous ammonia (ii) Liquor ammonia (iii) Liquid ammonia.

Liquor ammonia is a saturated solution of ammonia in water and is stored in tightly packed bottles in a cold place.

➤ **Lab Preparation of Ammonia :**

By warming any ammonium salt with an alkali.

$$2NH_4Cl + Ca(OH)_2 \xrightarrow{\text{Heat}} CaCl_2 + 2H_2O + \underset{\text{(amm. gas)}}{2NH_3\uparrow}$$

➤ **Physical Properties of Ammonia Gas :**

(i) It is colourless gas, that fumes strongly in air.

(ii) It has a sharp burning smell, which brings tears to the eyes. It increases the heart beat in large doses.

(iii) It is highly soluble in water. One volumes of water can dissolve 1300 volumes of gas, at a pressure of 76 nm of mercury.

(iv) It is lighter than air. Its V.D. is 8·5 as compared to air, which is 14·4. For this reason it is collected by the downward displacement of air. It cannot be collected over water because of its extreme solubility.

(v) However, being basic in character still it is non-poisonous in nature.

(vi) It damages respiratory system.

➤ **Chemical Properties of Ammonia Gas :**

(i) Ammonia changes red litmus solution to blue. But it does not affect a blue litmus solution.

(ii) Ammonia produces a pink colour with phenolphthalein indicator.

(iii) Ammonia reacts with acids to form corresponding ammonium salts.

$$NH_3 + HCl \longrightarrow \underset{\text{Ammonium chloride}}{NH_4Cl}$$

(iv) Ammonia dissolves in water to form an aqueous solution of NH_4OH

$$NH_3 + H_2O \rightleftharpoons \underset{\text{Ammonium hydroxide}}{NH_4OH}$$

➤ **Uses :**

(i) Liquid NH_3 is used in Solvay's Process as refrigerant in the ice plant.

(ii) Ammonia water is used to remove grease and ink stains. The solution of ammonia is also used for cleaning glass windows, tiles and porcelain.

(iii) Ammonia is used in the fertilizer industry for the manufacture of urea, Ammonium sulphate and ammonium nitrate.

(iv) Ammonia is used in the Ostwald's Process of manufacture of nitric acid.

➤ **Catalytic Oxidation of Ammonia (Ostwald Process) :**

$$4NH_3 + HO_2 \xrightarrow[800°C]{Pt} 6H_2O + 2NO$$

$$2NO + O_2 \xrightarrow{50°C} 2NO_2$$

$$4NO_2 + O_2 + 2H_2O \longrightarrow \underset{\text{Nitric acid}}{4HNO_3}$$

➤ **Nitric acid :** Molecular formola – HNO_3

Vapour Density – 1.5

Molecular weight – 63

Nature – Corrosive, acidic in Nature.

➤ **Occurence :** It is naturally found in combined state in compounds such as Bengal saltpeter or nitre (KNO_3) and chile saltpetre ($NaNO_3$). During thunder and lightening, traces of nitric acid are formed in air.

➤ **Laboratory preparation of Nitric acid :** By distillation of potassium nitrate crystals with conc. sulphuric acid.

$$KNO_3 + H_2SO_4 \xrightarrow[200°C]{\text{Less than}} \underset{\text{Potassium bisulphate}}{KHSO_4} + \underset{\text{Nitric acid}}{HNO_3}$$

$$KNO_3 + KHSO_4 \xrightarrow[200°C]{\text{More than}} \underset{\text{Potassium sulphate}}{K_2SO_4} + \underset{\text{Nitric acid}}{HNO_3}$$

By heating a mixture of sodium nitrate and concentrated sulphuric acid at 200°C.

$$NaNO_3 + H_2SO_4 \xrightarrow{< 200°C} NaHSO_4 + HNO_3$$

$$2NaNO_3 + H_2SO_4 \xrightarrow{> 200°C} Na_2SO_4 + 2HNO_3$$

The nitric acid distills over is brown due to the dissolved oxides of nitrogen from thermal decomposition of nitric acid; these may be removed by bubbling dry air or CO_2 through warm acid.

➤ **Physical Properties :**

(i) Pure nitric acid is a colourless hygroscopic liquid with choking smell.

(ii) Commercial nitric acid is yellowish brown in colour due to presence of dissolved nitrogen dioxide in it.

(iii) It has a sour taste and highly corrosive in nature. It stains skin yellow and causes very painful blisters.

(iv) Pure nitric acid boils at 86°C and its specific gravity is 1·54.

On cooling it freezes at –42°C to form white crystals.

(v) It is highly soluble in water.

➤ **Chemical Properties :**

(i) Pure nitric acid contained in the glass bottles, on exposure to light, slowly decomposes to liberate NO_2, which redissolves in it and makes it pale yellow.

(ii) Nitric acid turns blue litmus red.

(iii) Nitric acid decomposes metal carbonates and hydrogen carbonates to corresponding salts with liberation of CO_2.

(iv) Nitric acid with bases, forms salts and water and neutralises alkalis.

(v) Nitric acid oxidises carbon, sulphur, phosphorus to their highest oxides or oxy-acids like carbon dioxide, sulphuric acid and phosphoric acid.

➤ **Use of Nitric Acid :**
Manufacture of explosives, fertilizers, nitrates, sulphuric acid, dyes, perfumes, drugs, aqua-regia, etc.

➤ **Nitrates :** The salt of nitric acid with metals are called nitrates. Only non-metallic nitrate is ammonium nitrate.

Chapter 8. (c) Sulphuric Acid

➤ **Sulphuric Acid :**
Molecular formula – H_2SO_4
Vapour Density – 1.8
Molecular Mass – 98
Nature – Corrosive, Acidic in nature
Sulphuric acid acts as an acid because it reacts with metallic hydroxide to form salt and water only.

➤ **Manufacture of Sulphuric acid by Sulphur dioxide :** (Contact Process)

(i) Sulphur dioxide is produced by burning sulphur or by roasting iron pyrites.

$$S + O_2 \xrightarrow{\text{Heat}} SO_2$$

$$\underset{\text{Iron pyrites}}{4FeS_2} + 11O_2 \xrightarrow{\text{Heat}} \underset{\text{Ferric oxide}}{2Fe_2O_3} + 8SO_2$$

(ii) Sulphur dioxide obtained above is oxidised to sulphur trioxide in presence of vanadium pentoxide (V_2O_5) as catalyst.

$$2SO_2 + O_2 \xrightarrow{V_2O_5} 2SO_3$$

The above reaction is exothermic. The favourable conditions are :
(a) Temperature, 450 to 500°C.
(b) Pressure, about 2 atmosphere.
The reaction gases SO_2 and O_2 must be free from all impurities as they poison the catalyst used.

(iii) Sulphur trioxide is then absorbed in sulphuric acid to get oleum which on dilution with water gives sulphuric acid of desired concentration.

$$SO_3 + H_2SO_4 \longrightarrow \underset{\text{Oleum}}{H_2S_2O_7}$$

$$H_2S_2O_7 + H_2O \longrightarrow 2H_2SO_4$$

Sulphur trioxide is not directly dissolved in water as it forms a mist of fine drops of sulphuric acid which fills the factory.
Platinized asbestos, silica gel, and V_2O_5 are all catalysts used in contact process because the rate of production of SO_3 is increased.

➤ **Physical properties :**
(i) Colourless, odourless, syrupy liquid.
(ii) Density : $1{\cdot}84 g\ cm^{-3}$.
(iii) It is hygroscopic in nature.
(iv) Highly corrosive, and poor conductor of electricity.

➤ **Chemical properties :**
(i) Chemical nature of H_2SO_4 changes with its concentrations because dil. H_2SO_4 shows acidic properties while conc. H_2SO_4 shows oxidising property.
(ii) It changes blue litmus into red colour.
(iii) It changes methyl orange into red.
(iv) Sulphuric acid decolourises the pink colour of alkaline solution of phenolphthalein.
(v) Concentrated H_2SO_4 is a non volatile dibasic acid.

$$NaCl + H_2SO_4 \longrightarrow NaHSO_4 + HCl$$

(vi) Oxidation of Carbon and Sulphur.

$$C + \underset{\text{(Conc.)}}{2H_2SO_4} \longrightarrow H_2CO_3 + H_2O + 2SO_2\uparrow$$

$$S + \underset{\text{(Conc.)}}{2H_2SO_4} \longrightarrow 3SO_2\uparrow + 2H_2O$$

Chapter 9. Organic Chemistry

- **Organic Chemistry :** The branch of chemistry dealing with carbon and its compounds other than oxides, carbonates and carbides.
- **Organic Compound :** The chemical compounds which are present in the living organisms, plants and animals.
- **Catenation :** This property of self linking of carbon atoms through covalent bonds to form long, straight or branched chain and rings of different sizes Carbon atoms possess this property to a great extent and thus, forms large number of compounds.
- **Molecular Formula :** The chemical formula of a compound, which tells the kind of atom present and their actual number in one molecule of a compound.
- **Structural Formula :** A formula which tells the arrangement of various atoms in one molecule of a chemical compound.

  Methane and ethane have the following structural formulae :

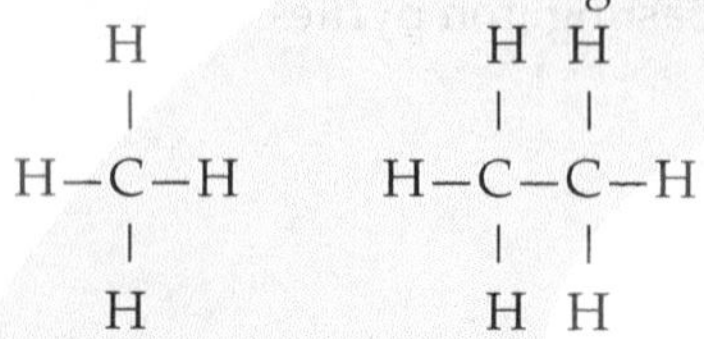

Methane Ethane

- **Hydrocarbon :** Compounds which are made up of only carbon and hydrogen, e.g., CH_4, C_2H_5.

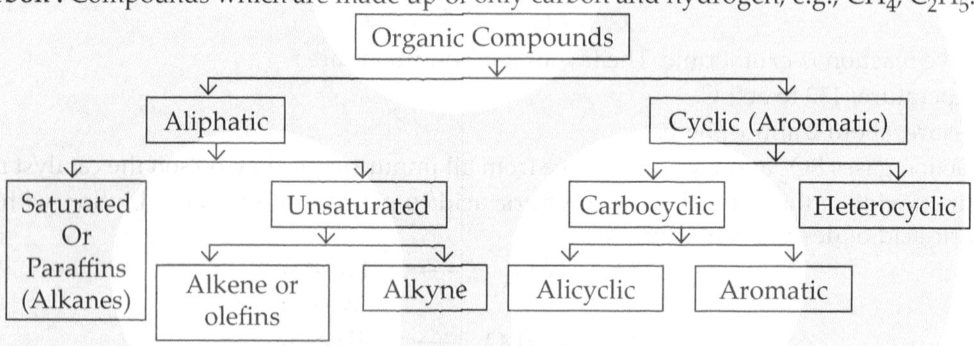

- **Condensed Formula :** Structural formula which indicates the group of atoms joined together to each of the carbon atom, in straight line or branched chains.

  Molecular formula : C_4H_{10}

  Condensed formula : $CH_3-CH_3-CH_2-CH_3$

  Or

  $CH_3-CH(CH_3)-CH_3$

- **Homologous Series :** A series of compounds of the same family in which each member differs from its adjacent member by one CH_2 unit.
- **IUPAC Nomenclature :** The system of assigning a proper name to an organic compound was evolved by the International Union of Pure and Applied Chemistry and the scheme is referred as IUPAC nomenclature. This scheme is based on the following parameters :

  (i) Number of carbon atoms present in one molecule of the compound.

  (ii) Kinds of covalent bonds between carbon atoms.

  (iii) Arrangement of carbon atoms in space, i.e. straight chain, branched chain or closed chain.

Organic compound	General Formula	Symbol	Suffix	Examples
Alkanes	C_nH_{2n+2}	C – C	– ane	Methane CH_4, Ethane CH_3-CH_3

Alkene	C_nH_{2n}	C = C	– ene	Ethene $CH_2 = CH_2$ Propene $CH_3 - CH = CH_2$
Alkyne	C_nH_{2n-2}	C ≡ C	– yne	Ethyne CH ≡ CH Propyne CH_3_C ≡ CH
Alkyl	C_nH_{2n+1}	C –	– yl	Methyl – CH_3– Ethyl – C_2H_5–
Alcohols	$C_nH_{2n+1} - OH$	– OH	– ol	Methanol $CH_3 - OH$ Ethanol $CH_3 - CH_2 - OH$
Aldehyde	C_nH_{2n+1} CHO	– CHO	– al	Methanal HCHO Ethanal CH_3 – CHO
Carboxylic Acids	C_nH_{2n+1} – COOH	– COOH	– oic acid	Methanoic Acid HCOOH Ethanoic Acid CH_3 – COOH

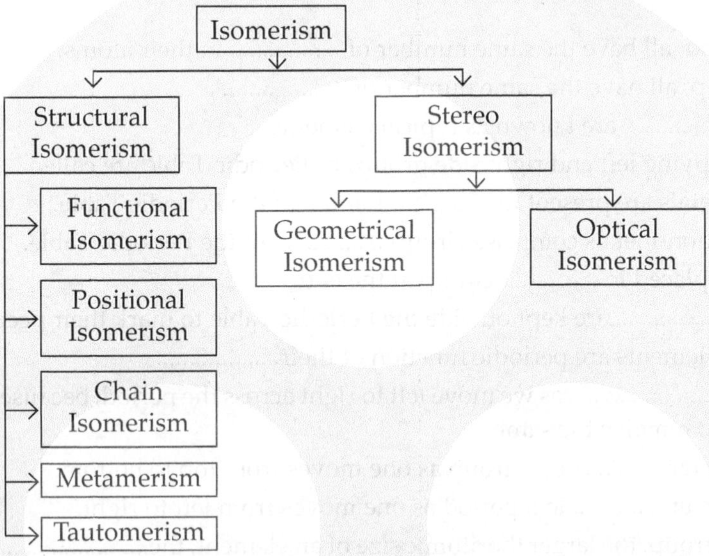

Fill in the Blanks

Set 2

Fill in the blanks in the following statements, with suitable words :

Chapter 1. Periodic Properties and Variations of Properties

1. The modern Periodic Table has .... periods.
2. Each .... in the Periodic Table is comprised of elements having the same number of electron shells.
3. Elements in a period, all have the same number of .... in their atoms.
4. Elements in a group, all have the same number of ....
5. The elements of .... are known as typical elements.
6. The elements occupying left and right side groups of Periodic Table are called .... elements.
7. The most active metals are present in .... 1 and 2 of the Periodic Table.
8. The most reactive non-metals comprise Group .... of the Periodic Table.
9. The rare gases are placed in .... Group at the end.
10. The actinides and .... are kept outside the Periodic Table to mark their peculiar properties.
11. The properties of elements are periodic function of their..................
12. The atomic size .... as we move left to right across the period, because the .... increases but the .... remains the same.
13. The metallic character .... in a group as one moves from top to bottom.
14. The metallic character .... in a period as one moves from left to right.
15. In a period or in a group, the larger the atomic size of an element, the........................ metallic is the element.
16. Moving across a .... of the periodic table, the elements show increasing .... character.
17. The energy required to remove an electron from a neutral isolated gaseous atom and convert it into a positively charged gaseous ion is called............
18. The metallic and non-metallic character depends upon the..................and................. of the elements.
19. The amount of energy involved in the reaction $X + \text{Energy} \rightarrow X^+ + e^-$ is known as the ....of the element X.
20. Across a period, the ionization potential.................
21. Down the group, electron affinity..............
22. The higher the electron affinity of a non-metal, ... chemically reactive the non-metal is.
23. The tendency to gain an electron .... on moving down a group and .... on moving across a period in the Periodic Table.
24. Elements having high ionization potential have .... electron affinities.
25. The electronegativity of elements .... across the period and .... down the group.
26. In general, non-metals are .... electronegative than metals.
27. On moving from left to right in a given period, the number of shells ....
28. Element X belongs to group 2 and period 3 of the Periodic Table. It has .... electrons in the outer most shell.
29. Each period except period 1 in the Periodic Table begins with .... and ends up with a ....
30. Actinides are the elements from atomic number .... to .... and are radioactive.
31. Elements from atomic number 57 to .... occupy same place in Periodic Table. These elements are called ....

Ans. 1. 7
2. period
3. electron shells
4. outermost valence electrons
5. period 3
6. representative
7. Groups
8. 17
9. zero
10. lanthanides
11. atomic number
12. decrease, nuclear charge, number of shells
13. increases
14. decreases
15. more
16. period, non-metallic
17. ionization potential
18. atomic size, ionization potential
19. ionization potential
20. increases
21. decreases
22. more
23. decrease; increase
24. high
25. increases; decreases
26. more
27. remains the same
28. 2
29. an alkali metal; noble gas
30. 89, 103
31. 71, lanthanides

Chapter 2. Chemical Bonding

1. Ionic compounds consist of .... charged ions.
2. Ionic compounds have .... melting points due to .... ionic bonds.
3. The physical state of ionic compounds is ....
4. Ionic compounds conduct electricity in their ....
5. The nature of bond in compounds of alkali and alkaline earth metals is ....
6. .... compounds have low boiling points because of .... intermolecular forces.
7. In covalent compounds, the bond is formed due to the..............of electrons.
8. A solution of a .... electrolyte will contain both ions and molecules of the solute.
9. In the sodium chloride crystal, each Na^+ ion is surrounded by .... Cl^- ions and each Cl^- ion by .... Na^+ ions.
10. In sodium chloride, the Na^+ ion has the .... configuration and the Cl^- ion has the .... configuration.
11. A compound conducted electricity in the fused state. The compound is made-up of ....
12. The duplet or octet structure of valence shell makes .... of an element chemically ....
13. The atomic number of oxygen is 8. The total number of electrons in the O^{2-} ion is ....
14. The atomic number of oxygen is 8. The number of protons in the O^{2-} ion is ....
15. Two nitrogen atoms join together by sharing .... pairs of electrons.
16. In NH_4^+ all the four bonds are..................... .
17. The water molecules easily break the .... bonds between the oppositly charged ions. The ions drift in water in all possible directions and hence electrovalent compounds .... in water.
18. Metals are good ....because they are electron...............
19. Electrovalent compounds have..............melting points.
20. Melting and boiling points of covalent compounds are generally...........
21. Covalent bond is rigid and directional, it is responsible for ....
22. Most covalent compounds have density .... than that of water.
23. The electronegativities of atoms giving covalent molecules are generally ....
24. Atoms of the same element combine to form molecules by means of .... bonds.
25. A molecule of..............contains a triple bond.
26. .... are identical to a normal covalent compound.

Ans. 1. oppositely
2. high, strong
3. solid state
4. fused state as well as in their aqueous solution
5. ionic bond
6. Covalent, weak
7. sharing
8. weak
9. six, six
10. neon, argon
11. ions
12. atom, inactive
13. 10
14. 8
15. three
16. identical
17. ionic, dissolve
18. reducing agents, donors
19. high
20. low
21. definite shape of the molecule
22. less
23. high
24. covalent
25. nitrogen
26. Coordinate bond

Chapter 3. Study of Acids, Bases and Salts

1. (1) An acid is a compound which when dissolved in water gives .... ions as the only .... ions.

(2) A .... acid undergoes almost complete dissociation on dissolving in water.

(3) An example of mineral acid is ....

(4) Vinegar contains .... acid.

(5) Sodium acetate on hydrolysis forms sodium hydroxide and ....

(6) The basicity of acetic acid is ....

(7) Metal A reacts with cold water forming hydrogen and new compound B. The anion combined with A in a substance B is named ....

(8) The solution of sodium carbonate is alkaline due to ....

(9) An alkali which completely dissociates into ions is ....*

(ammonium hydroxide, calcium hydroxide, lithium hydroxide)

(10) As the pH of solution decreases, its acidic strength progressively ....

(11) As the pH of solution increases, the .... strength of the solution progressively decreases.

(12) pH value of lemon juice is .... than 7.

(13) Higher the pH value of a solution, the more...........it is.

(14) Ammonium hydroxide is a .... base.

(15) Methyl orange turns .... in acidic solution.

(16) The formation of a salt from an acid and a base involves the combination of.................. from the acid and the..................from the...................to form...................... molecules.

(17) The number of H^+ ions of an acid which react completely with one molecule of a base is called its ....

(18) When a metallic oxide is dissolved in water, the solution formed has high concentration of.............. ions.

(19) A chemical reaction between hydronium ions of an acid and OH^- ions of a base to form unionised water is called ....

(20) Ammonium chloride is a .... salt.

(21) Sodium hydrogen sulphate is .... salt.

(22) The salt of the acid HNO_2 are called......................and those of the acid HNO3 are called ....

(23) Sodium sulphite reacts with dilute sulphuric acid to form sodium sulphate, ... gas and water.

(24) Copper (II) chloride reacts with sodium hydroxide to form .... copper hydroxide and sodium chloride.

(25) The salt prepared by the method of direct combination is ....*

(iron (II) chloride/iron (III) chloride).

*** Frequently asked previous years Board Exam Questions.**

(26) A salt prepared by displacement reaction is ....*

(ferric chloride, ferrous chloride, silver chloride)

(27) Dissolving of aluminium foil in HCl as well as NaOH shows that it is .... in nature.

(28) The metal whose oxide, which is amphoteric, is reduced to metal by carbon reduction............

(29) The metal which does not react with water or dilute H_2SO_4 but reacts with concentrated H_2SO_4 is...............

(30) The divalent metal whose oxide is reduced to metal by electrolysis of its fused salt is............

(31) Superphosphate is an example of a compound called........... .

(32) A salt which absorbs moisture from the air, but does not change in physical state in called .... salt.

(33) An example of a deliquescent salt is ....

Ans. (1) hydronium, positive
(2) strong
(3) HCl
(4) acetic
(5) acetic acid
(6) 1
(7) hydroxyl ion
(8) hydrolysis
(9) calcium hydroxide
(10) increases
(11) acidic
(12) less
(13) alkaline
(14) weak
(15) pink
(16) H^+, OH^-, base, water
(17) acidity
(18) OH^-
(19) neutralisation
(20) normal
(21) acidic
(22) nitrites, nitrates
(23) sulphur dioxide
(24) insoluble
(25) iron (III) chloride
(26) silver chloride
(27) amphoteric
(28) Pb
(29) Cu
(30) Mg
(31) acid salt
(32) hygroscopic
(33) anhydrous ($CaCl_2$)

2. A solution X turns blue litmus red, so it must contain : ....(1)....... ions. Another solution Y turns red litmus blue and therefore, must contain ....(2)....... ions. When solutions X and Y are mixed together, the products will be a ....(3)....... and ....(4)....... If a piece of magnesium was put into solution X. .... (5)....... gas would be evolved.

Ans. (1) hydronium, (2) hydroxide, (3) salt, (4) water, (5) Hydrogen.

3. Salts of strong acids and strong bases in water produce ....(1)....... solution. Solution of sodium carbonate in water is ....(2)....... in nature, while the solution of ferric chloride is ....(3).......in nature.

Ans. (1) neutral, (2) basic, (3) acidic.

4. In making copper sulphate crystals, a black powder called ....(1)....... is added to dilute ....(2)....... acid and the mixture is warmed. Excess black powder is added to a small volume of acid so as to ......(3)....... The excess insoluble black powder is got rid by ....(4)....... The resulting clear solution looks ....(5)....... in colour. This solution is heated to get rid of some of the ....(6)....... . Slow cooling of the remaining solution ....(7)....... of pure copper sulphate.

Ans. (1) copper oxide, (2) sulphuric, (3) neutralise the excess of the acid. (4) filtration, (5) blue, (6) copper sulphide as impurity, (7) gives well-defined crystals.

5. Deliquescence takes place because of the difference in (1) ....of atmospheric (2) .... and that of the deliquescent substance. The vapour pressure of atmospheric humidity is (3) ....than that of the deliquescent substance. The substances which take up water on exposure to atmosphere, without forming a solution are known as (4) .... substances.

Ans. (1) vapour pressure, (2) humidity, (3) higher, (4) hygroscopic.

6. Complete the following by selecting the correct option from the choices given :*

(i) pH of acetic acid is greater than dilute sulphuric acid. So acetic acid contains .... concentration of H^+ ions. (greater, same, low)

* Frequently asked previous years Board Exam Questions.

(ii) The indicator which does not change colour on passage of HCl gas is .... (methyl orange, moist blue litmus, phenolphthalein)

(iii) The acid which cannot act as an oxidizing agent is ....(conc. H_2SO_4, conc. HNO_3, conc. HCl)

Ans. (i) low

(ii) phenolphthalein

(iii) conc. HCl

Chapter 4. Analytical Chemistry

1. Salts of normal elements [1 (IA) to 17 (VIIA)] are generally.....................

2. Ferrous salts are...........in colour.

3. An example of weak alkali solution is......................

4. Both ammonium and sodium hydroxide are used in analytical chemistry for identifying....... of salts.

5. Zinc chloride solution reacts with ammonium hydroxide solution to give a .... coloured precipitate.

6. Calcium salts with sodium hydroxide give.........precipitates.

7. Ammonium hydroxide is a weak alkali which dissociates partially to furnish.........OH^- ions which precipitate...........metal hydroxides.

8. Sodium zincate and water is obtained on reaction of ....with concentrated caustic soda.

9. .... and ....salt dissolve in sodium hydroxide.

10. Oxides and hydroxides of certain metals such as ...., ..., ... are amphoteric in nature.

11. Amphoteric oxides and hydroxides react with an.........to give salt and water only.

12. ...., a white precipitate is soluble in excess NH_4OH.

Ans.

1. colourless
2. light green
3. ammonium hydroxide
4. cations
5. white
6. white
7. insufficient, soluble
8. zinc oxide
9. Zinc, lead
10. zinc, lead, aluminium
11. acid as well an alkali
12. AgCl

Chapter 5. Mole Concept and Stoichiometry

1. An...........is the smaller unit of matter, which may or may not have an independent existence, but always takes part in a chemical reaction.

2. The relative molecular mass is a number that represents how many times one......... of a substance is heavier than $\frac{1}{12}$ th mass of carbon ${}_6C^{12}$.

3. The temperature at which all molecular motion ceases is...........

4. Whenever the gases react chemically, they do so in........... which bear a simple ratio to each other and to the products, if gaseous, provided the temperature and pressure of reacting gases and products remains the same.

5. Equal volumes of all.........under similar conditions of temperature and pressure, contain equal number of molecules.

6. The mass of substance containing particles equal to Avogadro's number is called.........

7. The number of atoms present in one molecule of a substance is its.........

8. The ratio of certain mass of a gas or vapour to the mass of same volume of hydrogen is its..........

9. The volume occupied by one mole of a gas at S.T.P is.........

Ans.

1. atom
2. molecule
3. absolute zero
4. volumes
5. gases
6. mole

7. atomicity
8. vapour density
9. gram molecular volume

Chapter 6. Electrolysis

1. Molten lead bromide conducts electricity. It is called an ....(1)....... . It is composed of lead ions and bromide ....(2)....... The lead ....(3)....... are ....(4)....... charged and are called ....(5)....... . The bromide ....(6)....... are ....(7)....... charged and are called ....(8)....... . During electrolysis, the lead ....(9)....... are attracted to the ....(10)....... charged electrode which is called ....(11)....... and the bromide. ....(12)....... are attracted towards the ....(13)....... charged electrode which is called ....(14)....... .

Ans. (1) electrolyte (2) ions (3) ions (4) positively (5) cations (6) ions (7) negatively (8) anions (9) ions (10) negatively (11) cathode (12) ions (13) positively (14) anode

2. Solid lead(II) bromide will not conduct an ....(1).......The ....(2).......are held in a rigid crystal ....(3)....... and are not free to move to the ....(4)....... . When the solid is ....(5)......., it ....(6).......and allows the passage of an electric current ....(7)....... is liberated at the cathode and ....(8).......at the anode. The decomposition of an electrolyte by an electric current is called ....(9)....... .

Ans. (1) Electric current (2) ions (3) lattice (4) electrodes (5) heated (6) melts (7) lead (8) bromine (9) electrolysis

3. For electroplating an article with nickel require an ....(1)....... which must be a solution containing ....(2)....... ions. The article to be plated is placed as the ....(3)....... of the cell on which the plating is carried out. The ....(4)....... of the cell is made from pure nickel. The ions which are attracted to the negative electrode and discharged are called ....(5)....... .

Ans. (1) electrolyte (2) nickel (3) cathode (4) anode (5) cations

4. (1) Electrolysis is the passage of .... through a liquid or a solution accompanied by a .... change.
(2) An electrically charged atom is called........... .
(3) The solution of a substance which conducts electricity is called ....
(4) An electrolyte is a.................. .
(5) A weak electrolyte is one which..................... .
(6) A strong electrolyte is one which...................... .
(7) Sodium chloride is................ .
(8) Water is........................ .
(9) .... is a non-electrolyte.
(10) Pure water consists almost entirely of ....
(11) We can expect that pure water.........normally conduct electricity.
(12) .... lead bromide does not conduct electricity.
(13) Hydrogen chloride ....when dissolved in water.
(14) A solution of hydrogen chloride gas in water conducts electricity because.................. but a solution of hydrogen chloride gas in toluene does not conduct an electric current because.............................. .
(15) Negative electrode is called the.......... .
(16) Cations migrate to............during electrolysis.
(17) During electrolysis, anions undergo.......................at the.................
(18) The reactions occurring at the cathode during electrolysis involve.......................
(19) Elements liberated at the anode during electrolysis are said to be.................
(20) In solution or in molten state a .... electrolyte consists almost entirely of ions.
(21) The................... the concentration of an ion in a solution, the greater is the probability of its being discharged at its appropriate electrode.

(22) In the electrolysis of acidulated water, oxygen is produced by the discharge of..............ions at the anode.

(23) Ionization is a .... process.

(24) The gas given off at cathode during the electrolysis of acidulated water is................ .

(25) With platinum electrodes, hydrogen is liberated at the .... and oxygen at the........... during the electrolysis of acidified water.

(26) As we descend the electrochemical series containing cations, the tendency of the cations to get .... at the cathode increases.

(27) The electrode where the current enters the electrolyte is called the..........

Ans. (1) electricity, chemical
(2) an ion
(3) an electrolyte
(4) good conductor of electricity
(5) is feebly ionized in the solution
(6) is completely ionized in the solution
(7) strong electrolyte
(8) a weak electrolyte
(9) Sugar
(10) molecules
(11) will not
(12) Solid
(13) ionizes
(14) it ionizes in solution to form free ions; it remains as a single molecule in solution.
(15) cathode
(16) cathode
(17) oxidation, anode
(18) reduction
(19) electronegative
(20) strong
(21) higher
(22) OH^-
(23) reversible
(24) hydrogen
(25) cathode, anode
(26) reduced
(27) anode

Chapter 7. Metallurgy

1.

	X	Y
Normal Electronic Configuration	2, 8, 7	2, 8, 2
Nature of oxide	Dissolves in water and turns blue litmus red	Very low solubility in water. Dissolves in hydrochloric acid
Tendency for oxidising and reducing reactions	Tends to oxidise elements and compounds	Tends to act as a reducing agent
Electrical and Thermal conductivity	Very poor electrical conductor Poor thermal conductivity	Good electrical conductor good thermal conductor
Tendency to form alloys and amalgams	No tendency to form alloys	Forms alloys

Using the information above, complete the following :

(1) ....is the metallic element.

(2) Metal atoms tend to have a maximum of .... electrons in the outermost energy level.

(3) Non-metallic elements tend to form ....oxides while metals tend to form ....oxides.

(4) Non-metallic elements tend to be ....conductors of heat and electricity.

(5) Metals tend to ....electrons and act as ....agents in their reactions with elements and compounds.

Ans. (1) Y
(2) three
(3) acidic, basic
(4) very poor
(5) lose, reducing

2. Fill in the blanks with a suitable word in the following paragraph :

In the smelting, the ore is heated ....(1)....... the melting point either along or with some ....(2)....... . In calcination, the ore is heated ....(3)....... the melting point and thus no ....(4)....... occurs in this process.

Similar to calcination, ...(5)....... involves heating at high temperature but chemical change occurs here. The ....(6)....... is used only when ore or impurity is ....(7)....... in nature.

Ans. (1) above (4) chemical change (7) magnetic.
(2) flux (5) roasting
(3) below (6) magnetic separation

3. X is an element in the form of a powder. X, burns in oxygen and the product is soluble in water. The solution is tested with litmus. Write down only the word, which will correctly complete each of the following sentences.

(1) If X is a metal, then the litmus will turn............
(2) If X is a non-metal, then the litmus will turn.............
(3) If X is a reactive metal, then..........will be evolved when X reacts with dilute sulphuric acid.
(4) If X is a metal it will form........oxide, which will form.........solution with water.

Ans. (1) blue (3) hydrogen
(2) red (4) basic, alkaline.

4. (1) The metal other than aluminium present both in magnalium and duralumin is.........
(2) The ore from which aluminium is extracted must first be treated with .... so that pure aluminium oxide can be obtained.
(3) Aluminum is an important constituent metal in duralumin since it is...........
(4) In a thermetic mixture, aluminium...........iron(III) oxide.
(5) Pure aluminium oxide is dissolved in .... to make a conducting solution.
(6) ....is a dark coloured crystalline solid.
(7) The divalent metal whose oxide is reduced to metal by electrolysis of its fused salt is.........
(8) Pine oil used in froth floatation process act as a ....
(9) In dry cells, the zinc container acts as an ....
(10) An .... is a homogeneous mixture of two or more metals or a metal and a ....
(11) The properties of an alloy are not necessarily .... between those of its .... The M.P. of an alloy is always less than the melting point of its constituent ....
(12) An alloy in which .... is present as one of the constituents is called ferrous alloy.
(13) The alloy that contains lead is ....
(14) The alloy of nickel and iron is known as...........
(15) An alloy which is sonorous is.........
(16) An alloy used for making cases for cartridges is ....
(17) The alloy used for making magnets is ....
(18) .... is a ferrous alloy.
(19) Brass is an alloy of ....
(20) Bell metal is an alloy of ....
(21) Type metal is an alloy of ....
(22) Addition of Tin to.........lowers the melting point of alloy solder.
(23) Alnico is a mixture of ....
(24) ....is used in machine parts due to its..............tensile strength.
(25) The non-metallic component in stainless steel is ....
(26) Carbon content of steel is ....
(27) Stainless steel contains ....

Ans. (1) magnesium (6) Iodine
(2) sodium hydroxide solution (7) magnesium
(3) light weight (8) water repellant
(4) reduces (9) anode
(5) cryolite (10) alloy, non-metal.

(11) intermediate, constituents, metals.
(12) iron
(13) solder
(14) invar
(15) bell metal
(16) brass
(17) alnico
(18) Manganese steel
(19) Cu and Zn
(20) Cu and Sn
(21) Pb, Sn and Sb
(22) lead
(23) Al, Ni and Co
(24) Steel, high
(25) carbon
(26) (0.5 to 1.5)% .
(27) Cr, Ni and C

Chapter 8. (a) Study of Compounds : Hydrogen Chloride

1. Hydrogen chloride occurs in a ....
2. Molecular formula of Hydrogen chloride is ....
3. HCl has the .... of 36.5.
4. It is a colourless .... gas at room temperature.
5. HCl is .... times heavier than air.
6. In laboratory HCl is prepared by the action of .... on H_2SO_4.
7. NaCl is used in preparation of HCl because it is ....
8. Formula weight of HCl is ....

Ans. 1. Free state
2. HCl
3. Molecular mass
4. Pungent
5. 1.28
6. NaCl
7. Abundant
8. 36.46

Chapter 8. (b) Study of Compounds : Ammonia and Nitric Acid

1. Most of the nitric acid today is manufactured by Ostwald's process. In this process a mixture of pure dry ammonia and air in the ratio of ....(1)....... volume is first compressed and then passed over ....(2)....... at about ....(3).......°C. This results in the oxidation of ammonia into ....(4)....... which combines with ....(5)....... from the air to give ....(6)....... . This is an acidic gas from which nitric acid can be obtained by simply dissolving in ....(7)....... .

Ans. (1) 10 : 1
(2) platinum gauze
(3) 800
(4) nitric oxide
(5) oxygen
(6) nitrogen dioxide
(7) water

2. Nitrogen and hydrogen combine in the presence of a catalyst to give ....(1)....... gas. When the above mentioned gas is passed through water it forms a solution which will be ....(2)....... in nature and the solution contains ....(3)....... ions and ....(4)....... ions. The above solution when added to iron(II) sulphate solution, gives a ....(5)....... coloured precipitate of iron(II) hydroxide.

Ans. (1) ammonia
(2) alkaline
(3) ammonium
(4) hydroxyl
(5) dirty green.

3. (1) Ammonia in liquid form is ....
(2) Ammonia gas is collected by ....
(3) Nitric acid is also called as..............
(4) The alkaline behaviour of liquor ammonia is due to the presence of ....
(5) ....is used as a catalyst during preparation of ammonia.
(6) Liquor ammonia fortis is a saturated solution of ....
(7) ....turns turmeric paper brown.
(8) The salt solution which gives white ppt. on the addition of ammonium hydroxide solution is ....
(9) Excess of ammonia reduces chlorine to ....

(10) The gas most difficult to liquefy is..................

(11) When ammonium chloride is heated, it undergoes ....

(12) Heating ammonium chloride with sodium hydroxide produces ....

(13) Heating a solution of ammonium chloride and sodium nitrite produces ....

(14) Cold, dilute nitric acid reacts with copper to form...........

(15) The reaction of ....with dil. nitric acid is an example of neutralisation reaction.

(16) Lead nitrate is a...........salt of nitric acid.

(17) The mineral acid obtained from conc. nitric acid on reaction with a non-metal is.........

(18) A nitrate which leaves a black residue on heating is...........

(19) The nitrate which on heating melts and liberates only one neutral gas is ....

(20) The oxidised product obtained on reaction with H_2S gas and dil. HNO_3 is ....

(21) Ammonium nitrate is used in preparing ....

(22) Nitric acid is manufactured by .... process.

Ans. (1) neutral
(2) a downward displacement of air
(3) aqua fortis
(4) OH^- ions
(5) Molybdenum
(6) NH_4OH
(7) Ammonia
(8) magnesium chloride
(9) NH_4Cl
(10) hydrogen
(11) thermal dissociation
(12) ammonia
(13) nitrogen
(14) nitric oxide
(15) CaO
(16) normal
(17) H_2SO_4
(18) $Cu(NO_3)_2$
(19) $NaNO_3$
(20) sulphur
(21) explosives
(22) Ostwald's

Chapter 8. (c) Study of Compounds : Sulphuric Acid

1. The conversion of sulphur dioxide into sulphur trioxide is an ....(1)....... therefore, ...(2)....... during the reaction. The reaction is ....(3)....... by ....(4)....... . At low temperature, the reaction is ....(5)....... so an ....(6)....... of 500° C and a catalyst is needed, the reaction is accompanied by ....(7)....... in volume and ....(8)....... pressure ....(9).......the yield of sulphur trioxide.

Ans. (1) exothermic reaction
(2) heat is evolved
(3) favoured
(4) low temperature
(5) slow
(6) optimum temperature
(7) decrease
(8) high
(9) favours.

2. (1) Sulphuric acid is a.........acid.

(2) The physical property of concentrated sulphuric acid that makes it valuable in preparing other acids is its ....

(3) Sulphuric acid absorbs sulphur trioxide and forms..............

(4) The acid anhydride of sulphuric acid is ....

(5) When hot concentrated sulphuric acid is added to sulphur, it gets oxidised to...............

(6) The reduced product obtained when hydrogen sulphide reacts with conc. sulphuric acid is...........

(7) The salt which reacts with dil. sulphuric acid to form insoluble ppt is...........

(8) The type of salt formed when excess of caustic soda reacts with sulphuric acid is...........

(9) The gas produced by the action of concentrated sulphuric acid on sodium chloride is............

(10) The dehydrated product obtained when sugar reacts with conc. sulphuric acid is........

(11) The chemical formula of oxalic acid is...........

(12) The sulphur compound used for the identification of several cations is ....

(13) An explosive prepared by using sulphuric acid is.........

(14) Sulphuric acid used for making fertilizers is made by the .... process.

Ans. (1) non-volatile (2) high boiling point (3) pyrosulphuric acid (4) sulphur trioxide (5) sulphur dioxide, SO_2 (6) sulphur (7) Lead nitrate $[Pb(NO_3)_2]$ (8) normal salt (9) HCl (10) carbon (11) $C_2H_2O_4$ (12) hydrogen sulphide (13) trinitrotoluene (14) lead-chamber

Chapter 9. Organic Chemistry

1. The alkanes form an ....(1)....... series with the general formula ....(2)............ . The alkanes are ....(3)............. ....(4).......... which generally undergo ....(5).............. reactions.

Ans. (1) homologous (2) C_nH_{2n+2} (3) saturated (4) hydrocarbons (5) substitution

2. (1) .... was the first organic compound prepared in laboratory.
(2) Organic compounds are generally insoluble in ....
(3) The compounds of carbon and hydrogen are called ....
(4) The hydrocarbon containing only single bonds are known as ....
(5) Alkanes are .... hydrocarbons.
(6) Alkanes are open-chain hydrocarbons in which the carbon atoms are joined by.................. only.
(7) The four C—H bonds in the methane molecule are directed towards the corners of a regular ....
(8) Methane can be prepared by heating .... with solid ....
(9) When methane reacts with excess of chlorine, the product obtained is .... of the molecular formula ....
(10) Ethane is an ....
(11) Ethene reacts with chlorine to give.............. products.
(12) ....does not react with hydrogen.
(13) When methane is heated to a high temperature in the absence of oxygen, it yields .... and .... This reaction is known as .... or ....
(14) The general formula C_nH_{2n} represents ....
(15) Ethene is prepared by the .... of ethyl alcohol by heating it with ....
(16) Ethene reacts with chlorine to give .... the molecular formula of which is ....
(17) .... has two carbon atoms joined by a triple covalent bond.
(18) Ethene and ethyne are................... hydrocarbons.
(19) .... reacts with bromine to give two different addition products.
(20) Ethyne is prepared by the action of .... on ....
(21) When ethyne is hydrogenated, the first product is .... and the final product is ....
(22) The conversion of ethene to ethane is an example of ....
(23) The catalyst used in the conversion of ethene to ethane is commonly ....
(24) The product formed when ethene gas reacts with water in the presence of sulphuric acid is...........
(25) The product of the dehydration of ethyl alcohol is ....
(26) Converting ethanol to ethene requires the use of ....
(27) .... is commonly called as wood spirit.
(28) When acetaldehyde is oxidized with acidified potassium dichromate, it forms......
(29) Ethanoic acid reacts with ethanol in presence of concentrated H_2SO_4, so as to form a compound and water. The chemical reaction which takes place is called.........
(30) The ability of carbon atom to link with other carbon atom is known as ....
(31) Compounds represented by a single molecular formula but having different structural formulae are called .... and this phenomenon is known as ....
(32) The name of the compound CH_2Cl_2 is ....
(33) The compound formed when ethene reacts with hydrogen is............
(34) Conversion of ethene to ethane is an example of............

Ans. (1) Urea
(2) water
(3) hydrocarbons
(4) alkanes
(5) saturated
(6) single covalent bonds
(7) tetrahedron
(8) sodium acetate, sodium hydroxide
(9) carbon tetrachloride, CCl_4
(10) alkane
(11) addition
(12) Ethane
(13) carbon, hydrogen, pyrolysis, cracking
(14) alkene
(15) dehydration, concentrated sulphuric acid
(16) ethene dichloride, $C_2H_4Cl_2$
(17) Ethyne
(18) unsaturated
(19) Ethyne
(20) water, calcium carbide
(21) ethene, ethane
(22) hydrogenation
(23) nickel
(24) ethanol
(25) ethene
(26) concentrated sulphuric acid
(27) Methyl alcohol or methanol
(28) acetic acid
(29) esterification
(30) catenation
(31) isomers, isomerism
(32) dichloromethane
(33) ethane
(34) hydrogenation

6. Copy and complete the following paragraph using the options given in brackets :*
Alkenes are a homologous series of (i) .... (saturated / unsaturated) hydrocarbons characterised by the general formula (ii) .... (C_nH_{2n+2}/C_nH_{2n}). Alkenes undergo (iii) .... (addition/ substitution) reactions and also undergo (iv) .... (hydrogenation / dehydrogenation) to form alkanes.

Ans. (i) unsaturated (ii) C_nH_{2n} (iii) addition (iv) hydrogenation

❑

* **Frequently asked previous years Board Exam Questions.**

Match the Column

Set 3

Match the column A with column B :

Chapter 1. Periodic Properties and Variations of Properties

1.

Column A	Column B
(1) Proton	An alkaline earth metal
(2) Sodium	Halogen
(3) Barium	Noble gas
(4) Chlorine	An alkali metal
(5) Electron	Responsible for nuclear charge
(6) Completed shell	Occupied sub-shell.

Ans.

Column A	Column B
(1) Proton	Responsible for nuclear charge
(2) Sodium	An alkali metal
(3) Barium	An alkaline earth metal
(4) Chlorine	Halogen
(5) Electron	Occupied sub-shell
(6) Completed shell	Noble gas

Chapter 2. Chemical Bonding

1.

Column A	Column B
(1) Metal	Methane (CH_4)
(2) Covalent compound	Sodium chloride (NaCl)
(3) Non-polar	Selenium (Se)
(4) Polar	Chloride ion (Cl^-)
(5) Non-metal	Aluminium (Al)
(6) Electrovalent compound	Carbon dioxide (CO_2)
(7) Anion	Hydrogen chloride (HCl)
(8) Cation	Potassium ion (K^+)

Ans.

Column A	Column B
(1) Metal	Aluminium (Al)
(2) Covalent compound	Carbon dioxide (CO_2)
(3) Non-polar	Methane (CH_4)
(4) Polar	Hydrogen chloride (HCl)
(5) Non-metal	Selenium (Se)
(6) Electrovalent compound	Sodium chloride (NaCl)
(7) Anion	Chloride ion (Cl^-)
(8) Cation	Potassium ion (K^+)

Chapter 3. Study of Acids, Bases and Salts

1.

Column A	Column B
(1) Acid salt	Ferrous ammonium sulphate
(2) Double salt	Contains only ions
(3) Ammonium hydroxide solution	Sodium hydrogen sulphate
(4) Dilute hydrochloric acid	Contains only molecules
(5) Carbon tetrachloride	Contains ions and molecules

Ans.

Column A	Column B
(1) Acid salt	Sodium hydrogen sulphate
(2) Double salt	Ferrous ammonium sulphate
(3) Ammonium hydroxide solution	Contains ions and molecules
(4) Dilute hydrochloric acid	Contains only ions
(5) Carbontetrachloride	Contains only molecules

2.

Column A	Column B
(1) Acid salt	Sodium potassium carbonate
(2) Mixed salt	Alum
(3) Complex salt	Sodium carbonate
(4) Double salt	Sodium zincate
(5) Normal salt	Sodium hydrogen carbonate

Ans.

Column A	Column B
(1) Acid salt	Sodium hydrogen carbonate
(2) Mixed salt	Sodium potassium carbonate
(3) Complex salt	Sodium zincate
(4) Double salt	Alum
(5) Normal salt	Sodium carbonate

Chapter 4. Analytical Chemistry

1.

Column A	Column B
(1) Copper(II) nitrate	Green
(2) Iron(II) sulphate	White
(3) Magnesium chloride	Pink
(4) Cobalt chloride	Orange
(5) Iron(III) chloride	Blue
(6) Potassium dichromate	Black
(7) Copper(II) oxide	Yellow

Ans.

Column A	Column B
(1) Copper (II) nitrate	Blue
(2) Iron (II) sulphate	Green
(3) Magnesium chloride	White
(4) Cobalt chloride	Pink
(5) Iron (III) chloride	Yellow
(6) Potassium dichromate	Orange
(7) Copper (II) oxide	Black

2.

Column A	Column B
(1) Pb^{2+}	Reddish brown
(2) Fe^{2+}	White insoluble in excess
(3) Zn^{2+}	Dirty green
(4) Fe^{3+}	White soluble in excess
(5) Cu^{2+}	White insoluble in excess
(6) Ca^{2+}	Blue

Ans.

Column A	Column B
(1) Pb^{2+}	White insoluble in excess
(2) Fe^{2+}	Dirty green
(3) Zn^{2+}	White soluble in excess
(4) Fe^{3+}	Reddish brown
(5) Cu^{2+}	Blue
(6) Ca^{2+}	White insoluble in excess

3.

Column A	Column B
(1) $Pb(NO_3)_2$ from PbO	Simple displacement
(2) $MgCl_2$ from Mg	Titration
(3) $FeCl_3$ from Fe	Neutralization
(4) $NaNO_3$ from NaOH	Precipitation
(5) $ZnCO_3$ from $ZnSO_4$	Combination

Ans.

Column A	Column B
(1) $Pb(NO_3)_2$ from PbO	Neutralization
(2) $MgCl_2$ from Mg	Simple displacement
(3) $FeCl_3$ from Fe	Combination
(4) $NaNO_3$ from NaOH	Titration
(5) $ZnCO_3$ from $ZnSO_4$	Precipitation

Chapter 7. Metallurgy

1.

Column A	Column B
(1) Duralumin	Electrical fuse
(2) Solder	Surgical instrument
(3) Brass	Aircraft body
(4) Stainless Steel	Decorative articles

Ans.

Column A	Column B
(1) Duralumin	Aircraft body
(2) Solder	Electrical fuse
(3) Brass	Decorative articles
(4) Stainless Steel	Surgical instruments

Multiple Choice Questions | Set 4 |

Choose the correct answer out of the four available choices (a), (b), (c) and (d) given under each question :

Chapter 1. Periodic Properties and Variations of Properties

1. In the periodic table, alkali metals are placed in the group :
(a) 1 (b) 11 (c) 17 (d) 18

2. An alkaline earth metal is :
(a) Palarium (b) Calcium (c) Lead (d) Copper

3. On moving from left to right across a period of the periodic table, the atomic size :
(a) Decreases
(b) Increases
(c) Remains the same
(d) Sometimes increases and sometimes decreases

4. On moving from left to right across a period of the periodic table, the non-metallic character of the elements :
(a) Decreases (c) Remains the same
(b) Increases (d) Depends on the period

5. On moving from left to right across a period of the periodic table, the ionization potential :
(a) Goes up and down (c) Increases
(b) Decreases (d) Remains the same

6. Ionisation potential increases over a period from left to right because the :
(a) Atomic radius increases and nuclear charge increases
(b) Atomic radius decreases and nuclear charge decreases
(c) Atomic radius increases and nuclear charge decreases
(d) Atomic radius decreases and nuclear charge increases

7. On moving from left to right across a period of the periodic table, the electron affinity of the elements in groups 1 to 7 :
(a) Goes up and then down (c) Increases
(b) Decreases and then increases (d) Decreases

8. An element in period 3 whose electron affinity is zero :
(a) Neon (b) Sulphur (c) Sodium (d) Argon

9. Among the elements of period 2 the element which has high electron affinity is :
(a) Lithium (b) Carbon (c) Chlorine (d) Fluorine

10. On moving from left to right across a period of the periodic table, the electronegativity :
(a) Depends on the number of valence electrons (c) Decreases
(b) Remains the same (d) Increases

11. Among the elements given below, the element with the least electronegativity is :
(a) Lithium (b) Carbon (c) Boron (d) Fluorine

12. The element with highest ionization potential, is :*
(a) Hydrogen (b) Caesium (c) Radon (d) Helium

13. The most electronegative element from the following elements is :*

* **Frequently asked previous years Board Exam Questions.**

(a) Magnesium (b) Chlorine (c) Aluminium (d) Sulphur

14. The number of electrons present in the valence shell of halogen is :
(a) 1 (b) 3 (c) 5 (d) 7

15. If an element A belongs to period 3 and Group II then it will have :
(a) 3 shells and 2 valence electrons
(c) 3 shells and 3 valence electrons
(b) 2 shells and 3 valence electrons
(d) 2 shells and 2 valence electrons

16. An element with the atomic number 19 will most likely combine chemically with the element whose atomic number is :
(a) 17 (b) 11 (c) 18 (d) 20

Ans. (1) (a) (4) (b) (7) (c) (10) (d) (13) (b) (16) (a)
(2) (b) (5) (c) (8) (d) (11) (a) (14) (d)
(3) (a) (6) (d) (9) (d) (12) (d) (15) (a)

Chapter 2. Chemical Bonding

1. The number of electrons lost or gained by an atom refers to :
(a) Electrovalency
(b) Covalency
(c) Donation
(d) Acceptance

2. The capacity of an atom to attract the shared pair of electrons towards itself is called :
(a) Electronegativity
(c) Sharing of electrons
(b) Electron affinity
(d) Electron donation

3. Sodium and sodium ions :
(a) Are chemically same
(c) Have same number of protons
(b) Have same number of electrons
(d) None of these

4. The most ionic compound of periodic table is :
(a) Sodium chloride
(c) Magnesium chloride
(b) Potassium chloride
(d) Caesium chloride

5. A compound having one lone pair of electrons :
(a) Water
(c) Ammonia
(b) Methane
(d) Hydrogen sulphide

6. A compound X consists of only molecules. Hence X will have :
(a) A crystalline hard structure
(b) A low melting point and low boiling point
(c) An ionic bond
(d) A strong force of attraction between its molecules

7. Molecular reactions which are generally slow reactions are shown by :
(a) Covalent compounds
(c) Coordinate compounds
(b) Ionic compounds
(d) Both ionic and covalent compounds

8. Which of the following doesn't represent oxidation ?
(a) Loss of electrons
(c) Increase in oxidation number
(b) Addition of oxygen
(d) Addition of hydrogen

9. A compound which contains all the three types of bonds *i.e.*, ionic, covalent and coordinate is :
(a) Sodium chloride
(c) Calcium chloride
(b) Ammonia
(d) Ammonium chloride

10. The molecule containing a triple covalent bond is :
(a) Ammonia
(c) Water
(b) Methane
(d) Nitrogen

11. Bonding in this molecule can be understood to involve coordinate bonding :

(a) Carbon tetrachloride
(b) Hydrogen
(c) Hydrogen chloride
(d) Ammonium chloride

12. Which of the following is a common characteristic of a covalent compound ?

(a) High melting point
(b) Consists of molecules
(c) Always soluble in water
(d) Conducts electricity when it is in the molten state

Ans. (1) (a) (2) (a) (3) (c) (4) (d) (5) (c) (6) (b) (7) (d) (8) (d) (9) (d) (10) (d) (11) (d) (12) (b)

Chapter 3. Study of Acids, Bases and Salts

1. A particular solution contains molecules and ions of the solute so it is a :

(a) Weak acid (b) Strong acid (c) Strong base (d) Salt solution

2. Select the acid which contains four hydrogen atoms in it :

(a) Formic acid (b) Sulphuric acid (c) Nitric acid (d) Acetic acid

3. An organic weak acid is :

(a) Formic acid
(b) Sulphuric acid
(c) Nitric acid
(d) Hydrochloric acid

4. The aqueous solution of the following compounds which contains both ions and molecules is :

(a) Sulphuric acid
(b) Hydrochloric acid
(c) Nitric acid
(d) Acetic acid

5. An acid which is not a hydracid is :

(a) H_2S (b) H_2SO_3 (c) HBr (d) HCl

6. Which one of the following will not produce an acid when made to react with water ?

(a) Carbon monoxide
(b) Carbon dioxide
(c) Nitrogen dioxide
(d) Sulphur trioxide

7. To increase the pH value of neutral solution, we should add :

(a) An acid (b) An acid salt (c) An alkali (d) A salt

8. The metal oxide which can react with acid as well as alkali is :

(a) Silver oxide
(b) Copper(II) oxide
(c) Aluminium oxide
(d) Calcium oxide

9. During ionisation metals lose electrons, this change can be called :

(a) Oxidation (b) Reduction (c) Redox (d) Displacement

10. An example of a complex salt is :

(a) Zinc sulphate
(b) Sodium hydrogen sulphate
(c) Iron(II) ammonium sulphate
(d) Tetrammine copper(II) sulphate

11. The salt which in solution gives a pale green precipitate with sodium hydroxide solution and a white precipitate with barium chloride solution is :

(a) Iron(III) sulphate
(b) Iron(II) sulphate
(c) Iron(II) chloride
(d) Iron(III) chloride

12. The acid which can produce carbon from cane sugar, is :*

(a) Concentrated Hydrochloric acid
(b) Concentrated Nitric acid
(c) Concentrated Sulphuric acid
(d) Concentrated Acetic acid

Ans. (1) (a) (2) (d) (3) (a) (4) (d) (5) (b) (6) (a) (7) (c) (8) (c) (9) (a) (10) (d) (11) (b) (12) (c)

* Frequently asked previous years Board Exam Questions.

Chapter 4. Analytical Chemistry

1. Salts of which elements are generally coloured :
(a) Transition (b) Normal (c) Lanthanides (d) Inner-transition.
2. Which one of the following salt solutions on reaction with excess of ammonium hydroxide solution gives a deep blue solution ?
(a) $FeCl_3$ (aq) (b) $CuSO_4$ (aq) (c) $Al_2(SO_4)_3$ (aq) (d) $ZnSO_4$ (aq)
3. Which one of the following salt solutions on reaction with excess sodium hydroxide solution gives a clear solution finally ?
(a) $Pb(NO_3)_2$ (aq) (b) $CuSO_4$ (aq) (c) $FeCl_3$ (aq) (d) $ZnSO_4$ (aq)
4. The precipitate of which of the following compounds is soluble in excess of ammonia solution ?
(a) Iron(II) chloride (c) Copper(II) sulphate
(b) Magnesium chloride (d) Lead nitrate
5. Which one of the following salt solutions on reaction with excess of ammonium hydroxide solution results finally in dissolution of the precipitate first formed ?
(a) $AlCl_3$ (aq) (b) $FeSO_4$ (aq) (c) $Fe(SO_4)_3$ (aq.) (d) $ZnSO_4$ (aq.)
6. Hydroxide of this metal is soluble in sodium hydroxide solution :
(a) Magnesium (b) Lead (c) Silver (d) Copper
7. The hydroxide which is soluble in excess of NaOH is :
(a) $Zn(OH)_2$ (b) $Fe(OH)_2$ (c) $Fe(OH)_3$ (d) $Al(OH)_3$
8. Name the reagent from the following which can be used to distinguish zinc nitrate solution from magnesium nitrate.
(a) NH_4OH (aq) (b) NaOH (aq) (c) $BaCl_2$ (d) H_2SO_4
9. The oxide and hydroxide of which metal is amphoteric :
(a) Zinc (b) Copper (c) Iron (d) Manganese
10. Anhydrous iron(III) chloride is prepared by :
(a) Direct combination (c) Decomposition
(b) Simple displacement (d) Neutralization
11. A chloride which forms a precipitate that is soluble in excess of ammonium hydroxide is :
(a) Calcium chloride (c) Ferric chloride
(b) Ferrous chloride (d) Copper chloride

Ans. (1) (a) (3) (a) (5) (d) (7) (a) (9) (a) (11) (d)
(2) (b) (4) (c) (6) (b) (8) (a) (10) (a)

Chapter 5. Mole Concept and Stoichiometry

1. The temperature at which all molecular motion ceases is :
(a) 0°C (b) 100°C (c) Zero power (d) Absolute zero
2. The volume occupied by 1 mole of gas at STP is :
(a) 2·24 litres (b) 22·4 litres (c) 2·42 litres (d) 2·44 litres
3. The volume occupied by 7 g of nitrogen at STP is :
(a) 5·4 litres (b) 4·8 litres (c) 5·6 litres (d) 4·6 litres
4. Which law states that "under the same conditions of temperature and pressure, equal volumes of all gases contain the same number of molecule."
(a) Avogadro's law (c) Mole law
(b) Gay-Lussac's law (d) Law of conservation of mass
5. What is the value of Avogadro number ?
(a) $6{\cdot}022 \times 10^{23}$ (b) $6{\cdot}022 \times 10^{-23}$ (c) $6{\cdot}22 \times 10^{23}$ (d) $6{\cdot}22 \times 10^{-23}$
6. One atomic mass unit is how much part the mass of carbon-12 atoms ?
(a) $\frac{1}{4}$ (b) $\frac{1}{12}$ (c) $\frac{1}{8}$ (d) $\frac{1}{16}$

7. The number of atoms present in one molecule of an element is called its :
   (a) Molecular number
   (b) Atomic number
   (c) Avogadro's number
   (d) Atomicity
8. The vapour density of carbon dioxide [C = 12, O = 16] is :
   (a) 12 (b) 16 (c) 44 (d) 22
9. The empirical formula of hexane is :
   (a) C_2H_7 (b) C_5H_8 (c) C_3H_7 (d) C_4H_7
10. If empirical formula of an organic compound is CH_2O then its molecular formula can be :
   (a) $C_2H_2O_2$ (b) C_2H_4O (c) C_3H_6O (d) $C_6H_{12}O_6$
11. Which of the following would weight the least?
   (a) 2g atoms of nitrogen
   (b) 1 mole of silver
   (c) 22.4 litres of oxygen gas at 1 atmospheric pressure and 273 K
   (d) $6.02 \times 10_{23}$ atoms of carbon
   [Atomic masses : Ag = 108, N = 14, O = 16, C = 12]
12. The ratio between the number of molecules in 2 g of hydrogen and 32 g of oxygen is :
   (a) 1 : 2 (b) 1 : 0.01 (c) 1 : 1 (d) 0.01 : 1
   Given that H = 1, O = 16]

Ans. (1) (d) (3) (c) (5) (a) (7) (d) (9) (c) (11) (d)
(2) (b) (4) (a) (6) (b) (8) (d) (10) (d) (12) (c)

Chapter 6. Electrolysis

1. Identify the weak electrolyte from the following :
   (a) Sodium chloride solution
   (b) Dilute hydrochloric acid
   (c) Dilute sulphuric acid
   (d) Aqueous acetic acid
2. The particles present in strong electrolytes are :
   (a) only molecules
   (b) mainly ions
   (c) ions and molecules
   (d) only atoms
3. Which of these will act as a non-electrolyte ?
   (a) Liquid carbon tetrachloride
   (b) Acetic acid
   (c) Sodium hydroxide aqueous solution
   (d) Potassium chloride aqueous solution
4. During ionisation metals lose electrons, this change can be called :
   (a) Oxidation (b) Reduction (c) Redox (d) Displacement
5. The metallic electrode which does not take part in an electrolytic reaction ? (Inert electrode)
   (a) Cu (b) Ag (c) Pt (d) Ni
6. When dilute sodium chloride is electrolysed using graphite electrodes, the cation is discharged at the cathode most readily.
   (a) Na^+ (b) OH^- (c) H^+ (d) Cl^-
7. During electrolysis of NaCl, the gas discharged at the anode is :
   (a) Chlorine
   (b) Oxygen
   (c) Hydrogen
   (d) None of the above
8. A compound which liberates reddish brown gas around the anode during electrolysis in its molten state is :
   (a) Sodium chloride
   (b) Copper (II) oxide
   (c) Copper(II) sulphate
   (d) Lead (II) bromide
9. When fused lead bromide is electrolysed we observe :
   (a) a silver grey deposit at anode and a reddish brown deposit at cathode
   (b) a silver grey deposit at cathode and a reddish brown deposit at anode

(c) a silver grey deposit at cathode and reddish brown fumes at anode
(d) silver grey fumes at anode and reddish brown fumes at cathode.

10. The vessel in which electrolysis of lead bromide is carried out is :
(a) Clay crucible (c) Silica crucible
(b) Glass vessel (d) Aluminium vessel

11. The ion which is discharged at the cathode during the electrolysis of copper sulphate solutions using copper electrodes as anode and cathode ?
(a) Cu^{2+} (b) OH^- (c) SO_4^{2-} (d) H^+

12. An aqueous electrolyte consists of the ions mentioned in the list, the ion which could be discharged most readily during electrolysis ?
(a) Fe^{2+} (b) Cu^{2+} (c) Pb^{2+} (d) H^+

13. During the electrolysis of molten lead bromide which of the following takes place :
(a) Bromine is released at the cathode (c) Bromine ions gain electrons
(b) Lead is deposited at the anode (d) Lead is deposited at the cathode

14. An electrolyte which completely dissociates into ions is :*
(a) Alcohol (c) Sucrose
(b) Carbonic acid (d) Sodium hydroxide

15. During silver plating of an article using potassium argentocyanide as an electrolyte, the anode material should be :
(a) Cu (b) Ag (c) Pt (d) Fe

16. The electrolysis of acidified water is an example of :*
(a) Reduction (b) Oxidation (c) Redox reaction (d) Synthesis

Ans. (1) (d) (4) (a) (7) (a) (10) (c) (13) (d) (16) (c)
(2) (b) (5) (c) (8) (d) (11) (a) (14) (d)
(3) (a) (6) (c) (9) (c) (12) (b) (15) (b)

Chapter 7. Metallurgy

1. An element of an inorganic compound found in nature is known as :
(a) Mineral (b) Ore (c) Both (a) and (b) (d) None of these

2. Which one of the following is not true for metals ?
(a) Metals are good conductors of electricity.
(b) Metals are malleable and ductile.
(c) Metals form non-polar covalent compounds.
(d) Metal will have 1 or 2 or 3 electrons in their valence shell.

3. Which ore is metalloid ?
(a) C (b) Ge (c) Si (d) Sn

4. A chemical process of extracting a metal from its ore is known as :
(a) Mineralogy (b) Metallurgy (c) Liquation (d) None of these

5. A mineral from which the metal is extracted economically is known as :
(a) Matrix (b) Gangue (c) Ore (d) None of these

6. An unwanted earthly material associated with tin ore as impurity is known as :
(a) Gangue (b) Flux (c) Froth (d) None of these

7. The process of heating the ore strongly in excess of air so that the volatile impurities are removed and the ore is changed to oxide is known as :
(a) Calcination (b) Roasting (c) Froth floatation (d) Leaching

8. Heating an ore in a limited supply of air or in the absence of air at a temperature just below its melting point is known as :
(a) Smelting (b) Ore dressing (c) Calcination (d) Bessemerisation

***** **Frequently asked previous years Board Exam Questions.**

9. Smelting is carried out in :
   (a) Blast furnace
   (b) Muffle furnace
   (c) Open Heat furnace
   (d) Electric furnace
10. The salt which is least likely to be found in minerals is :
   (a) Chloride (b) Sulphate (c) Sulphite (d) Nitrate
11. The commonest method of extraction of metals from oxide ores involves :
   (a) Reduction with carbon
   (b) Reduction with aluminium
   (c) Reduction with hydrogen
   (d) Electrolytic method
12. Froth floatation process for the concentration of ores is an illustration of the practical application of :
   (a) Adsorption (b) Absorption (c) Sedimentation (d) Coagulation
13. In the froth floatation process for the purification of ores, the ore particles float because :
   (a) They are light
   (b) Their surface is not easily wetted by water
   (c) They bear electrostatic charge
   (d) They are insoluble
14. Froth floatation method may be used to increase the concentration of mineral in :
   (a) Chalcopyrites (b) Bauxite (c) Haematite (d) Calamine
15. Sulphide ore is generally concentrated by :
   (a) Roasting
   (b) Froth floatation process
   (c) Reduction by carbon
   (d) Tempering
16. The main ore used for the extraction of iron is :
   (a) Haematite (b) Calamine (c) Bauxite (d) Cryolite
17. The process used to convert impure alumina to pure alumina is :
   (a) Roasting
   (b) Electrolytic refining
   (c) Purification
   (d) Bayer's process
18. The part of fluorspar (CaF_2) which is added in small quantities in the electrolytic reduction of alumina dissolved in fused cryolite (Na_3AlF_6) is :
   (a) As a catalyst
   (b) To make the fused mixture very conducting
   (c) To lower the temperature of the melt
   (d) To decrease the rate of oxidation of carbon at the anode
19. The electrolyte used for electroplating an article with silver is :
   (a) Silver nitrate solution
   (b) Silver cyanide solution
   (c) Sodium argentocyanide solution
   (d) Nickel sulphate solution
20. Aluminium powder is used in thermite welding because :
   (a) It is a strong reducing agent
   (b) It is a strong oxidising agent
   (c) It is corrosion resistant
   (d) It is a good conductor of heat
21. The metals zinc and tin are present in the alloy :
   (a) Solder (b) Brass (c) Bronze (d) Duralumin
22. In electrolytic refining of metals the impure metal is made :
   (a) Cathode (b) Salt bridge (c) Anode (d) Electrode
23. Stainless steel does not contain :
   (a) Cr (b) Al (c) C (d) Ni
24. This is not an alloy of copper :
   (a) Brass (b) Bronze (c) Solder (d) Duralumin
25. The two main metals in bronze are :
   (a) Copper and zinc
   (b) Copper and lead
   (c) Copper and nickel
   (d) Copper and tin
26. The reason for using aluminium in the alloy duralumin is :*
   (a) Aluminium is brittle.
   (b) Aluminium gives strength.
   (c) Aluminium brings lightness.
   (d) Aluminium lowers melting point.

* **Frequently asked previous years Board Exam Questions.**

Ans. (1) (a) (2) (c) (3) (b) (4) (b) (5) (c) (6) (a) (7) (b) (8) (c) (9) (a) (10) (d) (11) (a) (12) (a) (13) (b) (14) (a) (15) (b) (16) (a) (17) (d) (18) (c) (19) (c) (20) (a) (21) (c) (22) (c) (23) (b) (24) (c) (25) (d) (26) (c)

Chapter 8. (a) Study of Compounds : Hydrogen Chloride

1. An aqueous solution of HCl gas is named :
   (a) Aqua fortis (b) Aqua regia (c) Oil of vitriol (d) Muriatic acid
2. An acid which is not a monobasic acid ?
   (a) HNO_3 (b) HCOOH (c) H_2SO_4 (d) HCl
3. An acid which is not an oxidising agent ?
   (a) H_2SO_4 (b) HCl (c) HNO_3 (d) CH_3COOH
4. Hydrogen chloride gas being highly soluble in water is dried by :
   (a) Anhydrous calcium chloride (c) Quick lime
   (b) Phosphorus penta oxide (d) Concentrated sulphuric acid
5. A substance which reacts with conc. HCl to liberate chlorine :
   (a) PbO (b) $PbCl_2$ (c) PbO_2 (d) $Pb(NO_3)_2$
6. A metal which reacts with dil HCl to liberate hydrogen ?
   (a) Zn (b) Cu (c) Ag (d) Hg
7. Gas liberated when dil. HCl gas is added to iron(II) sulphide.
   (a) Hydrogen gas (c) Hydrogen sulphide gas
   (b) Chlorine gas (d) Carbon dioxide gas
8. Gas liberated when hydrochloric acid reacts with galena.
   (a) Sulphur dioxide (c) Nitrogen
   (b) Carbon dioxide (d) Hydrogen sulphide
9. The gases which react chemically to form a solid ?
   (a) $H_{2(g)}$ and Cl_{2} (g) (c) $CO_{2(g)}$ and $SO_{2(g)}$
   (b) $NH_{3(g)}$ and $HCl_{(g)}$ (d) $NO_{2(g)}$ and $CO_{2(g)}$
10. Constant boiling mixtures are known as :
   (a) Constant compounds (c) Distillators
   (b) Woulfe's compound (d) Azeotropes
11. Aqua regia is a mixture of :
   (a) Dilute hydrochloric acid and concentrated nitric acid
   (b) Concentrated hydrochloric acid and dilute nitric acid
   (c) Concentrated hydrochloric acid [1 part] and concentrated nitric acid [3 parts]
   (d) Concentrated hydrochloric acid [3 parts] and concentrated nitric acid [1 part]
12. The aim of the fountain experiment is to prove that :
   (a) HCl turns blue litmus red (c) HCl is highly soluble in water
   (b) HCl is denser than air (d) HCl fumes in moist air
13. The drying agent used to dry HCl gas is :*
   (a) Conc. H_2SO_4 (b) ZnO (c) Al_2O_3 (d) CaO

Ans. (1) (d) (2) (c) (3) (b) (4) (d) (5) (c) (6) (a) (7) (c) (8) (d) (9) (b) (10) (d) (11) (d) (12) (c) (13) (a)

* Frequently asked previous years Board Exam Questions.

Multiple Choice Questions

Chapter 8. (b) Study of Compounds : Ammonia and Nitric Acid

1. Ammonia can be obtained by adding water to :
(a) Ammonium chloride
(b) Ammonium nitrite
(c) Magnesium nitride
(d) Magnesium nitrate

2. Ammonia is soluble in water because it is :
(a) A polar molecule
(b) An acid
(c) A base
(d) A simple covalent compound

3. Nitrogen gas can be obtained by heating :
(a) Ammonium nitrate
(b) Ammonium nitrite
(c) Magnesium nitride
(d) Ammonium chloride

4. Rotten egg smell is due to the liberation of :
(a) HCl gas (b) Cl_2 gas (c) H_2S gas (d) SO_2 gas

5. The temperature at which catalytic oxidation of ammonia is carried out :
(a) 200°C (b) 800°C (c) 1000°C (d) 500°C

6. With excess of chlorine, NH_3 forms :
(a) NH_4Cl (b) NCl_3 (c) NOCl (d) N_2Cl

7. Liquid NH_3 is employed in refrigerator, because :
(a) It is more basic
(b) It is stable hydride
(c) It has a high dipole moment
(d) It has high heat of evaporation

8. Ammonia nitrate is used in :
(a) Lining vessels
(b) Paints
(c) Textile industry
(d) Preparing explosives

9. Which acid is prepared by Ostwald's process :
(a) Nitric acid
(b) Oleic acid
(c) Sulphuric acid
(d) Tartaric acid

10. The nitrate which evolves laughing gas on decomposition ?
(a) Ferric nitrate
(b) Calcium nitrate
(c) Sodium nitrate
(d) Ammonium nitrate

11. The brown ring test is used for the detection of :
(a) CO_3^{2-} (b) NO_3^- (c) SO_3^{2-} (d) Cl^-

12. The composition of brown ring formed at the junction is :
(a) $FeSO_4.OH$ (b) $FeSO_4.NO$ (c) $FeSO_4.H_2O$ (d) $FeSO_4$

Ans. (1) (c) (3) (b) (5) (b) (7) (d) (9) (a) (11) (b)
(2) (a) (4) (c) (6) (b) (8) (d) (10) (d) (12) (b)

Chapter 8. (c) Study of Compounds : Sulphuric Acid

1. The catalyst used in the Contact Process is :*
(a) Copper
(b) Iron
(c) Vanadium pentoxide
(d) Manganese dioxide

2. The molecular formula of epsom salt is :
(a) $MgSO_4 \cdot 5H_2O$
(b) $FeSO_4 \cdot 7H_2O$
(c) $MgSO_4 \cdot 7H_2O$
(d) $MgSO_4 \cdot 6H_2O$

3. Nitroglycerine and nitrotoluene are :
(a) Fertilizers (b) Explosives (c) Fibres (d) Detergents

4. Pyrosulphuric acid is the chemical name of :
(a) Green vitriol (b) White vitriol (c) Oleum (d) Gypsum

5. In the given equation identify the role played by concentrated sulphuric acid

$$S + 2H_2SO_4 \longrightarrow 3SO_2 + 2H_2O$$

* Frequently asked previous years Board Exam Questions.

Multiple Choice Questions

(a) Non-volatile acid
(b) Oxidising agent
(c) Dehydrating agent
(d) None of these

6. When dilute sulphuric acid reacts with iron sulphide, the gas evolved is :
(a) Hydrogen sulphide
(b) Sulphur dioxide
(c) Sulphur trioxide
(d) Vapour of sulphuric acid

7. Dilute sulphuric acid will produce a white precipitate when added to a solution of :
(a) Copper nitrate (b) Zinc nitrate (c) Lead nitrate (d) Sodium nitrate

8. Corrosive action of H_2SO_4 on skin is due to :
(a) Exothermic nature
(b) Volatile nature
(c) Dehydrating character
(d) Oxidising nature

9. When sulphuric acid is added to sodium carbonate brisk effervesence is produced which is due to :
(a) Evolution of H_2S gas
(b) Evolution of Cl_2 gas
(c) Evolution of CO_2 gas
(d) Evolution of O_2 gas

10. When hot concentrated sulphuric acid is added to sulphur it gets oxidised to :
(a) H_2S (b) SO_2 (c) O_3 (d) HSO_4

11. When concentrated sulphuric acid reacts with NaOH, the product formed is :
(a) Sodium hydroxide
(b) Sodium sulphate
(c) Sodium carbonate
(d) Sodium hydrogen sulphate

Ans. (1) (c) (2) (c) (3) (b) (4) (c) (5) (b) (6) (b) (7) (c) (8) (c) (9) (c) (10) (b) (11) (b)

Chapter 9. Organic Chemistry

1. The property of carbon of form chains and rings is called :
(a) Catenation (b) Polymerisation (c) Cracking (d) Hydrogenation

2. Two neighbours of a homologous series differ by :
(a) CH (b) CH_2 (c) CH_3 (d) CH_4

3. Heating sodium acetate with soda lime produces :
(a) Methane (b) Ethane (c) Ethene (d) Ethyne

4. The number of C–H bonds in ethane molecule are :
(a) Four (b) Six (c) Eight (d) Ten

5. Halogenation of alkane can be carried out in :
(a) Dark
(b) Bright light
(c) UV light
(d) Diffused sunlight

6. The unsaturated hydrocarbons undergo :
(a) a substitution reaction
(b) an oxidation reaction
(c) an addition reaction
(d) None of these

7. Identify the statement which does not describe the property of alkenes :
(a) They are unsaturated hydrocarbons
(b) They decolourise bromine water
(c) They can undergo addition as well as substitution reactions
(d) They undergo combustion with oxygen forming carbon dioxide and water

8. Dehydrohalogenation of X with alcoholic KOH produces ethene. X is :
(a) Ethyl chloride
(b) Methyl chloride
(c) Ethylene dichloride
(d) None of these

9. If the molecular formula of an organic compound is $C_{10}H_{18}$ it is :
(a) Alkene
(b) Alkane
(c) Alkyne
(d) Not a hydrocarbon

10. The I.U.P.A.C. name of acetylene is :

(a) Propane (b) Propyne (c) Ethene (d) Ethyne

11. Acetylene polymerises into benzene by joining :

(a) 3 molecules (b) 2 molecules (c) 4 molecules (d) 6 molecules

12. An organic compound undergoes addition reactions and gives a red colour precipitate with ammoniacal cuprous chloride. Therefore, the organic compound could be :

(a) Ethane (b) Ethene (c) Ethyne (d) Ethanol

13. The organic compound mixed with ethanol to make it spurious is :

(a) Methanol (b) Methanoic acid (c) Methanal (d) Ethanoic acid

14. Which compound is used as an antifreeze agent ?

(a) Propene (b) Acetylene (c) Ethylene glycol (d) Methanol

15. The functional group present in acetic acid is :

(a) Ketonic C = O
(b) Hydroxyl —OH
(c) Aldehydic —CHO
(d) Carboxyl —COOH

16. An acid used for removing ink stains :

(a) Acetic acid (b) Oxalic acid (c) Malic acid (d) Formic acid

17. The resulting ester, when ethyl alcohol and acetic acid are mixed together :

(a) CH_3COOCH_3 (b) $C_2H_5COOC_2H_5$ (c) $CH_3COOC_2H_5$ (d) $C_2H_5COOCH_3$

18. A compound with low boiling point, is :*

(a) Sodium chloride
(b) Calcium chloride
(c) Potassium chloride
(d) Carbon tetrachloride

19. The organic compound having a triple carbon-carbon covalent bond, is :*

(a) C_3H_4 (b) C_3H_6 (c) C_3H_8 (d) C_4H_{10}

20. A hydrocarbon which is a greenhouse gas is :*

(a) Acetylene (b) Ethylene (c) Ethane (d) Methane

21. The organic compound which undergoes substitution reaction is :*

(a) C_2H_2 (b) C_2H_4 (c) $C_{10}H_{18}$ (d) C_2H_6

22. The IUPAC name of dimethyl ether is :*

(a) Ethoxy methane
(b) Methoxy methane
(c) Methoxy ethane
(d) Ethoxy ethane

Ans.

(1) (a)	(5) (c)	(9) (c)	(13) (a)	(17) (c)	(21) (d)
(2) (b)	(6) (c)	(10) (d)	(14) (c)	(18) (d)	(22) (b)
(3) (a)	(7) (c)	(11) (a)	(15) (d)	(19) (a)	
(4) (b)	(8) (a)	(12) (c)	(16) (b)	(20) (d)	

❑

* **Frequently asked previous years Board Exam Questions.**

One Word or Chemical Term

Set 5

Chapter 1. Periodic Properties and Variations of Properties

1. The law that groups elements in family of three.
2. The law where every 8th element repeats properties.
3. The law where repetition of properties varies periodically.
4. Horizontal arrangement of elements.
5. Vertical arrangement of elements.
6. The elements present in the first period.
7. The transition element whose electronic configuration does not agree with the general pattern.
8. The element with the least ionization potential in the third period.
9. The amount of energy released when an atom in the gaseous state accepts an electron to form an anion.
10. The element which has the highest electron affinity in the third period.
11. The electrons present in the outermost shell of an atom.
12. A positive charge present in the nucleus of an atom which is equal to the number of protons in the nucleus.
13. Elements with 1, 2 or 3 electrons in the last shell.
14. Elements with 4, 5, 6 or 7 electrons in the last shell.
15. Elements with 8 electrons in the last shell.
16. Elements where the penultimate shell is being filled with electrons.
17. A non-metal which has three electrons in the outermost shell.
18. The most metallic element in the third period.
19. The most non-metallic element in the third period.
20. The element which is the most electronegative in the third period.
21. Most active non-metal.
22. A metal which is a liquid at room temperature.
23. A liquid non-metal.
24. An alkaline earth metal present in group 2 and period 3.*
25. The combining capacity of the atom of an element.
26. Elements lose electrons to complete their octet.

Ans.

1. Law of triads
2. Newland's law of Octaves
3. Mendeleev's Periodic law
4. Periods
5. Groups
6. Hydrogen and Helium
7. Palladium
8. Sodium
9. Electron affinity
10. Chlorine
11. Valence electrons
12. Nuclear charge
13. Metals
14. Non-metals
15. Rare gases
16. Transitional metals
17. Gallium
18. Sodium
19. Chlorine
20. Chlorine
21. Fluorine
22. Mercury
23. Bromine
24. Magnesium or Mg
25. Valency
26. Reducing agents

* **Frequently asked previous years Board Exam Questions.**

Chapter 2. Chemical Bonding

1. A bond formed by the transfer of electrons.
2. A bond formed by the sharing of electrons.
3. A bond formed by a shared pair of electrons with both electrons coming from the same atom.
4. Bond present in metallic chloride and between two chlorine atoms.
5. Electrons lost or gained by an atom in compound formation.
6. Ions formed by the loss and gain of electrons respectively.
7. The process in which an atom lose electrons.
8. The process in which an atom gain electrons.
9. An inert gas which contain triple bond.
10. A molecule with little affinity.
11. Two molecules, which are polar.
12. A polar covalent molecule.
13. A non-polar covalent compound.
14. A covalent molecule which on dissolving form ions.
15. A molecule with one lone pair and three bond pairs.
16. A molecule with two lone pairs and two bond pairs.

Ans. 1. Electrovalent bond
2. Covalent bond
3. Coordinate bond
4. Electrovalent, covalent.
5. Electrovalency
6. Cation and anion respectively
7. Oxidation
8. Reduction
9. Nitrogen
10. Methane
11. Hydrogen chloride, Ammonia.
12. Hydrogen chloride (HCl)
13. Carbon dioxide (CO_2)
14. HCl
15. Ammonia (NH_3)
16. Water (H_2O)

Chapter 3. Study of Acids, Bases and Salts

1. Name the acids, which are used in food industry.
2. Two monobasic acids containing nitrogen.
3. Two dibasic acids containing sulphur.
4. Strong acid containing chlorine.
5. The substance that releases hydronium ion as the only positive ion when dissolved in water.*
6. Basic anhydride of calcium hydroxide.
7. Product obtained when lime water dries up.
8. Three common indicators to indicate presence of acids and alkalies.
9. A salt formed by incomplete neutralisation of an acid by a base.
10. This salt gives nitrogen dioxide on heating.
11. This compound is an acid salt.
12. A compound whose aqueous solution is neutral in nature.
13. Three metals that can form bases when treated with water.
14. Acid anhydride of sulphuric acid.
15. The process in which a substance absorbs moisture from the atmospheric air to become moist, and ultimately dissolves in the absorbed water.
16. The property by which certain hydrated salts, when left exposed to atmosphere, lose their water of crystallization and crumble into powder.*
17. A compound which is deliquescent.
18. Name an acidic hygroscopic substance.

* **Frequently asked previous years Board Exam Questions.**

19. Name a basic hygroscopic substance.

20. Name any two efflorescent substances.

Ans.
1. Citric acid, tartaric acid and acetic acid
2. Nitric acid and nitrous acid
3. Sulphuric acid and sulphurus acid
4. Hydrochloric acid
5. Protic acid
6. Calcium oxide
7. Calcium oxide and carbon dioxide
8. Litmus, methyl orange, phenolphthalein
9. Acid salt
10. $Pb(NO_3)_2$
11. $NaHSO_4$
12. Sodium chloride
13. Sodium, calcium and potassium
14. Sulphur trioxide
15. Deliquescence
16. Efflorescence or Efflorescent
17. Caustic soda (NaOH)
18. Sulphuric acid, or oil of vitriol
19. Calcium oxide
20. Copper sulphate (Blue vitriol) and Washing soda (Sodium carbonate)

Chapter 4. Analytical Chemistry

1. Name two coloured and two colourless metal ions.

2. A salt of zinc which is efflorescent.

3. The ion responsible for the blue colour of an aqueous solution of a salt.

4. One metal which forms more than one type of positive ions.

5. A yellow monoxide that dissolves in hot and concentrated caustic alkali.

6. A chloride which gives reddish brown precipitate with sodium hydroxide solution.

7. A metallic hydroxide which is insoluble in caustic soda but soluble in excess of NH_4OH.

8. A metal that evolves a gas which burns with a pop sound when boiled with alkali solutions.

9. An oxide which dissolves in both acid as well as base.

10. A metallic hydroxide soluble in excess of ammonium hydroxide solution.

Ans.
1. Coloured-Cu^{2+}, Fe^{2+} colourless—Mg^{2+}, Pb^{2+}
2. Zinc sulphate ($ZnSO_4$)
3. Copper
4. Iron
5. Lead monoxide, PbO
6. Ferric chloride
7. Copper hydroxide, $Cu(OH)_2$
8. Zinc, Zn
9. Lead oxide, PbO
10. Zinc hydroxide, $Zn(OH)_2$

Chapter 5. Mole Concept and Stoichiometry

1. The law which states that–``under similar conditions of temperature and pressure the volume of gases taking part in a chemical reaction bear a simple whole number ratio".

2. The volume occupied by one mole of a gas at STP is.

3. Equal volumes of gases under similar conditions of temperature and pressure contain equal number of molecules.

4. The number of atoms present in 12 g of carbon $\left[{}^{12}_{6}C\right]$.

5. It is a number that represents how many times an atom of an element is heavier than $\frac{1}{12}$ th mass of carbon atom.

6. The molecular weight of an element expressed in grams.

7. The ratio of the mass of a certain volume of gas to the mass of an equal volume of hydrogen under the same conditions of temperature and pressure.

8. A formula of a chemical substance which tells the actual number of atoms in one molecule of a substance.

9. A formula which shows the simplest whole number ratio.

Ans.
1. Gay Lussac's law
2. 22·4 litres.
3. Avogadro's law
4. $6{\cdot}023 \times 10^{23}$ atoms
5. Atomic weight
6. Gram-molecular weight
7. Vapour density
8. Molecular formula
9. Empirical formula

Chapter 6. Electrolysis

1. The fundamental particles of the atoms that move when electric current is passed through a metal wire.
2. The type of substances which undergo decomposition in aqueous solution form when electric current passes through them.
3. The type of materials which neither in solution nor in molten state allows an electric current to pass through it.
4. The process by which a particle gains or loses electrons.
5. The process by which an electrovalent substance breaks up into free mobile ions in the molten or aqueous form.
6. An acid which is added during electrolysis of water.
7. The acid used in electrolysis of water.
8. Electrodes used in electrolysis of copper sulphate.
9. A material used for making attackable electrodes.
10. Liquid which does not conduct electricity.
11. Electrolytic deposition of a superior metal on a baser metal.
12. Electrolyte used in the process of silver plating.
13. Electrolyte used in the process of nickel plating.
14. A suitable cathode used in the electrorefining of copper.
15. Impurities which remain behind at the bottom of the cell.
16. Process used for obtaining highly electropositive metals.

Ans. 1. Electron
2. Strong electrolytes
3. Non-electrolytes
4. Ionization
5. Dissociation
6. Conc. H_2SO_4
7. Sulphuric acid
8. Platinum electrodes
9. Platinum
10. Non-electrolyte
11. Electroplating
12. Potassium argento cyanide
13. Nickel ammonium sulphate
14. Sheets of pure copper metal
15. Anode mud
16. Electrometallurgy

Chapter 7. Metallurgy

1. Name the metals which can be extracted from the following ores :
(1) Haematite
(2) Malachite
(3) Cinnabar
(4) Barytes
(5) Cuprite
(6) Bauxite
(7) Galena
(8) Zinc blende
(9) Iron pyrites
(10) Epsom
(11) Calamine
(12) Dolomite

Ans.
(1) Iron
(2) Copper
(3) Mercury
(4) Barium
(5) Copper
(6) Aluminium
(7) Lead
(8) Zinc
(9) Iron
(10) Magnesium
(11) Zinc
(12) Magnesium

2. Name an ore which is :
(1) An oxide
(2) A hydrated oxide
(3) A carbonate
(4) A sulphate
(5) A sulphide
(6) A nitrate
(7) A phosphate
(8) A chloride

Ans. (1) Haematite (Fe_2O_3)
(2) Bauxite ($Al_2O_3.2H_2O$)
(3) Magnetite ($MgCO_3$)
(4) Gypsum ($CaSO_4.2H_2O$)
(5) Galena (PbS)
(6) Chile salt-peter ($NaNO_3$)
(7) Rock phosphate [$Ca_3(PO_4)_2$]
(8) Horn silver (AgCl)

3. Name the property of a metal by virtue of which it :

(1) Can be beaten into thin sheets
(2) Can be drawn into wires
(3) Possesses tensile strength
(4) Is good conductor of heat
(5) Is good conductor of electricity
(6) Dissolves into another metal
(7) Is liberated at cathode
(8) Can cut a non-metal
(9) Is heavier than a non-metal
(10) Acts as a reducing agent.

Ans.

(1) Malleability
(2) Ductility
(3) Durability
(4) Thermal conductivity
(5) Electrical conductivity
(6) Alloy formation
(7) Electropositive nature
(8) Hardness
(9) Density
(10) Donor of electrons

4. (1) Name a metal which is found abundantly in the earth's crust.
(2) Name two metals which have a high degree of malleability.
(3) Name any two metals which are both malleable and ductile.
(4) Name the metal which is a good conductor of both heat and electricity.
(5) Name a metal which can be cut even with a knife.
(6) Name a metal which is not ductile, malleable and tenacious.
(7) Name a metal which is a poor conductor.
(8) A metal which exists in liquid state at room temperature.
(9) Name the metal which floats in water.
(10) Name the metal which is stored in kerosene oil.
(11) Name a metal which has a low melting point.
(12) Name a metal which has acidic oxide.
(13) Name the metal that burns in air with a golden flame.
(14) Name a metal which forms anions which are liberated at anode.
(15) The metal which combines directly with sulphur on heating.
(16) The burning metal which combines directly with nitrogen.
(17) The molten metal which gives white fumes while reacting with chlorine.
(18) A metal which reacts reversibly with steam.
(19) Metal which is rendered passive on reaction with concentrated nitric acid.
(20) Name a metal which is used in accumulators or car batteries.
(21) Name a metal which is used for galvanizing iron.
(22) Name a metal which is used as silver paper.
(23) Name a metal which is used in a torch cell.
(24) Name a metal which is used in flashlight photography.
(25) Name a metal which is used in vapour lamps.
(26) Name a non-metal which has shiny appearance.
(27) Name a non-metal which is the hardest substance known.
(28) Name a non-metal which has high melting and boiling points.
(29) Name a non-metal which is a good conductor of heat and electricity.
(30) Name a non-metal which forms alloys with metals.
(31) Name a non-metal which is electropositive in nature.
(32) Name a non-metal which forms a neutral oxide.
(33) Name a non-metallic element which forms both acidic and neutral oxides.
(34) Name a non-metal which posses metallic lustre and sublimes on heating.
(35) Name a non-metallic element which is a liquid at ordinary temperatures.
(36) The non-metal, which forms two compounds, while reacting with chlorine.
(37) The process of removal of gangue from ore.
(38) Name the sulphide ore of mercury.
(39) The most common ore of aluminium.

(40) Name the process used for the enrichment of sulphide ore.
(41) The formula of slag.
(42) The chemical name of slag.
(43) The major impurity associated with iron obtained from blast furnace.
(44) The process of coating thin layer of zinc over the surface of iron.
(45) The process by which zinc is purified.
(46) Gas obtained when zinc blende is roasted.
(47) Name one alloy each of aluminium and iron.
(48) Name the alloy of zinc used in simple voltaic cells.
(49) The elements added to iron to form stainless steel.
(50) Name the metal which is alloyed with zinc to form brass.
(51) Name an element from which fencing wire is made.
(52) Name the purest form of iron.
(53) Two elements, whose hydroxides are easily soluble in water and form alkaline solutions.
(54) One metal, which forms more than one type of positive ions.
(55) One ion responsible for blue colour of an aqueous solution of copper sulphate.
(56) The alloy of steel with a minimum of 10·5% chromium content by mass.
(57) Name the solution used to react with Bauxite as a first step in obtaining pure aluminium oxide in the Baeyer's process.
(58) Name the compound added to pure alumina to lower the fusion temperature during the electrolytic reduction of alumina.
(59) The process of coating of iron with zinc.
(60) An alloy of lead and tin that is used in electrical circuits.
(61) An ore of zinc containing its sulphide.
(62) A metal oxide that can be reduced by hydrogen.

Ans. (1) Aluminium
(2) Silver and aluminium
(3) Gold and silver
(4) Aluminium
(5) Sodium
(6) Arsenic
(7) Bismuth
(8) Mercury
(9) Sodium
(10) Sodium
(11) Sodium
(12) Antimony oxide (Sb_2O_3)
(13) Sodium
(14) Manganese in the form of permanganate ions [MnO_4^-]
(15) Iron
(16) Magnesium
(17) Sodium
(18) Iron
(19) Iron
(20) Lead
(21) Zinc
(22) Aluminium
(23) Zinc
(24) Magnesium
(25) Sodium
(26) Iodine
(27) Diamond
(28) Carbon
(29) Graphite
(30) Carbon
(31) Hydrogen
(32) Hydrogen (forms neutral oxide, water)
(33) Carbon
(34) Iodine
(35) Bromine
(36) Phosphorus
(37) Concentration
(38) Cinnabar
(39) Bauxite
(40) Froth floatation process
(41) $CaSiO_3$
(42) Calcium silicate
(43) Carbon
(44) Galvanization
(45) Distillation
(46) Sulphur dioxide
(47) Stainless steel is an alloy of iron and duralium is an alloy of aluminium
(48) Amalgamated zinc
(49) Chromium (Cr) and Nickel (Ni)
(50) Copper

(51) Soft-iron
(52) Wrought-iron
(53) Sodium and Potassium
(54) Iron which forms Fe^{2+} and Fe^{3+}
(55) Cupric ion (Cu^{2+})
(56) Stainless steel
(57) Sodium hydroxide (NaOH)
(58) Cryolite
(59) Galvanisation
(60) Solder
(61) Zinc blende
(62) Copper oxide

Chapter 8. (a) Study of Compounds : Hydrogen Chloride

1. The gas obtained when rock salt reacts with conc. sulphuric acid.
2. Give the name of a chemical, which on being added to sodium chloride, will produce hydrogen chloride gas.
3. Name the experiment use for the density of hydrogen chloride.
4. Drying agent for hydrogen chloride gas.
5. Compound which cannot be used to dry hydrogen chloride gas.
6. A polar covalent compound which on dissolving in water produces ions.
7. The experiment which demonstrates extreme solubility of hydrogen chloride gas.
8. A compound formed when iron reacts with hydrogen chloride.
9. Name one lead compound that can be used to oxidise HCl to chlorine.
10. The gas evolved when Mangnese(IV) oxide and concentrated hydrochloric acid are heated.
11. Two colourless gases, when mixed, form white dense fumes. Name the two gases and white dense fumes.
12. A gas evolved when HCl is added to red lead.
13. Two compounds of lead which combine with conc. HCl to liberate chlorine.
14. A substance which reacts with conc. HCl to liberate chlorine.
15. The colour of a gas evolved when conc. hydrochloric acid is heated with manganese dioxide.
16. A gas evolved when metal carbonates reacts with hydrogen chloride.
17. An acid which is not an oxidising agent.
18. A mixture of three parts of conc. HCl and one part of conc. nitric acid.
19. Name chemical in which gold can be dissolved.

Ans. 1. Hydrogen chloride gas
2. Conc. sulphuric acid
3. Fountain experiment
4. Concentrated sulphuric acid
5. Quick lime, CaO
6. Hydrogen chloride gas
7. Fountain experiment
8. Iron(II) chloride
9. Lead dioxide or Red lead
10. Chlorine (Cl_2).
11. Two gases : hydrogen chloride and ammonia. White dense fumes : ammonium chloride
12. Chlorine
13. Lead dioxide and red lead (Pb_3O_4)
14. PbO_2
15. Greenish yellow
16. Carbon dioxide, CO_2
17. Hydrogen chloride, HCl
18. Aqua regia
19. Aqua regia

Chapter 8. (b) Study of Compounds : Ammonia and Nitric Acid

1. An element that acts as a promoter in the industrial preparation of ammonia.
2. A compound used to dry ammonia.
3. A catalyst used during Haber's process.
4. Products obtained by the catalytic oxidation of ammonia.
5. Ammonium salt used in the preparation of alum.
6. A fertilizer that reacts slowly with moisture to give ammonia.
7. A fertilizer made by combining ammonia and carbon dioxide.
8. The product of oxidation of ammonia in Ostwald process.
9. A compound which is manufactured by the oxidation of ammonia.

10. Experiment which demonstrates the extreme solubility of ammonia.
11. The fertilizer formed when nitrogen is reacted with calcium carbide at high temperature.
12. A white ppt. formed when an aqueous solution of the compound is added to a solution of barium chloride
13. A colourless gas which becomes reddish brown when it comes in contact with atmosphere.
14. The common name of carbamide.
15. An explosive formed when ammonia and chlorine react together.
16. The compound known by the name, Aqua fortis.
17. A catalyst used in the manufacture of nitric acid by Ostwald's process.
18. The process by which nitric acid is obtained on a large scale.
19. Salt used in the laboratory to prepare nitric acid.
20. The compound that makes laboratory acid yellow.
21. Organic liquid that bursts into flames when sprayed into highly conc. nitric acid.
22. Gas obtained by treating copper with dilute nitric acid.
23. Products obtained by heating concentrated nitric acid.
24. Insoluble component present in super phosphate of lime.
25. The gas produced by the action of dil. HNO_3 on copper.
26. A compound formed when nitric acid and hydrochloric acid reacts together.
27. The property of nitric acid which allows it to react with copper.
28. A concentrated acid which oxidies sulphur directly to H_2SO_4.
29. The gas produced when copper reacts with conc. HNO_3.
30. A gas which possess the smell of rotten eggs.
31. The test carried out with freshly prepared ferrous sulphate.
32. A nitrate which leaves behind no residue on heating.
33. A solution which gives nitrogen dioxide with copper.
34. Two gases–one of which is basic and which combine to give a solid
35. The compound responsible for the brown ring during the brown ring test of nitrate ion.

Ans. 1. Molybdenum
2. Quick lime (CaO)
3. Finely divided iron
4. Nitric oxide and water vapour
5. Ammonium sulphate
6. Calcium cyanamide
7. Urea (NH_2CONH_2)
8. Nitrogen(II) oxide (nitric oxide) and steam
9. Nitric acid
10. Fountain experiment
11. Nitrogen
12. Ammonium sulphate
13. Nitric oxide
14. Urea
15. NCl_3
16. Nitric acid
17. Platinum
18. Ostwald's process
19. Sodium or potassium nitrate
20. Nitrogen dioxide (NO_2)
21. Turpentine
22. Nitric oxide
23. Water, nitrogen dioxide and oxygen
24. Calcium sulphate
25. Nitric oxide
26. Nitrosyl chloride
27. Oxidising agent
28. Conc. HNO_3
29. NO_2
30. H_2S
31. Brown ring test
32. Ammonium nitrate
33. Conc. HNO_3
34. Ammonia and hydrogen chloride
35. Nitroso Iron(II) sulphate

Chapter 8. (c) Study of Compounds : Sulphuric Acid

1. Acid commonly known as oil of vitriol.
2. Catalyst used during contact process.
3. Promoter used in contact process.

4. Atomicity of sulphur.
5. Basicity of sulphuric acid.
6. Acid obtained by dissolving sulphur dioxide in water.
7. A black coloured substance formed when sugar is dehydrated.
8. Bleaching agent used to bleach delicate articles.
9. Name the solution which turns colourless when sulphur dioxide gas is passed through it?
10. Name an acid that cannot be used to obtain sulphur dioxide, from sulphites.
11. Give the chemical name and common name of a reagent used in photographic work.
12. The process involved during the bleaching action of sulphur dioxide.
13. A neutral oxide of carbon formed when formic acid reacts with concentrated sulphuric acid.
14. Name the chemical used to test hydrogen sulphide.
15. Name an orange colour solution which turns in green colour when sulphur dioxide gas is passed through it.
16. An organic acid which on reaction with concentrated sulphuric acid, produces two oxides of carbon ?
17. The solution which turns black on coming in contact with hydrogen sulphide?
18. Green coloured compound formed when SO_2 is passed through acidified $K_2Cr_2O_7$.
19. Process by which sulphur is boiled with rubber.

Ans. 1. Sulphuric acid
2. V_2O_5 or platinised asbestos
3. Potassium oxide
4. Eight
5. Dibasic
6. Sulphurus acid
7. Carbon or sugar charcoal
8. Moist sulphur dioxide
9. Potassium permanganate
10. Nitric acid
11 Sodium thiosulphate
12. Reduction
13. Carbon monoxide
14. Lead acetate
15. Potassium dichromate
16. Oxalic acid
17. Lead acetate, lead nitrate solution
18. Chromium sulphate
19. Vulcanization

Chapter 9. Organic Chemistry

1. Name the organic compound which was first synthesized.
2. A simplest hydrocarbon.
3. The shape of methane molecule.
4. Hydrocarbons containing a $-\overset{\overset{\displaystyle O}{\|}}{C}-$ functional group.
5. An organic compound whose functional group is carboxyl.
6. The compound with –OH as the part of its structure.
7. The compound with –COOH as the part of its structure.
8. The compounds having same molecular formula but different structural formula.
9. Homologue of homologous series with general formula C_nH_{2n}.
10. The next higher homologue of methyl alcohol.
11. The type of reactions alkenes undergo.
12. The type of reactions alkanes undergo.
13. The hydrocarbon which contributes towards greenhouse effect.
14. A reaction in which hydrogen of an alkane is replaced by a halogen.
15. Process by which ethene is obtained from ethanol.
16. A hydrocarbon which on catalytic hydrogenation gives a saturated hydrocarbon.
17. The catalyst used during hydrogenation of alkene.
18. A white solid which on treatment with water liberate acetylene.
19. The chemical name of the gas, which evolves in the marshy lands in the form of bubbles.

20. The solution which detects the presence of unsaturation in the given hydrocarbon?
21. An organic compound used as a thermometric liquid.
22. Hydrocarbon, which increases the rate of fruit ripening.
23. A compound which is used in making denatured spirit.
24. A gas which forms explosive mixture with air.
25. A solution used for storing biological specimens.
26. An unsaturated hydrocarbon used for welding purposes.
27. Process by which ethane is obtained from ethene.
28. A hydrocarbon which contributes towards the greenhouse effect.
29. Distinctive reaction that takes place when ethanol is treated with acetic acid.
30. The property of elements by virtue of which atoms of the element can link to each other in the form of a long chain or ring structure.
31. Reaction when an alkyl halide is treated with alcoholic potassium hydroxide.
32. The catalyst used in the conversion of ethyne to ethane.
33. Name of the first member of alkyne series.

Ans. 1. Urea
2. Methane
3. Tetrahedral
4. Carbonyl compounds alkanones
5. Acetic acid
6. Ethanol
7. Ethanoic acid
8. Isomer
9. Ethene
10. Ethyl alcohol, C_2H_5OH
11. Addition reaction
12. Substitution reaction
13. Methane
14. A substitution reaction—halogenation
15. Dehydration
16. Ethene
17. Nickel
18. Calcium carbide
19. Methane gas
20. Bromine solution in carbon tetrachloride or Alkaline potassium permanganate
21. Ethanol
22. Ethene
23. Methyl alcohol
24. Acetylene
25. Formalin
26. Ethyne
27. Hydrogenation
28. Methane (CH_4)
29. Esterification
30. Catenation
31. Dehydrohalogenation
32. Lindlar catalyst
33. Acetylene (Ethyne) C_2H_2.

Chapter 10. Practical Chemistry

1. A molecule of diatomic gas with a triple bond.
2. A compound with six water of crystallisation.
3. A black coloured reducing agent.
4. A trivalent metal.
5. A mixed salt.
6. A stable carbonate.
7. A green coloured carbonate.
8. A blue coloured nitrate.
9. A nitrate which does not leave behind solid residue on heating.
10. The nitrate which gives a black residue on heating.
11. A black metallic oxide which dissolves in nitric acid to give greenish blue solution.
12. A salt soluble in hot water but insoluble in cold water.
13. A chloride which dissolves in ammonia to give a complex.
14. A solution which can absorb oxygen.
15. A double salt used for the sediment fo muddy water.

16. A hydrated salt that is blue in colour.

17. The salt formed when aluminium react with potassium hydroxide.

18. The salt which gives a yellow residue on heating.

19. This gas turns moist lead acetate paper black.

20. The compound which is responsible for the green colouration when sulphur dioxide is passed through acidified potassium dichromate solution.

21. On heating, this salt changes from green to black.

22. On treating with concentrated sulphuric acid, this salt changes from blue to white.

23. A compound which is insoluble in cold water, but soluble in hot water.

Ans. 1. N_2 molecule
2. Cobalt chloride/Copper (II) Sulphate
3. Charcoal
4. Aluminium
5. Bleaching powder
6. Sodium carbonate
7. Copper carbonate
8. Copper nitrate
9. Ammonium nitrate
10. Copper nitrate
11. Copper oxide
12. Lead chloride
13. Silver chloride
14. Alkaline pyrogallol
15. Potash alum
16. Lead nitrate
17. Potassium aluminate
18. $CuSO_4 \cdot 5H_2O$
19. Hydrogen sulphide
20. Chromium sulphate
21. $CuCO_3$
22. $CuSO_4.5H_2O$
23. Lead(II) chloride

❑

Identification of Gases

Set 6

Name the gas evolved in the following cases :

Chapter 8. (b) Study of Compounds : Ammonia and Nitric Acid

1. The gas produced when excess ammonia reacts with chlorine.
2. The gas produced when dry ammonia and oxygen are ignited.
3. The gas produced when ammonium nitrite undergoes a thermal decomposition.
4. The gas evolved when dry ammonia and oxygen are passed over heated platinum and then allowed to cool.
5. The gas produced when ammonia gas is passed over heated litharge.
6. The gas produced when ammonium nitrate is heated.
7. The gas produced when copper reacts with concentrated nitric acid.
8. The gas produced when magnesium reacts with dil. nitric acid.
9. The gas produced when sodium nitrate is heated.
10. Sodium nitrite and ammonium chloride.
11. The gas produced when sodium carbonate reacts with dilute nitric acid.
12. The gases produced when silver nitrate is heated.
13. The gas produced when ammonia burns in an atmosphere of oxygen without any catalysts.
14. Calcium hydroxide and ammonium chloride.

Ans. 1. Nitrogen
2. Nitrogen
3. Nitrogen
4. Nitrogen dioxide
5. Nitrogen
6. Nitrous oxide (Laughing gas N_2O)
7. Nitrogen dioxide
8. Hydrogen
9. Oxygen
10. Nitrogen gas
11. Carbon dioxide
12. Nitrogen dioxide and oxygen
13. Nitrogen
14. Ammonia gas

Chapter 8. (c) Study of Compounds : Sulphuic Acid

1. The gas which turns acidified potassium dichromate clear green.
2. The gas produced on reaction of dilute sulphuric acid with a metallic sulphide.
3. The gas released when sodium carbonate is added to a solution of sulphur dioxide.
4. The gas produced by the action of concentrated sulphuric acid on sodium chloride.
5. The gas evolved when dilute sulphuric acid is added to sodium sulphite.

Ans. 1. Sulphur dioxide
2. Hydrogen sulphide
3. Carbon dioxide
4. Hydrogen chloride
5. Sulphur dioxide

Chapter 10. Practical Chemistry

1. A colourless gas which forms an explosive mixture with air and water is the only product of combustion.
2. A colourless gas which rekindles glowing splint but cannot be used for breathing.
3. A colourless gas which burns with a pale blue flame, forming carbon dioxide gas as the only product.
4. A colourless gas having a sharp pungent smell, which gives dense white fumes with HCl.
5. A colourless gas having a choking smell, which causes coughing.
6. An extremely soluble gas in water, which forms dense white fumes with ammonia solution.
7. A colourless gas having a disagreeable smell which it turns lead acetate paper black.
8. A colourless gas which is neither combustible nor does it support combustion. It does not turn lime water milky.
9. A colourless gas which on coming in contact with air, forms a reddish-brown gas.
10. A reddish-brown gas obtained when lead nitrate crystals are strongly heated.
11. A colourless gas which bleaches moist coloured flowers.
12. Dilute hydrochloric acid reacts with iron(II) sulphide.
13. Two gases which react with ammonia to form dense white fumes.
14. The gas evolved on reaction of aluminium with boiling concentrated caustic alkali solution.

Ans. 1. Hydrogen gas
2. Nitrous oxides (N_2O) gas
3. Carbon monoxide gas
4. Ammonia gas
5. Sulphur dioxide gas
6. Hydrochloric acid gas
7. Hydrogen sulphide gas
8. Nitrogen gas
9. Nitric oxide (NO) gas
10. Nitrogen dioxide gas
11. Sulphur dioxide gas/Chlorine Gas
12. Hydrogen sulphide
13. HCl gas and chlorine gas
14. Hydrogen gas

❑

State the Observation

Set 7

Chapter 4. Analytical Chemistry

What do you observe when (write equations if necessary) :

1. Sodium hydroxide solution is slowly added and then in excess to zinc sulphate solution.
2. To a solution of lead nitrate small amount of sodium hydroxide is added and then excess of sodium hydroxide is added.
3. When sodium hydroxide is added to a solution of ferric chloride write equation for the reaction taking place.
4. Ammonium hydroxide solution is added to Iron(II) sulphate solution.
5. Ammonium hydroxide is added to iron(III) sulphate solution.
6. Ammonium hydroxide solution is slowly added and then in excess to copper sulphate solution.
7. Ammonium hydroxide is added to zinc sulphate solution. Write the name and the formula of the final product.
8. Ammonium hydroxide solution is added to copper(II) nitrate solution in small quantities and then in excess.
9. Ammonium hydroxide solution is added to zinc nitrate solution in minimum quantities and then in excess.
10. Ammonium hydroxide solution is added to iron(III) chloride solution.

Ans. 1. Sodium hydroxide solution when slowly added to zinc sulphate solution, a white precipitate of zinc hydroxide is obtained, which is soluble in excess of sodium hydroxide to form a clear solution of sodium zincate.

$$ZnSO_4 + 2NaOH \longrightarrow \underset{\text{White ppt.}}{Zn(OH)_2} \downarrow + Na_2SO_4$$

$$Zn(OH)_2 + 2NaOH \longrightarrow \underset{\text{Sodium zincate}}{Na_2ZnO_2} \downarrow + 2H_2O$$

2. A curdy white precipitate of lead hydroxide is formed which dissolves in excess NaOH giving a colourless solution.

$$Pb(NO_3)_2 + 2NaOH \longrightarrow Pb(OH)_2 \downarrow + 2NaNO_3$$

$$Pb(OH)_2 + 2NaOH\ (\text{Excess}) \longrightarrow \underset{\text{Sodium Plumbate}}{Na_2PbO_2}$$

3. When sodium hydroxide solution is added to ferric chloride, a reddish brown precipitate of ferric hydroxide, insoluble even in the excess of the sodium hydroxide, is obtained.

$$FeCl_3 + 3NaOH \longrightarrow \underset{\text{(Reddish brown)}}{Fe(OH)_2} \downarrow + 3NaCl$$

4. A dirty green ppt. of ferrous hydroxide $Fe(OH)_2$, insoluble in excess of NH_4OH, is formed.

$$FeSO_4 + 2NH_4OH \longrightarrow Fe(OH)_3 \downarrow (NH_4)_2SO_4$$

5. A reddish brown precipitate of iron (III) hydroxide is formed.

$$Fe_2(SO_4)_3 + 6NH_4OH \longrightarrow 2Fe(OH)_3 \downarrow + 3(NH_4)_2SO_4$$

6. Ammonium hydroxide solution, when slowly added to copper sulphate solution, light blue precipitate of copper hydroxide is obtained, which is soluble in excess of ammonium hydroxide to form an intense deep blue solution of tetra amine cupric sulphate.

$$CuSO_4 + 2NH_4OH \longrightarrow \underset{\text{Pale blue ppt.}}{Cu(OH)_2} \downarrow + (NH_4)_2SO_4$$

$$CuSO_4 + 4NH_4OH \longrightarrow [Cu(NH_3)_4]SO_4 + 4H_2O$$

Tetra amine cupric sulphate (soluble complex) Deep blue solution.

7. White ppt. of $Zn(OH)_2$ formed which redissolves to give a colourless solution of tetraamine zinc sulphate $[Zn(NH_3)_4]SO_4$.

$$ZnSO_4 + 2NH_4OH \longrightarrow Zn(OH)_2 + (NH_4)_2SO_4$$

Colourless — white, gelatinous ppt. — colourless

with excess of NH_4OH, ppt. dissolves

$$Zn(OH)_2 + (NH_4)_2SO_4 + 2NH_4OH \longrightarrow [Zn(NH_3)_4]SO_4 + 4H_2O$$

(excess) — Tetrammine zinc (II) sulphate (colourless)

8. Initially a light blue ppt. is formed which on addition of excess of ammonium hydroxide dissolves and a deep inky blue solution is formed.

$$CuSO_4 + 2NH_4OH \longrightarrow Cu(OH)_2 \downarrow + (NH_4)_2SO_4$$

Blue with — Pale blue ppt. — Colourless in solution

excess of NH_4OH, ppt. dissolves

$$Cu(OH)_2 + (NH_4)_2SO_4 + 2NH_4OH \longrightarrow [Cu(NH_3)_4]SO_4 + 4H_2O$$

Excess — Tetrammine copper (II) sulphate

9. Initially a white ppt. is formed which disappears (dissolves) in excess of ammonium hydroxide.

$$Zn(NO_3)_2 + 2NH_4OH \longrightarrow Zn(OH)_2 \downarrow + 2NH_4NO_3$$

White ppt.

$$Zn(OH)_2 + NH_4OH \longrightarrow [Zn\ (NH_3)_4]^{2+}$$

Colourless solution, soluble

10. A reddish brown ppt. of ferric hydroxide insoluble in excess of NH_4OH is formed.

$$FeCl_3 + 3NH_4OH \longrightarrow 3NH_4Cl + Fe(OH)_3 \downarrow$$

white ppt.

Chapter 8. (a) Study of Compounds : Hydrogen Chloride

1. Hydrogen chloride gas is passed through silver nitrate solution.
2. Hydrogen chloride gas comes in contact with ammonia solution.
3. Hydrogen chloride gas is passed through lead nitrate solution and the product thus formed is heated.
4. A few drops of dilute hydrochloric acid are added to silver nitrate solution followed by addition of ammonium hydroxide solution.
5. Dilute hydrogen chloride is added to lead nitrate solution.
6. Copper sulphide is treated with dilute hydrochloric acid.
7. Manganese dioxide is treated with concentrated hydrochloric acid.
8. Copper oxide is treated with conc. hydrochloric acid.
9. Magnesium strip is dropped in dil. hydrochloric acid.
10. When platinum is added to a solution of aqua regia.
11. Dilute hydrochloric acid is added to copper carbonate.
12. Dilute hydrochloric acid is added to sodium thiosulphate.

Ans. 1. A white precipitate is formed which is soluble in excess of ammonium hydroxide.
2. Dense white fumes are observed.
3. A white precipitate is formed which gets dissolved on heating.
4. A white precipitate of silver chloride ($AgCl_2$) is formed which dissolves in excess of NH_4OH.
5. A white precipitate of lead chloride ($PbCl_2$) is obtained. It is insoluble in cold water but soluble in hot water.
6. Rotten egg smell of hydrogen sulphide (H_2S) gas will be given out.

7. Black coloured manganese dioxide on reacting with conc. hydrochloric acid gives light brown coloured solution with the evolution of greenish yellow coloured gas.
8. Black coloured copper oxide on reacting with conc. hydrochloric acid gives blue coloured solution with the evolution of greenish yellow coloured gas which turns moist starch iodide paper blue black, turns moist blue litmus to red and finally bleaches it to white.
9. Magnesium metal slowly dissolves with the evolution of a colourless and odourless gas which burns off with a popping sound.
10. Platinum dissolves in the solution of aqua regia.
11. When dilute hydrochloric acid is added to copper carbonate, a brisk effervescence is seen due to the evolution of CO_2, with the formation of copper chloride.

$$CuCO_3(s) + 2HCl(aq) \longrightarrow CuCl_2(aq) + H_2CO_3(aq) \longrightarrow CuCl_2(aq) + H_2O(l) + CO_2(g)$$

12. Sodium thiosulphate reacts with dilute hydrochloric acid to produce sodium chloride, gas of sulphur dioxide, water and sulphur in a yellow solid form.

$$Na_2S_2O_3 + 2HCl \longrightarrow 2NaCl + S + H_2O + SO_2$$

Chapter 8. (b) Study of Compounds : Ammonia and Nitric Acid

1. A piece of moist red litmus paper is placed in a gas jar of ammonia.
2. Solution of ammonium chloride and sodium hydroxide are mixed and heated.
3. Ammonia is dissolved in water.
4. When a glass rod dipped in hydrochloric acid is introduced into a gas jar full of ammonia, dense white fumes are produced.
5. An aqueous solution of ammonia is added to a solution of ferric chloride.
6. Excess of ammonia is passed through an aqueous solution of lead nitrate.
7. Ammonia is passed through excess chlorine.
8. Ammonia is passed through black copper oxide.
9. Ammonia is passed through yellow lead oxide.
10. Ammonia gas is burnt in an atmosphere of oxygen in the absence of catalyst.
11. When a jar filled with ammonia gas is inverted in a trough filled with water.
12. A fertilizer containing an ammonium salt gets spoiled, if accidentally mixed with slaked lime.
13. Copper turnings are heated with concentrated nitric acid.
14. Nitric acid reacts with glycerol.
15. Nitric acid reacts with phenol in presence of conc. sulphuric acid.
16. When a few drops of concentrated nitric acid are dropped on hot saw dust.
17. When a little water is added to a yellow coloured nitric acid.
18. Concentrated nitric acid is added to copper.
19. Action of nitric acid on limestone.
20. Action of heat on concentrated nitric acid.
21. Nitric acid comes in contact with skin.
22. An aqueous solution of KI(aq) is added to HNO_3.
23. Hot and conc. nitric acid is added to sulphur.

Ans. 1. A piece of moist red litmus paper turns blue when it is placed in a gas jar of ammonia. This is because of the basic nature of ammonia gas.

2. Ammonia gas is produced when ammonium chloride (NH_4Cl) and sodium hydroxide (NaOH) is heated.

$$NH_4Cl + NaOH \longrightarrow NaCl + H_2O + NH_3 \uparrow$$

3. Ammonia dissolves in water to form a weak base, NH_4OH.

$$NH_3 + H_2O \rightleftharpoons \underset{\text{Ammonium hydroxide}}{NH_4OH}$$

4. Hydrochloric acid present on the glass rod reacts with ammonia to form dense white fumes of ammonium chloride.

$$NH_3 + HCl \longrightarrow NH_4Cl$$

5. When an aqueous solution of ammonia is added to ferric chloride solution, a reddish brown precipitate of ferric hydroxide is produced which is insoluble even in the excess of ammonium hydroxide.

$$FeCl_3 + 3NH_4OH \longrightarrow \underset{\substack{\text{Ferric hydroxide}\\\text{(Reddish brown ppt.)}}}{Fe(OH)_3\downarrow} + 3NH_4Cl$$

6. A white coloured precipitate of lead hydroxide [$Pb(OH)_2$] is formed which is insoluble in an excess of ammonia solution.
7. When ammonia is passed through excess of chlorine, a highly explosive yellow liquid of nitrogen trichloride is obtained.

$$NH_3 + 3Cl_2 \longrightarrow \underset{\substack{\text{Nitrogen trichloride}\\\text{(highly explosive)}\\\text{(yellow liquid)}}}{NCl_3} + 3HCl$$

8. When ammonia is passed through black copper oxide, it changes to red copper.
9. When ammonia is passed through yellow lead oxide, it changes to silvery white lead.

$$\underset{\substack{\text{Lead}\\\text{Oxide}\\\text{(yellow)}}}{3PbO} + 2NH_3 \xrightarrow{\Delta} \underset{\substack{\text{Lead}\\\text{(silvery}\\\text{white)}}}{3Pb} + N_2 + 3H_2O$$

10. A greenish yellow flame is observed when ammonia gas is burnt in an atmosphere of oxygen in the absence of catalyst.
11. Ammonia is highly soluble in water. When a jar of ammonia is inverted in a trough of water, a low pressure is created in a jar and the water rushes up, as the gas dissolves in water.
12. Slaked lime is a strong alkali, which reacts with ammonium salt present in fertilizer to liberate ammonia gas.

$$(NH_4)_2SO_4 + Ca(OH)_2 \longrightarrow CaSO_4 + 2NH_3 + 2H_2O$$

13. When copper turnings are heated with concentrated nitric acid, copper nitrate, water and nitrogen dioxide gas (which has reddish brown colour and pungent smell) is evolved.

$$Cu + 4HNO_3 \longrightarrow Cu(NO_3)_2 + 2H_2O + 2NO_2$$

14. Nitric acid react with glycerol in presence of conc. H_2SO_4 below 25°C to form glycerol tri-nitrate.

$$\underset{\text{Glycerol}}{\begin{array}{l}CH_2OH\\|\\CHOH\\|\\CH_2OH\end{array}} + 3HNO_3 \xrightarrow[H_2SO_4]{\text{Conc.}} \underset{\text{Nitroglycerine}}{\begin{array}{l}CH_2ONO_2\\|\\CHONO_2\\|\\CH_2ONO_2\end{array}} + 3H_2O$$

15. Nitric acid reacts with phenol in presence of conc. sulphuric acid to form orange coloured compound, picric acid.

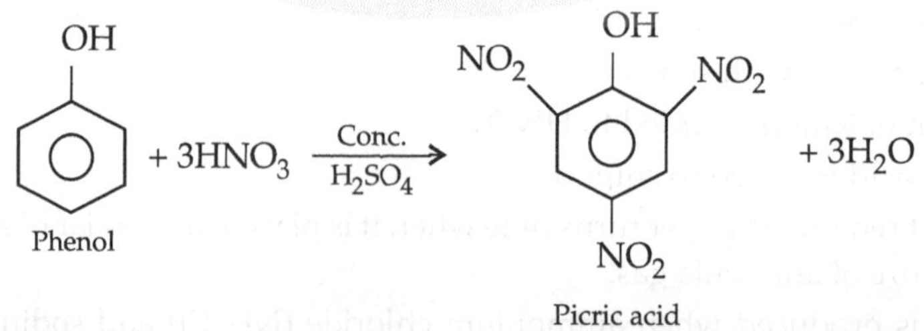

Phenol

Picric acid
(orange colour)

16. Hot saw dust catches fire. This is because concentrated nitric acid decomposes to give oxygen at a higher temperature.

$$4HNO_3 \longrightarrow 4NO_2 + 2H_2O + O_2$$

Oxygen gas is a supporter of combustion and it supports the combustion of hot saw dust, which catches fire.

17. The yellow colouration of nitric acid is due to the decomposition of nitric acid into nitrogen dioxide and other oxides of nitrogen. When a little water is added, all these oxides of nitrogen are dissolved and thus the acid becomes colourless.
18. When conc. HNO_3 is added to copper, a reddish brown gas (NO_2) having pungent smell is evolved.
19. Carbon dioxide gas is liberated when nitric acid reacts with limestone.

$$CaCO_3 + 2HNO_3 \longrightarrow Ca(NO_3)_2 + H_2O + CO_2 \uparrow$$

20. When conc. HNO_3 is heated, it rapidly decomposes to reddish brown nitrogen dioxide gas alongwith the formation of water and oxygen gas

$$4HNO_3 \xrightarrow{\Delta} 2H_2O + 4NO_2 \uparrow + O_2 \uparrow$$

21. The acid produces yellow stain on skin and wood, since skin and wood contain protein. The protein reacts with nitric acid to produce xanthoproteic complex compound. This compound is yellow in colour.
22. When HNO_3 is added to KI(aq), it is oxidised to Iodine. During the reaction, violet colour vapours of iodine is observed.

$$2KI + 4HNO_3 \longrightarrow 2KNO_3 + 2H_2O + 2NO_2 + \underset{\text{violet colour vapours}}{I_2 \uparrow}$$

23. A dense brown fumes of nitrogen dioxide (NO_2) gas are evolved.

$$S + \underset{\text{Hot Conc.}}{6HNO_3} \xrightarrow{\Delta} H_2SO_4 + 2H_2O + 6NO_2 \uparrow$$

Chapter 8. (c) Study of Compounds : Sulphuric Acid

1. Concentrated sulphuric acid is added to a lump of blue vitriol.
2. Crystal of $FeSO_4.7H_2O$ (Green) is placed in conc. sulphuric acid.
3. Lead nitrate is treated with dilute sulphuric acid.
4. Dilute sulphuric acid is added to barium chloride.
5. Sodium chloride is heated with conc. sulphuric acid.
6. Dil. sulphuric acid is added to a solution of barium chloride and lead nitrate.
7. Dilute sulphuric acid is added to iron sulphide.
8. Bleaching powder is heated with dil. sulphuric acid.
9. Zinc metal is treated with dil sulphuric acid.
10. Carbon is heated with conc. sulphuric acid.
11. Sugar is heated with conc. sulphuric acid.

Ans. 1. When concentrated sulphuric acid is added to hydrated copper sulphate, it loses its water of crystallization and thus white annhydrous copper(II) sulphate is formed. In other words, The blue coloured hydrated(II) sulphate turns white due to the loss of water of crystallization.

$$\underset{\substack{\text{Hydrated copper sulphate}\\\text{(Blue crystals)}}}{CuSO_4.5H_2O} + \underset{\text{Conc.}}{[H_2SO_4]} \longrightarrow \underset{\substack{\text{Anhydrous copper}\\\text{sulphate}\\\text{(White crystals)}}}{CuSO_4} + [H_2SO_4.5H_2O]$$

2. When the green coloured hydrated salt $FeSO_4.7H_2O$ is placed in conc. H_2SO_4, it lose the water molecules and the solution becomes white. The colour and crystalline nature of hydrated salt is due to presence of water of crystallization in it.

$$\underset{\text{(Green)}}{FeSO_4.7H_2O} + H_2SO_4 \longrightarrow \underset{\substack{\text{Iron sulphate}\\\text{(White)}}}{FeSO_4} + 7H_2O$$

3. Lead nitrate solution reacts with dilute sulphuric acid to form a white precipitate of lead sulphate, which is insoluble in all mineral acids.

$$Pb(NO_3)_2 + H_2SO_4 \longrightarrow \underset{\text{White ppt.}}{PbSO_4 \downarrow} + 2HNO_3$$

4. Barium chloride solution when added to dilute sulphuric acid, a thick white precipitate of barium sulphate is obtained which is insoluble in all mineral acids. Hydrochloric acid is also formed.

$$BaCl_2 + H_2SO_4 \longrightarrow BaSO_4 \downarrow + 2HCl$$

5. Sodium chloride when heated with concentrated sulphuric acid, liberates pungent fumes of hydrogen chloride gas and sodium sulphate is formed.

$$2NaCl + H_2SO_4 \longrightarrow Na_2SO_4 + \underset{\text{Hydrochloric acid}}{2HCl}$$

6. When dil. sulphuric acid is added to a solution of barium chloride and lead nitrate it forms a white precipitate of barium sulphate and lead sulphate respectively.

$$BaCl_2 + \underset{\text{Dil.}}{H_2SO_4} \longrightarrow 2HCl + BaSO_4 \downarrow$$

$$Pb(NO_3)_2 + H_2SO_4 \longrightarrow PbSO_4 + 2HNO_3$$

7. Iron sulphide reacts with dilute sulphuric acid to form iron(II) sulphate with the evolution of hydrogen sulphide gas, which has a rotten egg smell.

$$FeS + H_2SO_4 \longrightarrow FeSO_4 + \underset{\text{Hydrogen sulphide}}{H_2S \uparrow}$$

8. When bleaching powder is heated with dilute sulphuric acid, chlorine gas is liberated and calcium sulphate is formed along with water.

$$CaOCl_2 + H_2SO_4 \longrightarrow CaSO_4 + H_2O + \underset{\text{Chlorine}}{Cl_2 \uparrow}$$

9. Zinc metal reacts with dilute sulphuric acid to liberate hydrogen gas and zinc sulphate is formed.

$$Zn + H_2SO_4 \longrightarrow \underset{\text{White ppt.}}{ZnSO_4 \downarrow} + \underset{\text{Hydrogen}}{H_2 \uparrow}$$

10. When carbon is heated with concentrated sulphuric acid, it is oxidised to carbon dioxide and the acid is reduced to sulphur dioxide.

$$C + 2H_2SO_4 \longrightarrow CO_2 + 2SO_2 + 2H_2O$$

11. Sulphuric acid acts as a dehydrating agent. When sugar is heated with concentrated sulphuric acid, charring of sugar takes place due to the loss of water molecules and black charred residue is formed.

$$\underset{\text{Sugar}}{C_{12}H_{22}O_{11}} + H_2SO_4 \longrightarrow \underset{\text{Sugar charcoal (black)}}{12C \downarrow} + 11H_2O$$

Chapter 9. Organic Chemistry

1. Methane gas is burnt in air.
2. A mixture of sodium acetate and soda lime is heated.
3. When a mixture of acetylene with twice its volume of hydrogen is passed over nickel as catalyst at about 200°C.
4. Ethylene is heated at 400°C under very high pressure in traces of oxygen.
5. A mixture of ethylene and hydrogen is passed over nickel at 150°C.
6. Ethylene combines with hydrogen chloride.
7. Ethylene reacts with chlorine.
8. Ethene is oxidized with alkaline $KMnO_4$ solution.
9. Ethene is burnt in excess of oxygen.
10. Ethene reacts with bromine.
11. Ethene is bubbled through a solution of bromine in tetrachloromethane (carbon tetrachloride).
12. Calcium carbide is boiled with water.
13. Acetylene is oxidized with alkaline $KMnO_4$ solution.
14. Ethyne is bubbled through a solution of bromine in carbon tetrachloride.

15. Ethyl alcohol is heated at 300°C temperature, in presence of copper catalyst.

16. Ethanol is burnt in air.

17. Acetic acid and ethyl alcohol react in the presence of sulphuric acid.

18. Sodium propionate is heated with soda lime.

Ans. 1. Methane gas burns in air with a blue flame and produces carbon dioxide and water. A large amount of heat energy is liberated during the reaction.

2. When a mixture of sodium acetate and soda lime is heated, colourless and odourless methane gas is liberated.

3. When mixture of acetylene and hydrogen is heated at 200°C temperature, ethane is formed.

$$C_2H_2 + 2H_2 \xrightarrow[\text{temp.}]{200°C} \underset{\text{Ethane}}{C_2H_6}$$

4. When ethylene is heated at 400°C under very high pressure in traces of oxygen, it polymerises to form polyethylene or polyethene

$$nC_2H_4 \xrightarrow[\text{High pressure}]{400°C} \underset{\text{Polyethylene}}{—(CH_2—CH_2—)_n}$$

5. When a mixture of ethylene and hydrogen is passed over nickel at 150°C it forms ethane.

$$\underset{\text{Ethylene}}{CH_2 = CH_2} + H_2 \xrightarrow[150°C]{Ni} \underset{\text{Ethane}}{CH_3—CH_3}$$

6. Ethylene (unsaturated) combines with hydrogen chloride and results in the formation of an addition product chloroethane.

$$CH_2 = CH_2 + HCl \longrightarrow \underset{\text{Chloroethane}}{CH_3—\underset{\substack{|\\ Cl}}{CH_2}}$$

7. When ethylene is treated with chlorine, it combines and results in the formation of saturated 1, 2-dichloro ethane.

$$\underset{\substack{\text{Ethylene}\\ \text{(unsaturated)}}}{CH_2 = CH_2} + Cl–Cl \longrightarrow \underset{\substack{\text{1, 2-dichloroethane}\\ \text{(saturated)}}}{\underset{\substack{|\\ Cl}}{CH_2}—\overset{\substack{Cl\\ |}}{CH_2}}$$

8. Ethene gets oxidized with $KMnO_4$ and produce glycol.

$$H_2C = CH_2 + [O] \xrightarrow[\text{Alkaline}]{KMnO_4} \underset{\text{Glycol}}{\begin{matrix} CH_2OH \\ | \\ CH_2OH \end{matrix}} + H_2O$$

9. A large amount of heat liberated with sooty flame is produced when, ethene is burnt in excess of oxygen.

$$H_2C = CH_2 + 3O_2 \longrightarrow 2CO_2 + 2H_2O + \text{Heat}$$

10. When ethene reacts with bromine, addition reaction takes place. A molecule of bromine is added to the molecule of ethene forming dibromoethane. The almond colour of bromine is discharged.

11. Colour of bromine gets decolourised.

12. Calcium carbide is boiled with water to form acetylene and calcium hydroxide.

$$CaC_2 + H_2O \longrightarrow \underset{\text{Acetylene}}{CH \equiv CH} + Ca(OH)_2$$

13. Acetylene gets oxidized with alkaline $KMnO_4$ solution to produce oxalic acid.

$$\begin{matrix} CH \\ \| \\ CH \end{matrix} + 4[O] \xrightarrow{KMnO_4} \underset{\text{Oxalic acid}}{\begin{matrix} COOH \\ | \\ COOH \end{matrix}}$$

14. Brown colour of bromine disappears due to the formation of ethylene tetrachloride.
15. Ethyl alcohol is heated at 300°C temp. in presence of copper catalyst to form acetaldehyde.

$$CH_3CH_2OH + 2H \xrightarrow[300°C]{Cu} CH_3-CHO$$

16. Ethanol burns in air with a blue flame, producing carbon dioxide and water. During the reaction, large amount of heat energy is liberated.
17. When acetic acid and ethyl alcohol react in the presence of sulphuric acid, esterification process takes place and ethyl acetate is formed.
18. Ethane is prepared by heating a dry mixture of sodium propionate and soda-lime.

$$C_2H_5COONa + NaOH \xrightarrow[\Delta]{CaO} \underset{\text{Ethane}}{C_2H_6} + Na_2O_3$$

Chapter 10. Practical Chemistry

1. Chlorine water is exposed to sunlight.
2. Piece of moist red litmus is placed in the gas jar of ammonia.
3. Piece of moist blue litmus paper is placed in the jar of chlorine.
4. Copper hydroxide is heated in a glass tube.
5. Lead nitrate is heated in a glass tube.
6. Copper carbonate is heated in a glass tube.
7. SO_2 is bubbled through acidified $KMnO_4$ solution.
8. AgCl is shaken with ammonium hydroxide.
9. Anhydrous calcium chloride is exposed to air.
10. Barium chloride solution is added to dil H_2SO_4.
11. Dry red rose petals are placed in the jar of sulphur dioxide.
12. Paper soaked in potassium permangnate solution is introduced into a gas jar of sulphur dioxide.
13. Ammonia gas is burnt in an atmosphere of oxygen in the absence of a catalysts.
14. Moist blue litmus is introduced into a gas jar of sulphur dioxide.
15. Ammonia gas is bubbled through red litmus solution.
16. Neutral litmus solution is added to sodium carbonate solution.
17. Hydrogen sulphide gas is passed through the lead acetate solution.
18. Glass rod dipped in ammonium hydroxide is brought near the mouth of the concentrated hydrochloric acid bottle.
19. A piece of iron is placed in copper sulphate solution (only colour change is recorded).
20. Barium chloride solution is mixed with sodium sulphate solution.
21. When powdered zinc is added to copper sulphate solution.
22. When powdered copper is added to zinc sulphate solution.
23. When a rod of zinc metal is dipped into the solution of copper sulphate.
24. When zinc nitrate crystals are strongly heated.
25. When a piece of calcium is dropped into a trough of water.
26. When a piece of sodium is added to water.
27. When hydrogen is passed over a heated metallic oxide.
28. When a strip of copper is kept immersed in the solution of silver nitrate.
29. Zinc metal is heated in air at 500°C.
30. Zinc metal is treated with ferric sulphate.
31. What happens to the (a) aluminium oxide, (b) iron(III) oxide, when bauxite is treated with sodium hydroxide solution.

32. When iron filings are strongly heated.

33. When a finely ground (powdered) mixture of iron filings and sulphur is heated.

Ans. 1. Bubbles of colourless gas oxygen are observed.

2. The red litmus paper turns blue.
3. The blue litmus turns red and finally white (due to bleaching).
4. Colourless odourless gas evolves. Residue turns black in colour.
5. A crackling sound is heard. Brown gas evolves and the residue turns lime yellow in colour.
6. Colourless odourless gas evolves. Residue turns black in colour.
7. The pink colour of potassium permangnate disappears.
8. The white precepitate dissolves to form colourless solution.
9. It absorbs moisture and forms saturated solution.
10. A white precipitate of barium sulphate is formed.
11. Red rose petals turn white.
12. Paper turns pink to white.
13. Ammonia burns with a green flame.
14. Litmus turns blue to red then gets bleached.
15. The red litmus solution turns blue.
16. Neutral litmus solution turns blue.
17. Solution turns silver black due to lead sulphide formation.
18. Dense white fumes are observed.
19. Pale blue solution slowly turns pale green.
20. When barium chloride is added to sodium sulphate, the products are sodium chloride (which remains in the solution) and barium sulphate precipitate (which settles down as a white precipitate).

$$BaCl_2 + Na_2SO_4 \longrightarrow \underset{\text{(white ppt.)}}{BaSO_4 \downarrow} + 2NaCl$$

21. Zinc displaces copper from copper sulphate solution and zinc sulphate (white gelatinous ppt.) is formed decolorising the blue solution.

$$Zn + CuSO_4 \longrightarrow ZnSO_4 + Cu$$

22. No reaction takes place because copper is less reactive than zinc.
23. Zinc displaces copper from the solution of copper sulphate, decolorising blue solution.
24. A precipitate of zinc oxide (ZnO) is formed with the evolution of NO_2 and O_2 gases. The precipitate (ZnO) formed is yellow when hot and white when cold.

    A reddish brown gas (NO_2) thus liberated has a pungent, irritating odour which turns potassium iodide paper brown.

    A colourless and odourless gas (O_2) thus liberated rekindles a glowing splinter.
25. When a piece of calcium is dropped into water, a vigorous reaction takes place and a colourless, odourless hydrogen gas is evolved, which can be collected in a test tube and water becomes alkaline due to the formation of calcium hydroxide.

$$Ca + 2H_2O \longrightarrow Ca(OH)_2 + H_2 \uparrow$$

26. When a small piece of sodium is dropped into water, producing hydrogen gas, which catches fire, the solution becomes alkaline due to the formation of sodium hydroxide.

$$2Na + 2H_2O \longrightarrow 2NaOH + H_2 \uparrow$$

27. When hydrogen is passed over heated metallic oxide, it is oxidized into water (its corresponding oxide) and the metallic oxide is reduced to free metal.

$$\underset{\text{Metallic oxide}}{MO} + H_2 \longrightarrow H_2O + \underset{\text{Free metal}}{M}$$

28. Copper is more reactive than silver. So it displaces silver from silver nitrate solution.
29. Zinc metal is heated in air at 500°C, the metal burns with bluish-white flame and forms philosopher's wool.

$$2Zn + O_2 \xrightarrow{500°C} ZnO$$

Zinc oxide

30. Zinc metal reduces ferric sulphate to ferrous sulphate. Zinc is a powerful reducing agent.

$$Fe_2(SO_4)_3 + Zn \longrightarrow 2FeSO_4 + ZnSO_4$$

Ferric sulphate — Ferrous sulphate

31. (a) Aluminium oxide dissolves in NaOH solution to form sodium meta-aluminate.
    (b) The iron(III) oxide remains undissolved in the sodium.
32. When iron filings are strongly heated in the presence of atmospheric oxygen, iron is oxidized to form reddish brown ferric oxide.

$$4Fe + 3O_2 \longrightarrow 2Fe_2O_3$$

Iron(III) oxide.

33. When the mixture of iron filings and sulphur is heated, iron combines with sulphur to form a black powder of iron sulphide. During heating the mixture begins to glow, as it is an exothermic reaction.

$$Fe + S \longrightarrow FeS$$

Iron (II) sulphide

❑

Define or Explain the Following

Set 8

Chapter 1. Periodic Properties and Variations of Properties

1. Typical elements.
2. Transition elements.
3. Actinides
4. Rare earths.
5. Transuranium elements.
6. Bridge elements
7. Ionization potential
8. Electron affinity

Ans. 1. The elements of the third period (Na, Mg, Al, Si, P, S, Cl, Ar) are called the typical elements. Each of these elements represent the general properties of all the elements of its group.

2. The elements which possesses partially filled *d*-orbital in its penultimate shell are called transition elements.

3. The 15 elements beginning with actinium and have atomic numbers 89 to 103 are called actinides.

4. The sixth period has 15 elements with atomic numbers 57 to 71 (La-Lu), beginning with lanthanum. These 15 elements are very much alike chemically and are known as rare earths or lanthanides.

5. The elements beyond uranium U, in the seventh period are called transuranium elements.These elements do not occur naturally. They all have been prepared artificially.

6. The elements of second period are called bridge elements because they connect their own group elements with the elements of next group lying diagonally to them in the periodic table *e.g.*, Li (Lithium) and Mg (Magnesium).

7. It is the amount of energy required to remove an electron from the outer most (valence) shell of an isolated gaseous atom.

8. It is the amount of energy released when an electron is added to the outer most (valence) shell of an isolated gaseous atom.

Chapter 2. Chemical Bonding

1. Chemical bond
2. Electrovalent bond
3. Electrovalency
4. Covalent bond
5. Polar covalent bond
6. Non-polar covalent bond
7. Coordinate bond.
8. Lone pair
9. Hydronium ion
10. Electropositive elements
11. Electronegative elements

Ans. 1. An attractive force which hold atoms, ions or molecules together, is known as a chemical bond or linkage.

2. A chemical bond formed by transfer of one or more electrons between two atoms, is known as electrovalent or polar ionic bond.

3. The number of electrons lost or gained by an atom, is called the electrovalency of that element.

4. The bond formed by equal contribution and mutual sharing of electrons between two atoms so that both the atoms acquire the stable nearest noble gas configuration *i.e.*, get their octet complete is called covalent bond.

5. When a covalent bond is formed between two atoms having different electronegativities, the shared pair of electrons is attracted more towards the atom having higher electronegativity. This atom acquires a partial negative charge and other atom acquires a partial positive charge.

6. When a covalent bond is formed between two identical atoms of the same element, the shared pair of electrons is attracted equally by both the bonded atoms. Such a covalent bond is called non-polar covalent bond. *e.g.*, Cl–Cl, H–H etc.

7. A coordinate bond is a special type of covalent bond which is formed by one side sharing of lone pair of electron between the two atoms.
8. Lone pair is defined as the pair of electrons which are not shared by any of the reacting atoms. Lone pair of electrons is actually responsible for its basic nature. Example of lone pairs are $\ddot{N}H_2, H_2\ddot{O}:$.
9. It is a hydrogen ion in association with a molecule of water. In a water molecule, the oxygen atom has two pairs of unshared electrons. The hydrogen ion does not have any electron in the valence shell. Hence, it shares a pair of electrons from the oxygen atom by means of coordinate valency to from an H_3O^+ ion.
10. Elements producing cations are known as electropositive elements, *e.g.* Na^+, K^+, etc.
11. Those elements which produce anions are called as electronegative elements, *e.g.*, Cl^-, N^{3-} etc.

Chapter 3. Study of Acids, Bases and Salts

1. An acid
2. A base
3. An alkali
4. Salts
5. Normal salt
6. Acid salt
7. Basic salt

Ans. 1. A compound which on dissolving in water furnishes a proton [Hydrogen ion (H^+)] as the only positively charged ion is called acid or it is a compound which dissolves in water to form hydronium ion, *i.e.*, HCl, H_2SO_4 etc.

2. A base is a compound which combines with the hydronium ion (H_3O^+) of an acid to form salt and water only, or it is the oxide or hydroxide of a metal which will neutralize the acid.
3. An alkali is a compound, which on dissolving in water furnishes OH^- ions as only negative ions, *e.g.*, NaOH, KOH etc. The soluble bases are called alkali.
4. A salt is an ionic compound which when dissolved in water, yields a positive ion other than hydrogen ion (H^+) and a negative ion other than hydroxyl ion (OH^-).
5. A normal salt is one which does not contain any of ionisable or replaceable hydrogen atoms in its molecule. For example : NaCl, Na_2SO_4, AgCl, $FeCl_3$ etc.
6. An acid salt contains one or more partially replaceable hydrogen ions (H^+) in it. For example : $NaHCO_3$, $KHSO_3$, $Ca(HCO_3)_2$ etc.
7. A basic salt is formed as a result of incomplete neutralisation of a base by an acid. It contains replaceable hydroxyl ions (OH^-). For example : [Cu(OH) Cl], Pb(OH)Cl etc.

Chapter 5. Mole Concept and Stoichiometry

1. Gay Lussac's law of gaseous volumes
2. S.T.P.
3. Absolute Zero
4. Avogadro's law
5. Atomic weight
6. Gram atom
7. Gram molecule
8. Molar volume of gas
9. Vapour Density
10. Torr
11. Atomicity of a gas
12. Molecular formula of a compound
13. Empirical formula of a compound

Ans. 1. It states that whenever gases react, they always do so in volumes which bear a simple whole number ratio to one another, and to the volume of gaseous products, provided that all the volumes are being measured under similar conditions of temperature and pressure.

2. S.T.P. stands for standard temperatutre and pressure. The standard temperature is 273K and the standard pressure is 1 atm pressure.
3. The hypothetical temperature (– 273°C) at which a gas will have zero volume is known as Absolute Zero. This temperature is not attainable in practice and all gases condense to liquid state before this temperature is attained.
4. It states that equal volumes of all gases under similar conditions of temperature and pressure contain equal number of molecules.

5. Atomic weight of an element is the ratio between the mass of 1 atom of that element and $\frac{1}{12}$ of the mass of C–12 isotope.

$$\text{Atomic weight} = \frac{\text{Mass of one atom of the element}}{1/12 \text{ of the mass of C-12 isotope}}$$

6. It is the quantity of an element whose weight in grams is numerically equal to the atomic weight of the element. It is expressed in a.m.u.
7. It is the quantity of a substance whose weight in grams is numerically equal to the molecular weight of the substance.
8. The volume occupied by one gram molecular weight of a gas at S.T.P. is always equal to 22·4 litres. This volume is called molar volume of a gas.
9. It is the ratio between the masses of equal volumes of a gas and hydrogen under the same conditions of temperature and pressure.
10. The pressure exerted by 1mm of Mercury (Hg) column is called one torr.

$$1 \text{ atm} = 760 \text{ mm Hg} = 760 \text{ torr} = 1{\cdot}013 \times 10^5 \text{ Pa}$$

11. Atomicity of a gas is the number of atoms present in one molecule of a gas. For example, a gas which contains only one atom is called monoatomic. Similarly, a gas which contains two atoms is called diatomic.
12. Molecular formula represents the actual number of atoms of various elements present in a molecule of a compound.
13. It is the simplest formula of a compound as it represents the simplest whole number ratio between the atoms of various elements present in a molecule of a compound.

Chapter 6. Electrolysis

1. Electrolysis
2. Electrolytes.
3. Strong electrolyte
4. Weak electrolytes
5. Non-electrolytes
6. Electrolytic cell
7. Electrodes
8. Cathode
9. Anode
10. Ions
11. Cations
12. Anions
13. Ionisation
14. Dissociation
15. Reduction
16. Oxidation
17. Ionic equation
18. Electroplating
19. Electrometallurgy

Ans. 1. The process due to which, a chemical compound in aqueous or fused state conducts direct electric current and at the same time undergoes chemical decomposition, due to the discharge of charged ions is called electrolysis.

2. An electrolyte is a substance, which in the fused state or in aqueous solution conducts an electric current and itself is decomposed as a result of electric current passing through it. For example sodium chloride, potassium chloride and acidulated water.
3. An electrolyte having high degree of dissociation, is termed as strong electrolyte and hence it is completely dissociated in solution.
4. The compounds which in their fused state or aqueous solution are feebly ionised and are poor conductors of electricity are called weak electrolytes. For example, acetic acid, oxalic acid.
5. Non-electrolytes are substances which do not allow the current to pass through them either in the molten state or in aqueous solution. Non electrolytes are non-ionic.
6. The vessel in which electrolysis is carried out consisting of two electrodes and an electrolyte is called an electrolytic cell or voltameter.
7. The graphite or metal plates or rods, through which current enters or leaves an electrolyte, are called electrodes.

8. The electrode which is connected to the negative terminal of a battery, or a cell, is called cathode. The current from electrolyte leaves through cathode.
9. The electrode which is connected to the positive terminal of a battery, or a cell, is called anode. The current enters the electrolyte from anode.
10. When a chemical compound in fused state or in aqueous solution, breaks up into electrically charged atoms or group of atoms, then they are collectively called ions. For example, K^+, Mg^{2+}, Ca^{2+}, F^-, I^-, S^{2-}
11. The positively charged ions, which migrate towards cathode on the passage of electric current, are called cations. The cations accept electrons from cathode and change to neutral atoms. For example, Na^+, Ca^{2+}, K^+.
12. The negatively charged ions, which migrate towards anode on the passage of electric current are called anions. The anions donate electrons to anode and change to neutral atoms. For example, Cl^-, I^-, OH^-.
13. The process of formation of positively and negatively charged ions from molecules which are not initially in the ionic state is called ionisation. *e.g.*, Polar covalent compound HCl.

$$HCl \rightleftharpoons H^+ + Cl^-$$

14. The process of separation of ions which are already present in an ionic compound is called electric dissociation. *e.g.*, electrovalent compounds, NaCl.

$$NaCl \rightleftharpoons Na^+ + Cl^-$$

15. A process of gaining of electrons by a cation to form neutral atom (or gaining of electrons by an electronegative atom to form an anion) is called reduction. Reduction takes place at cathode.
16. A process of loss of electrons by an atom to form a cation (or loss of electrons by an anion to form atom) is called oxidation. Oxidation takes place at anode.
17. A chemical equation which represents the actual atoms (unionised) or ions or radicals taking part in a chemical reaction by the actual exchange of electrons is called an ionic equation.
18. The process of depositing a metal, generally a superior metal like silver, gold, chromium or nickel over another metal or article with the help of electric current is called electroplating.
19. The process of extracting metals from their ores with the help of electrolysis is called electrometallurgy. Highly electronegative metals like Na, K, Mg, Ca and Al are extracted by electrolysis of their fused halides or oxides.

Chapter 7. Metallurgy

1. Metal
2. Metalloids
3. Metallurgy
4. Ore
5. Gangue or Matrix
6. Slag
7. Flux
8. Roasting
9. Calcination
10. Alloy
11. Galvanized iron
12. Passive iron
13. Hardening of steel
14. Tempering of steel
15. Case hardening of steel
16. Amalgams

Ans. 1. A metal is an element which forms a positive ion by the loss of electrons. They are hard, malleable, ductile, lustrous, sonorous and also good conductors of heat and electricity.
2. Elements which exhibit the properties of both metals and non-metals, are called metalloids. They are also known as semimetals. For example, Arsenic, Antimony and Bismuth.
3. The processes involved in the extraction of pure metals from their ore are collectively called metallurgy.
4. Those minerals from which, metals are extracted commercially at a comparatively low cost and with minimum effort, are called ores of the metals.
5. The unwanted impurities which are associated with ore are called gangue or matrix. *e.g.*, stone, clay etc.
6. A fusible mass produced by the combination of flux and gangue is called slag.
7. A flux is a substance which is added to refine metals by combining with impurities to form a molten mixture that can be readily removed.

8. The process of strongly heating the ore in excess of air is called roasting.
9. The process of heating the ore in a limited supply of air, such that temperature is not sufficient to melt the ore is called calcination.
10. An alloy is a homogeneous mixture of either two or more metals or a metal and a non- metal which are mixed together in definite proportion in their molten state.
11. Iron sheets with their surface covered with zinc coating either by electrolysis or by dipping them in molten zinc are known as galvanised iron sheets.
12. Iron dipped in conc. nitric acid and rendered unreactive due to the formation of the layer of ferric oxide, is known as passive iron.
13. When steel heated to a temperature of 750°C to 800°C and then suddenly plunged into cold water or cold oil, the process is called hardening of steel.
14. The process of heating the hardened steel to some fixed temperature and then cooling it slowly is called tempering or annealing of steel.
15. The process of heating mould steel with powdered carbon in clay moulds, such that its upper surface becomes very hard, but inner surface remains soft and spongy is called case hardening of steel.
16. An alloy of mercury with, one or more other metals is called an amalgam, *e.g.*, zinc amalgam, sodium amalgam, etc.

Chapter 9. Organic Chemistry

1. Catenation
2. Hydrocarbons
3. Functional group
4. Isomerism
5. Homologous series
6. Pyrolysis
7. Fermentation

Ans. 1. The property of self-linking of carbon atoms through covalent bonds to form long, straight or branched chain and rings of different sizes is called catenation.

2. Hydrocarbons are the compounds made up of only carbon and hydrogen *e.g.*, CH_4, C_2H_6.
3. An atom of compounds of the same family in which each member differ from its adjacent.
4. Compounds having the same molecular formula, but different structural formula, are called 'Isomers' of one another and this phenomenon is called 'Isomerism'.
5. A series of compounds of the same family in which each member differ from its adjacent member by one CH_2 unit, is called homologous series.
6. The process of decomposition of an organic compound into elements on heating is called pyrolysis.
7. The process of slow decomposition of complex organic compounds into similar substances, in presence of enzymes is called fermentation.

❑

Balancing or Writing the Chemical Equations

Set 9

Chapter 3. Study of Acids, Bases and Salts

1. Write correctly balanced equations for the following reactions :

(1) Molten sodium and chlorine.
(2) Nitrogen and oxygen, when lightning strikes.
(3) Iron and dilute sulphuric acid.
(4) Decomposition of hypochlorous acid in sunlight.
(5) Decomposition of potassium nitrate.
(6) Sodium thiosulphate is reacted with dilute hydrochloric acid.
(7) Calcium bicarbonate reacts with dilute hydrochloric acid.
(8) Dilute sulphuric acid is poured over sodium sulphate.
(9) Zinc is heated with sodium hydroxide solution.
(10) Zinc oxide dissolves in sodium hydroxide.
(11) Lead sulphate from lead nitrate solution and dilute sulphuric acid.
(12) Copper sulphate from copper and concentrated sulphuric acid.
(13) Lead chloride from lead nitrate solution and sodium chloride solution.
(14) Ammonium sulphate from ammonia and dilute sulphuric acid.
(15) Sodium chloride from sodium carbonate solution and dilute hydrochloric acid.
(16) Magnesium and dilute sulphuric acid.
(17) Zinc carbonate and dilute sulphuric acid.
(18) Copper oxide and dilute sulphuric acid.
(19) Ferric hydroxide reacts with nitric acid.

Ans. (1) $2Na + Cl_2 \longrightarrow 2NaCl$

(2) $N_2 + O_2 \longrightarrow 2NO$

(3) $Fe + H_2SO_4 \longrightarrow FeSO_4 + H_2$

(4) $2HClO \xrightarrow{\text{Sunlight}} 2HCl + O_2$

(5) $2KNO_3 \longrightarrow 2KNO_2 + O_2$

(6) $Na_2S_2O_3 + 2HCl \longrightarrow 2NaCl + SO_2 + S\downarrow + H_2O$

(7) $Ca(HCO_3)_2 + 2HCl \longrightarrow CaCl_2 + 2H_2O + 2CO_2\uparrow$

(8) $Na_2SO_3 + H_2SO_4\text{ (dilute)} \longrightarrow Na_2SO_4 + H_2O + SO_2\uparrow$

(9) $Zn + 2NaOH \longrightarrow Na_2ZnO_2 + H_2\uparrow$

(10) $ZnO + 2NaOH \longrightarrow Na_2ZnO_2 + H_2O$

(11) $Pb(NO_3)_2 + H_2SO_4 \longrightarrow PbSO_4\downarrow + 2HNO_3$

Lead nitrate	Sulphuric acid	Lead sulphate	Nitric acid

(12) $Cu + 2H_2SO_4 \longrightarrow CuSO_4 + 2H_2O + SO_2$

Copper	Sulphuric acid	Copper sulphate	Water	Sulphur dioxide

(13) $Pb(NO_3)_2 + 2NaCl \longrightarrow PbCl_2 + 2NaNO_3$

Lead nitrate	Sodium chloride	Lead chloride	Sodium nitrate

(14) $2NH_3 + H_2SO_4 \longrightarrow (NH_4)_2SO_4$

Ammonia	Sulphuric acid	Ammonium sulphate

(15) $Na_2CO_3 + 2HCl \longrightarrow NaCl + H_2O + CO_2$
Sodium carbonate, Hydrochloric acid, Sodium chloride, Water, Carbon dioxide

(16) $Mg + H_2SO_4 \text{ (dil.)} \longrightarrow MgSO_4 + H_2 \uparrow$

(17) $ZnCO_3 + H_2SO_4 \longrightarrow ZnSO_4 + H_2O + CO_2 \uparrow$

(18) $CuO + H_2SO_4 \text{ (dil.)} \longrightarrow CuSO_4 + H_2O$

(19) $Fe(OH)_3 + 3HNO_3 \longrightarrow Fe(NO_3)_3 + 3H_2O$

2. Write the equation for the laboratory preparation of the following salts.

(1) Iron(II) sulphate from iron.
(2) Copper sulphate from copper.
(3) Lead sulphate from lead nitrate.
(4) Sodium sulphate from sodium carbonate.
(5) Copper sulphate from copper(II) oxide.
(6) Iron(III) chloride from iron.
(7) Potassium sulphate from potassium hydroxide solution.
(8) Lead chloride from lead carbonate (two equations).
(9) Zinc sulphate from zinc.
(10) Sodium sulphate from sodium hydroxide.
(11) Lead(II) chloride from lead nitrate.
(12) Copper(II) sulphate from copper carbonate.
(13) Calcium carbonate from calcium chloride.
(14) Sodium sulphate from sodium carbonate.
(15) Zinc carbonate from zinc nitrate.

Ans. (1) $Fe + H_2SO_4 \xrightarrow{\text{dil.}} FeSO_4 + H_2 \uparrow$

(2) $Cu + 2H_2SO_4 \xrightarrow{\text{conc.}} CuSO_4 + 2H_2O + SO_2$

(3) $Pb(NO_3)_2 + H_2SO_4 \xrightarrow{\text{dil.}} PbSO_4 + 2HNO_3$

(4) $Na_2CO_3 + H_2SO_4 \longrightarrow Na_2SO_4 + H_2O + CO_2$

(5) $CuO + H_2SO_4 \longrightarrow CuSO_4 + H_2O$
Copper(II) Oxide, Copper sulphate

(6) $2Fe + 3Cl_2 \longrightarrow 2FeCl_3$
Iron Chlorine, Ferric chloride

(7) $2KOH + H_2SO_4 \xrightarrow{\text{Temp. above } 200°C} K_2SO_4 + 2H_2O$
Potassium hydroxide, Potassium sulphate

(8) $PbCO_3 + 2HNO_3 \longrightarrow Pb(NO_3)_2 + H_2O + CO_2 \uparrow$
Lead carbonate (insoluble salt), Lead nitrate (soluble salt)

$Pb(NO_3)_2 + 2HCl \longrightarrow PbCl_2 + 2HNO_3$
Lead nitrate, Lead chloride

(9) $Zn + H_2SO_4 \text{ (dil.)} \longrightarrow ZnSO_4 + H_2 \uparrow$

(10) $2NaOH + H_2SO_4 \text{ (dil.)} \longrightarrow Na_2SO_4 + 2H_2O$

(11) $Pb(NO_3)_2 + 2HCl \longrightarrow PbCl_2 + 2HNO_3$

(12) $CuCO_3 + H_2SO_4 \text{ (dil.)} \longrightarrow CuSO_4 + H_2O + CO_2$
Copper carbonate, Copper(II) sulphate

(13) $CaCl_2 + Na_2CO_3 \longrightarrow CaCO_3 + 2NaCl$

(14) $Na_2CO_3 + H_2SO_4 \longrightarrow Na_2SO_4 + H_2O + CO_2$
Sodium carbonate, dil. sulphuric acid, Sodium sulphate

(15) $Zn(NO_3)_2 + Na_2CO_3 \longrightarrow ZnCO_3 + 2NaNO_3$
Zinc nitrate, Sodium carbonate, Zinc carbonate, Sodium nitrate

3. Action of dil. acids on :
(1) Carbonates
(2) Bicarbonates
(3) Sulphites
(4) Sulphides

Ans. (1) $CaCO_3 + 2HCl \longrightarrow CaCl_2 + H_2O + CO_2$
(2) $NaHCO_3 + HNO_3 \longrightarrow NaNO_3 + H_2O + CO_2$
(3) $ZnSO_3 + 2HCl \longrightarrow ZnCl_2 + H_2O + SO_2$
(4) $FeS + H_2SO_4 \longrightarrow FeSO_4 + H_2S$

Chapter 4. Analytical Chemistry

1. Write balanced chemical equations to show the reactions of the following :
(1) Aluminium and caustic potash solution.
(2) Aluminium oxide and sodium hydroxide.
(3) Aluminium oxide and potassium hydroxide.
(4) Zinc oxide and potassium hydroxide.
(5) Zinc is heated with sodium hydroxide solution.
(6) Zinc oxide dissolves in sodium hydroxide.
(7) Caustic soda solution and zinc oxide.
(8) Caustic soda solution and aluminium oxide.
(9) Caustic soda solution and lead monoxide.
(10) Action of KOH on $CuSO_4$
(11) Action of KOH on $MgSO_4$
(12) Action of KOH on $ZnSO_4$
(13) Action of KOH on $Fe_2(SO_4)_3$.
(14) Action of sodium hydroxide on freshly precipitated aluminium hydroxide.

Ans. (1) $2Al + 2KOH + 2H_2O \longrightarrow 2KAlO_2 + 3H_2\uparrow$
Aluminium, Caustic potash, Potassium aluminate

(2) $Al_2O_3 + 2NaOH \longrightarrow 2NaAlO_2 + H_2O$
Aluminium oxide, Sodium hydroxide, Sodium aluminate

(3) $Al_2O_3 + 2KOH \longrightarrow 2KAlO_2 + H_2O$
Aluminium oxide, Potassium hydroxide, Potassium aluminate

(4) $ZnO + 2KOH \longrightarrow K_2ZnO_2 + H_2O$
Zinc oxide, Potassium hydroxide, Potassium zincate

(5) $Zn + 2NaOH \longrightarrow Na_2ZnO_2 + H_2\uparrow$
Sodium zincate

(6) $ZnO + 2NaOH \longrightarrow Na_2ZnO_2 + H_2O$
Sodium zincate

(7) $ZnO + 2NaOH \longrightarrow Na_2ZnO_2 + H_2O$
Zinc oxide, Caustic soda, Sodium zincate, Water

(8) $Al_2O_3 + 2NaOH \longrightarrow 2NaAlO_2 + H_2O$
Aluminium oxide, Caustic soda, Sodium meta aluminate, Water

(9) $PbO + 2NaOH \longrightarrow Na_2PbO_2 + H_2O$
Lead oxide, Caustic soda, Sodium plumbite, water

(10) $CuSO_4 + 2KOH \longrightarrow Cu(OH)_2\downarrow + K_2SO_4$

Copper sulphate (Blue solution) + Potassium hydroxide ⟶ Copper hydroxide (Blue ppt.) + Potassium sulphate

$Cu(OH)_2 + KOH \longrightarrow$ No reaction

(Excess)

(11) $MgSO_4 + 2KOH \longrightarrow Mg(OH)_2\downarrow + K_2SO_4$

Magnesium sulphate + Potassium hydroxide ⟶ Magnesium hydroxide (A white ppt.) + Potassium sulphate

$Mg(OH)_2 + KOH \longrightarrow$ No reaction

(Excess)

(12) $ZnSO_4 + 2KOH \longrightarrow Zn(OH)_2 + K_2SO_4$

Zinc sulphate (Colourless) + Potassium hydroxide ⟶ Zinc hydroxide (White ppt.) + Potassium sulphate

$Zn(OH)_2 + 2KOH \longrightarrow K_2ZnO_2 + 2H_2O$

(Excess) Potassium zincate (Water soluble)

(13) $Fe_2(SO_4)_3 + 6KOH \longrightarrow 2Fe(OH)_3 + 3K_2SO_4$

Ferric sulphate (Brown) + Potassium hydroxide ⟶ Ferric hydroxide (Reddish brown ppt.) + Potassium sulphate

$Fe(OH)_3 + KOH \longrightarrow$ No reaction

(Excess)

(14) $Al(OH)_3 + NaOH \xrightarrow{\Delta} NaAlO_2 + 2H_2O$

Aluminium hydroxide ⟶ Sodium meta-aluminate

Chapter 6. Electrolysis

1. Write equations for the reactions taking place at cathode and at anode during the electrolysis of :

(1) Acidified nickel sulphate solution with nickel electrode.
(2) Acidified copper sulphate solution with copper electrode.
(3) Acidified copper sulphate solution with platinum electrode.
(4) Acidulated water with inert electrode.
(5) Molten lead bromide with inert electrodes.
(6) Electroplating a spoon with silver.

Ans. (1) Cathode : $Ni^{2+} + 2e^- \longrightarrow Ni$

Anode : $Ni - 2e^- \longrightarrow Ni^{2+}$

(2) Cathode : $Cu^{2+} + 2e^- \longrightarrow Cu$

Anode : $Cu - 2e^- \longrightarrow Cu^{2+}$

(3) Cathode : $Cu^{2+} + 2e^- \longrightarrow Cu$

Anode : $OH^- - e^- \longrightarrow OH$

$4OH^- \longrightarrow 2H_2O + H_2$

(4) Cathode : $H^+ + e^- \longrightarrow [H]$

$2\,[H] \longrightarrow H_2$

Anode : $OH^- - e^- \longrightarrow OH$

$4OH \longrightarrow 2H_2O + O_2$

(5) Cathode : $Pb^{2+} + 2e^- \longrightarrow Pb$

Anode : $2Br^- - 2e^- \longrightarrow 2\,[Br]$

$2[Br^-] \longrightarrow Br_2$

(6) Cathode : $AgNO_3 \longrightarrow Ag^+ + NO_3^-$

$Ag^+ + e^- \longrightarrow Ag$

Anode : $NO_3^- - e^- \longrightarrow NO_3$

$Ag + NO_3 \longrightarrow AgNO_3$

2. Complete and balance the following equations :

(1) $CH_3COOH \rightleftharpoons$ ...... + ...

(2) $4H^+ +$ ...... $\longrightarrow 2H_2$

(3) $Ag^+ + \ldots\ldots \longrightarrow Ag$

(4) $Cl^- - 1e^- \longrightarrow \ldots\ldots$

(5) $CuSO_4(aq) \rightleftharpoons \ldots\ldots + \ldots\ldots$

(6) $H_2O \rightleftharpoons \ldots\ldots + \ldots\ldots$

(7) $AgNO_3 \rightleftharpoons \ldots\ldots + \ldots\ldots$

(8) $4OH^- - \ldots\ldots \longrightarrow 4OH$

(9) $Na[Ag(CN)_2] \rightleftharpoons \ldots\ldots + \ldots\ldots + \ldots\ldots$

(10) $NiSO_4 \rightleftharpoons \ldots\ldots + \ldots\ldots$

Ans. (1) $CH_3COOH \rightleftharpoons CH_3COO^- + H^+$

(2) $4H^+ + 4e^- \longrightarrow 2H_2$

(3) $Ag^+ + e^- \longrightarrow Ag$

(4) $Cl^- - e^- \longrightarrow Cl$

(5) $CuSO_4(aq) \rightleftharpoons Cu^{2+} + SO_4^{2-}$

(6) $H_2O \rightleftharpoons H^+ + OH^-$

(7) $AgNO_3 \rightleftharpoons Ag^+ + NO_3^-$

(8) $4OH^- - 4e^- \longrightarrow 4OH$

(9) $Na[Ag(CN)_2] \rightleftharpoons Na^+ + Ag^+ + 2CN^-$

(10) $NiSO_4 \rightleftharpoons Ni^{2+} + SO_4^{2-}$

3. Write equations for the reactions taking place at the two electrodes (mentioning clearly the name of the electrode) during the electrolysis of :
(1) Acidified copper sulphate solution with copper electrodes.
(2) Molten lead bromide with inert electrodes.

Ans. (1) Overall cell reaction $CuSO_4 \rightleftharpoons Cu^{2+} + SO_4^-$

∴ $CuSO_4$ undergoes dissociation reaction.

Cathode : $Cu^{2+} + 2e^- \longrightarrow Cu$

Anode : $Cu - 2e^- \longrightarrow Cu^{2+}$

(2) Overall cell reaction $PbBr_2 \rightleftharpoons Pb^{2+} + 2Br^-$

Cathode : $Pb^{2+} + 2e^- \longrightarrow Pb$

Anode : $2Br^- - 2e^- \longrightarrow Br_2 \uparrow$

Chapter 7. Metallurgy

1. Write balanced chemical equation :
(1) The reduction of metallic oxide inside the blast furnace.
(2) Formation of slag inside the blast furnace.
(3) Heating of aluminium hydroxide.
(4) Reaction of zinc with hot concentrated sodium hydroxide.
(5) Reduction of zinc oxide.
(6) Burning of aluminium in air.
(7) Reduction of ferric oxide by aluminium powder.
(8) Calamine is heated.
(9) Zinc placed in ferrous sulphate solution.
(10) Reduction of copper oxide by hydrogen.
(11) Reduction of iron(III) oxide by carbon monoxide.
(12) Reduction of lead(II) oxide by carbon.
(13) Action of steam on zinc.
(14) Zinc is treated with dilute sulphuric acid.
(15) Action of copper sulphate solution on zinc.
(16) Alumina undergoes electrolytic reduction.*

*** Frequently asked previous years Board Exam Questions.**

Ans. (1) $Fe_2O_3 + 3CO \longrightarrow 2Fe + 3O_2$

(2) $CaO + SiO_2 \longrightarrow CaSiO_3$

(3) $2Al(OH)_3 \xrightarrow{\Delta} Al_2O_3 + 3H_2O$

(4) $Zn + 2NaOH \longrightarrow Na_2ZnO_2 + H_2$

(5) $ZnO + C \longrightarrow Zn + CO$

(6) $4Al + 3O_2 \xrightarrow{\Delta} 2Al_2O_3$

$2Al + N_2 \xrightarrow{\Delta} 2AlN$

(7) $Fe_2O_3 + 2Al \longrightarrow Al_2O_3 + 2Fe$

(8) $ZnCO_3 \xrightarrow{\Delta} ZnO + CO_2$

(9) $Zn + FeSO_4 \longrightarrow ZnSO_4 + Fe$

(10) $CuO + H_2 \longrightarrow Cu + H_2O$

(11) $Fe_2O_3 + 3CO \longrightarrow 2Fe + 3CO_2$

(12) $PbO + C \longrightarrow Pb + CO$

(13) $Zn + \underset{\text{Steam}}{H_2O} \longrightarrow \underset{\text{Zinc oxide}}{ZnO} + H_2\uparrow$

(14) $Zn + \underset{\text{(Dil.)}}{H_2SO_4} \longrightarrow ZnSO_4 + H_2\uparrow$

(15) $Zn + CuSO_4 \longrightarrow ZnSO_4 + Cu.$

(16) Oxygen gas (O_2) is evolved when alumina undergoes electrolysis.

$$Al_2O_3 \rightleftharpoons 2Al^{3+} + 3O^{2-}$$

2. Complete and balance the following equations :

(1) $Zn + HCl \longrightarrow$

(2) $Ag_2O \longrightarrow$

(3) $Al_2O_3 + NaOH \xrightarrow{\Delta}$

(4) $Al(OH)_3 \xrightarrow{\Delta}$

(5) $ZnCO_3 \xrightarrow{\Delta}$

(6) $ZnO + H_2SO_4 \xrightarrow{\Delta}$

(7) $NaAlO_2 + H_2O \xrightarrow{\Delta}$

(8) $Al + N_2 \xrightarrow{\Delta}$

(9) $Al + NaOH + H_2O \xrightarrow{\Delta}$

(10) $ZnS + O_2 \xrightarrow{\Delta}$

(11) $Fe_2O_3 + CO \xrightarrow{\Delta}$

(12) $Al + Fe_2O_3 \xrightarrow{\Delta}$

Ans. (1) $Zn + \underset{\text{(Dil.)}}{2HCl} \longrightarrow ZnCl_2 + H_2\uparrow$

(2) $2Ag_2O \xrightarrow{\Delta} 4Ag + O_2\uparrow$

(3) $Al_2O_3 + 2NaOH \xrightarrow{\Delta} 2NaAlO_2 + H_2O$

(4) $2Al(OH)_3 \xrightarrow{\Delta} Al_2O_3 + 3H_2O$

(5) $ZnCO_3 \xrightarrow{\Delta} ZnO + CO_2\uparrow$

(6) $ZnO + H_2SO_4 \xrightarrow{\Delta} ZnSO_4 + H_2O$

(7) $NaAlO_2 + 2H_2O \xrightarrow{\Delta} NaOH + Al(OH)_3\downarrow$

(8) $2Al + N_2 \xrightarrow{\Delta} 2AlN$

(9) $2Al + 2NaOH + 2H_2O \xrightarrow{\Delta} 2NaAlO_2 + 3H_2\uparrow$

(10) $2ZnS + 3O_2 \xrightarrow{\Delta} 2ZnO + 2SO_2\uparrow$

(11) $Fe_2O_3 + 3CO \longrightarrow 2Fe + 3CO_2\uparrow$

(12) $2Al + Fe_2O_3 \xrightarrow{\Delta} Al_2O_3 + 2Fe$

Chapter 8. (a) Study of Compounds : Hydrogen Chloride

1. Write balanced equations for the reaction of dilute hydrochloric acid with each of the following :

Ans. (1) Iron
(2) Sodium hydrogen carbonate
(3) Iron(II) sulphide
(4) Sodium sulphite
(5) Sodium thiosulphate solution
(6) Calcium bicarbonate
(7) Calcium carbonate
(8) Sodium hydroxide
(9) Zinc metal
(10) Potassium permanganate
(11) Red lead heated
(12) Magnesium metal
(13) Ammonium hydroxide
(14) Magnesium sulphite.
(15) Sodium hydrogen sulphide.
(16) Manganese dioxide.

Ans. (1) $Fe + \underset{\text{(Dil.)}}{2HCl} \longrightarrow FeCl_2 + H_2$

(2) $NaHCO_3 + \underset{\text{(Dil.)}}{HCl} \longrightarrow NaCl + H_2O + CO_2$

(3) $FeS + \underset{\text{(Dil.)}}{2HCl} \longrightarrow FeCl_2 + H_2S$

(4) $Na_2SO_3 + \underset{\text{(Dil.)}}{2HCl} \longrightarrow 2NaCl + SO_2 + H_2O$

(5) $Na_2S_2O_3 + \underset{\text{(Dil.)}}{2HCl} \longrightarrow 2NaCl + H_2O + SO_2 + S\downarrow$

(6) $Ca(HCO_3)_2 + \underset{\text{(Dil.)}}{2HCl} \longrightarrow \underset{\text{Calcium chloride}}{CaCl_2} + 2H_2O + 2CO_2\uparrow$

(7) $CaCO_3 + 2HCl \longrightarrow CaCl_2 + H_2O + CO_2$

(8) $NaOH + HCl \longrightarrow NaCl + H_2O$

(9) $Zn + 2HCl \longrightarrow ZnCl_2 + H_2$

(10) $2KMnO_4 + \underset{\text{(Conc.)}}{16\ HCl} \longrightarrow 2KCl + MnCl_2 + 8H_2O + 5Cl_2\ (g)$

(11) $Pb_3O_4 + \underset{\text{(Conc.)}}{8HCl} \longrightarrow 3PbCl_2 + 4H_2O + Cl_2\ (g)$

(12) $Mg + 2HCl\ (g) \longrightarrow MgCl_2 + H_2(g)$

(13) $\underset{\text{Ammonium hydroxide}}{NH_4OH} + HCl \longrightarrow NH_4Cl + H_2O$

(14) $\underset{\text{Magnesium sulphite}}{MgSO_3} + 2HCl \longrightarrow MgCl_2 + SO_2 + H_2O$

(15) $\underset{\text{Sod. hydrogen sulphide}}{NaHS} + HCl \longrightarrow NaCl + H_2S\ (g)$

(16) $\underset{\text{Maganese dioxide}}{MnO_2} + \underset{\text{(Conc.)}}{4HCl} \xrightarrow{\text{Boiling}} MnCl_2 + 2H_2O + \underset{\text{Chlorine gas}}{Cl_2\uparrow}$

2. Write balanced equation for the reaction of hydrochloric acid with each of the following :

(1) Marble chips
(2) Calcium sulphite
(3) Lead nitrate solution.
(4) Mangnese oxide.

Ans. (1) $CaCO_3 + 2HCl \longrightarrow CaCl_2 + H_2O + CO_2\uparrow$

(2) $CaSO_3 + 2HCl \longrightarrow CaCl_2\downarrow + H_2O + SO_2\uparrow$

(3) $Pb(NO_3)_2 + 2HCl \longrightarrow PbCl_2\downarrow + 2HNO_3$

(4) $MnO_2 + 4HCl \longrightarrow MnCl_2 + 2H_2O + Cl_2$

3. Complete and balance the following equations :

(1) $\underset{\text{(Conc.)}}{KCl + H_2SO_4} \xrightarrow{< 200°C}$

(2) $\underset{\text{(Conc.)}}{KCl + H_2SO_4} \xrightarrow{> 200°C}$

(3) $\underset{\text{(Dil.)}}{Cu + HCl} \longrightarrow$

(4) $\underset{\text{(Dil.)}}{CuO + HCl} \longrightarrow$

(5) $\underset{\text{(Conc.)}}{CuO + [O] + HCl} \longrightarrow$

(6) $Cu(NO_3)_2 + HCl \longrightarrow$

(7) $K_2O + 2HCl \longrightarrow$

(8) $2HCl \xrightarrow{500°C}$

(9) $Zn + HCl \longrightarrow$

(10) $Na_2SO_3 + HCl \longrightarrow$

(11) $Mg(HCO_3)_2 + HCl \longrightarrow$

(12) $PbS + HCl \longrightarrow$

(13) $CaCO_3 + HCl \longrightarrow$

(14) $KHSO_3 + HCl \longrightarrow$

(15) $K_2Cr_2O_7 + HCl \longrightarrow$

(16) $P_2O_5 + HCl \longrightarrow$

(17) $MnO_2 + HCl \xrightarrow{\Delta}$

(18) $CaOCl_2 + HCl \xrightarrow{\Delta}$

(19) $HCl + HNO_3 \longrightarrow$

(20) $ZnS + HCl \longrightarrow$

Ans. (1) $\underset{\text{(Conc.)}}{KCl + H_2SO_4} \xrightarrow{< 200°C} KHSO_4 + HCl$

(2) $\underset{\text{(Conc.)}}{2KCl + H_2SO_4} \xrightarrow{> 200°C} K_2SO_4 + 2HCl$

(3) $\underset{\text{(Dil.)}}{Cu + 2HCl} \longrightarrow CuCl_2 + H_2 \uparrow$

(4) $\underset{\text{(Dil.)}}{CuO + 2HCl} \longrightarrow CuCl_2 + H_2O$

(5) $\underset{\text{(Conc.)}}{CuO + [O] + 2HCl} \longrightarrow CuCl_2 + 2H_2O + Cl_2$

(6) $Cu(NO_3)_2 + HCl \longrightarrow CuCl_2 + 2HNO_3$

(7) $K_2O + 2HCl \longrightarrow 2KCl + H_2O$

(8) $2HCl \xrightarrow{500°C} H_2 + Cl_2$

(9) $Zn + 2HCl \longrightarrow ZnCl_2 + H_2 \uparrow$

(10) $Na_2SO_3 + 2HCl \longrightarrow 2NaCl + H_2O + SO_2 \uparrow$

(11) $Mg(HCO_3)_2 + 2HCl \longrightarrow MgCl_2 + H_2O + CO_2 \uparrow$

(12) $PbS + 2HCl \longrightarrow PbCl_2 + H_2O \uparrow$

(13) $CaCO_3 + 2HCl \longrightarrow CaCl_2 + H_2O + CO_2 \uparrow$

(14) $KHSO_3 + HCl \longrightarrow KCl + H_2O + SO_2 \uparrow$

(15) $\underset{\text{(Conc.)}}{K_2Cr_2O_7 + 14HCl} \longrightarrow 2KCl + 2CrCl_3 + 7H_2O + 3Cl_2 \uparrow$

(16) $2P_2O_5 + 3HCl \longrightarrow POCl_3 + 3HPO_3$

(17) $\underset{\text{(Conc.)}}{MnO_2 + 4HCl} \xrightarrow{\Delta} MnCl_2 + 2H_2O + Cl_2 \uparrow$

(18) $$CaOCl_2 + \underset{\text{(dil.)}}{2HCl} \xrightarrow{\Delta} CaCl_2 + H_2O + Cl_2 \uparrow$$

(19) $$3HCl + \underset{\text{(Conc.)}}{HNO_3} \longrightarrow \underset{\text{(Conc.)}}{NOCl} + 2H_2O + 2[Cl]$$

(20) $$ZnS + 2HCl \longrightarrow ZnCl_2 + H_2S \uparrow$$

Chapter 8. (b) Study of Compounds : Ammonia and Nitric Acid

1. Write balanced chemcial equation for the following :

(1) Reaction of hydrogen chloride with ammonia.
(2) The preparation of ammonia from ammonium chloride and calcium hydroxide.
(3) Magnesium heated in nitrogen.
(4) Action of concentrated nitric acid an copper.
(5) Action of heat on silver nitrate.
(6) Reaction of nitric acid with sodium bicarbonate.
(7) Laboratory preparation of nitric acid.
(8) Chlorine reacts with excess of ammonia.
(9) Ferric hydroxide reacts with nitric acid.
(10) Action of heat on ammonium nitrate.
(11) Dilute nitric acid producing carbon dioxide.
(12) Reaction between copper and concentrated nitric acid.
(13) Burning of ammonia in oxygen.
(14) Catalytic oxidation of ammonia.
(15) Action of heat on ammonium chloride
(16) Copper and concentrated nitric acid.
(17) Copper oxide and dilute nitric acid.
(18) Action of conc. HNO_3 on sulphur.
(19) Action of conc. HNO_3 on phosphorus.
(20) Action of conc. HNO_3 on carbon.
(21) Action of conc. HNO_3 on iodine.
(22) Dilute nitric acid and copper.
(23) Ammonia with lead oxide.
(24) Ammonia with copper oxide.
(25) Ammonia with excess chlorine.
(26) Action of water in magnesium nitride.
(27) Burning of ammonia in oxygen.
(28) Catalytic oxidation of ammmonia.
(29) Aluminium nitride and water.
(30) Potassium nitrate and concentrated sulphuric acid.
(31) Concentrated nitric acid and copper.
(32) Action of heat on ammonium nitrate.
(33) Nitrogen monoxide and oxygen.
(34) Dilute nitric acid and sodium carbonate.
(35) Decomposition of nitric acid.

Ans. (1) $$HCl + NH_3 \longrightarrow NH_4Cl.$$

(2) $$2NH_4Cl + Ca(OH)_2 \longrightarrow \underset{\text{Calcium chloride}}{CaCl_2} + \underset{\text{Ammonia}}{2NH_3} + \underset{\text{Water}}{2H_2O}$$

(3) $$3Mg + N_2 \longrightarrow \underset{\text{Magnesium nitrite}}{Mg_3N_2}$$

(4) $$Cu + \underset{\text{Nitric acid}}{4HNO_3} \longrightarrow \underset{\text{Copper(II) nitrate}}{Cu(NO_3)_2} + 2H_2O + 2NO_2 \uparrow$$

(5) $2AgNO_3 \xrightarrow{\Delta} 2Ag + 2NO_2 + O_2$
Silver

(6) $NaHCO_3 + HNO_3 \longrightarrow NaOH + H_2O + CO_2\uparrow$

(7) $NaNO_3 + H_2SO_4 \xrightarrow{\Delta<200°C} NaHSO_4 + HNO_3$
Sodium nitrate, (Conc.), Sodium hydrogen sulphate, Nitric acid

(8) $3Cl_2 + 8NH_3 \longrightarrow 6NH_4Cl + N_2$
Ammonium chloride

(9) $Fe(OH)_3 + 3HNO_3 \longrightarrow Fe(NO_3)_3 + 3H_2O$
Iron(III) nitrate

(10) $NH_4NO_3 \xrightarrow{\Delta} N_2O + 2H_2O$
Ammonium nitrate, Nitrous oxide, Water

(11) $Na_2CO_3 + 2HNO_3 \longrightarrow 2NaNO_3 + H_2O + CO_2\uparrow$
Carbon dioxide gas

(12) $Cu + 4HNO_3 \longrightarrow Cu(NO_3)_2 + 2H_2O + 2NO_2$
(Conc.), Nitrogen dioxide

(13) $4NH_3 + 3O_2 \longrightarrow 6H_2O + 2N_2 + Heat$
Ammonia, Oxygen, Water, Nitrogen

(14) $4NH_3 + 5O_2 \xrightarrow{800°C} 6H_2O + 4NO + Heat$

$4NO + 2O_2 \rightleftharpoons 4NO_2\uparrow$
(Brown gas)

(15) $NH_4Cl \xrightarrow{\Delta} NH_3 + HCl$

(16) $Cu + 4HNO_3 \longrightarrow Cu(NO_3)_2 + 2H_2O + 2NO_2$
Nitric acid, Copper (II) nitrate

(17) $CuO + 2HNO_3 \longrightarrow Cu(NO_3)_2 + H_2O$
(Dil.)

(18) $S + 6HNO_3 \xrightarrow{\Delta} H_2SO_4 + 6NO_2 + 2H_2O$

(19) $2P + 10HNO_3 \longrightarrow 2H_3PO_4 + 10NO_2 + 2H_2O$

(20) $C + 4HNO_3 \xrightarrow{\Delta} H_2CO_3 + 4NO_2 + H_2O$
Carbonic acid

(21) $I_2 + 10HNO_3 \xrightarrow{\Delta} 2HIO_3 + 10NO_2 + 4H_2O$
(Conc.), Iodic acid

(22) $3Cu + 8HNO_3 \longrightarrow 3Cu(NO_3)_2 + 2NO + 4H_2O$

(23) $3PbO + 2NH_3 \longrightarrow 3Pb + 3H_2O + N_2$
Yellow, White

(24) $3CuO + 2NH_3 \longrightarrow 3Cu + 3H_2O + N_2\uparrow$
Black, Reddish

(25) $NH_3 + 3Cl_2 \longrightarrow NCl_3 + 3HCl$
(Excess)

(26) $Mg_3N_2 + 6H_2O \longrightarrow 3Mg(OH)_2 + 2NH_3\uparrow$

(27) $4NH_3 + 3O_2 \xrightarrow{\Delta} 2N_2 + 6H_2O$

(28) $4NH_3 + 5O_2 \xrightarrow[Pt]{800°C} 4NO\uparrow + 6H_2O$

(29) $AlN + 3H_2O \longrightarrow Al(OH)_3\downarrow + NH_3\uparrow$

(30) $KNO_3 + H_2SO_4 \xrightarrow{<200°C} KHSO_4 + HNO_3$
(Conc.)

(31) $Cu + 4HNO_3 \longrightarrow Cu(NO_3)_2 + 2H_2O + NO_2\uparrow$
(Conc.)

(32) $NH_4NO_3 \xrightarrow{\Delta} 2H_2O + \underset{\text{Laughing gas}}{N_2O}$

(33) $2NO + O_2 \xrightarrow{\Delta} 2NO_2\uparrow$

(34) $Na_2CO_3 + \underset{\text{(Dil.)}}{2HNO_3} \longrightarrow 2NaNO_3 + H_2O + CO_2\uparrow$

(35) $4HNO_3 \xrightarrow[\text{Sunlight}]{\Delta} 4NO_2\uparrow + 2H_2O + O_2\uparrow$

2. Complete and balance the following equations :

(1) $Mg_3N_2 + 6H_2O \longrightarrow$

(2) $2NH_3 + 3CuO \longrightarrow$

(3) $8NH_3 + 3Cl_2 \longrightarrow$

(4) $4NH_3 + 5O_2 \longrightarrow$

(5) $NH_4OH + H_2SO_4 \longrightarrow$

(6) $(NH_4)_2SO_4 + Ca(OH)_2 \longrightarrow$

(7) $NH_3 + O_2 \longrightarrow$

(8) $NH_3 + HCl \longrightarrow$

(9) $NH_3 + Cl_2 \longrightarrow$

(10) $NH_3 + \underset{\text{(Conc.)}}{H_2SO_4} \longrightarrow$

(11) $NH_3 + CaCl_2 \longrightarrow$

(12) $NH_3 + P_2O_5 + H_2O \longrightarrow$

(13) $Mg_3N_2 + H_2O \longrightarrow$

(14) $NH_3 + O_2 \xrightarrow[800°C]{Pt}$

(15) $Ca + N_2 \longrightarrow$

(16) $N_2 + H_2 \xrightarrow[Fe]{450°C}$

(17) $NH_3 + H_2SO_4 \longrightarrow$

(18) $FeCl_3 + NH_4OH \longrightarrow$

(19) $NH_3 + CO_2 \xrightarrow[\text{150 atm.}]{150°C}$

(20) $\underset{\text{(Excess)}}{NH_3} + Cl_2 \longrightarrow$

(21) $NH_4OH + \underset{\text{(Excess)}}{HNO_3} \longrightarrow$

(22) $NH_3 + O_2 \xrightarrow{\text{Pt.}}$

(23) $Na_2CO_3 + HNO_3 \longrightarrow$

(24) $C + \underset{\text{(Conc.)}}{HNO_3} \longrightarrow$

(25) $KI + \underset{\text{(Conc.)}}{HNO_3} \longrightarrow$

(26) $H_2S + \underset{\text{(Conc.)}}{HNO_3} \longrightarrow$

(27) $Pb + \underset{\text{(Conc.)}}{HNO_3} \longrightarrow$

(28) $\underset{\text{(Conc.)}}{HNO_3} + \underset{\text{(Conc.)}}{HCl} \longrightarrow$

(29) $SO_2 + HNO_3 \longrightarrow$

(30) $Ag + \underset{\text{(Conc.)}}{HNO_3} \longrightarrow$

Ans. (1) $Mg_3N_2 + 6H_2O \longrightarrow 3Mg(OH)_2 + 2NH_3\uparrow$

(2) $2NH_3 + 3CuO \longrightarrow N_2\uparrow + 3H_2O + 3Cu$

(3) $8NH_3 + 3Cl_2 \longrightarrow 6NH_4Cl + N_2\uparrow$

(4) $4NH_3 + 5O_2 \longrightarrow 4NO\uparrow + 6H_2O$

(5) $2NH_4OH + H_2SO_4 \longrightarrow (NH_4)_2SO_4 + 2H_2O$

(6) $(NH_4)_2SO_4 + Ca(OH)_2 \longrightarrow CaSO_4 + 2NH_3 + 2H_2O$

(7) $4NH_3 + 3O_2 \longrightarrow 2N_2 + 6H_2O$

(8) $NH_3 + HCl \longrightarrow NH_4Cl$

(9) $8NH_3 + 3Cl_2 \longrightarrow 6NH_4Cl + N_2$

(10) $2NH_3 + \underset{\text{(Conc.)}}{H_2SO_4} \longrightarrow (NH_4)_2SO_4$

(11) $2NH_3 + 3CaCl_2 \longrightarrow Ca_3N_2 + 3HCl$

(12) $6NH_3 + P_2O_5 + 3H_2O \longrightarrow 2(NH_4)_3PO_4$

(13) $Mg_3N_2 + 6H_2O \longrightarrow 3Mg(OH)_2 + 2NH_3\ (g)$

(14) $4NH_3 + 5O_2 \xrightarrow[800°C]{Pt} 4NO + 6H_2O$

(15) $3Ca + N_2 \longrightarrow Ca_3N_2$

(16) $N_2 + 3H_2 \xrightarrow[Fe]{450°C} 2NH_3$

(17) $2NH_3 + H_2SO_4 \longrightarrow (NH_4)_2SO_4$

(18) $FeCl_3 + 3NH_4OH \longrightarrow Fe(OH)_3\downarrow + 3NH_4Cl$

(19) $2NH_3 + CO_2 \xrightarrow[150\ atm.]{150°C} \underset{\text{Urea}}{NH_2CONH_2} + H_2O$

(20) $\underset{\text{Excess}}{8NH_3} + 3Cl_2 \longrightarrow N_2 + 6HN_4Cl$

(21) $NH_4OH + HNO_3 \longrightarrow NH_4NO_3 + H_2O$

(22) $4NH_3 + 5O_2 \xrightarrow{Pt.} 4NO\uparrow + 6H_2O$

(23) $Na_2CO_3 + 2HNO_3 \longrightarrow 2NaNO_3 + H_2O + CO_2$

(24) $C + \underset{\text{(Conc.)}}{4HNO_3} \longrightarrow CO_2\uparrow + 4NO_2\uparrow + 2H_2O$

(25) $6KI + \underset{\text{(Conc.)}}{8HNO_3} \longrightarrow 6KNO_3 + 2NO\uparrow + 3I_2\uparrow + 4H_2O$

(26) $3H_2S + \underset{\text{(Conc.)}}{2HNO_3} \longrightarrow 2H_2O + 2NO\uparrow + 3S\downarrow$

(27) $Pb + \underset{\text{(Conc.)}}{4HNO_3} \longrightarrow Pb(NO_3)_2 + 2NO_2\uparrow + 2H_2O$

(28) $\underset{\text{(Conc.)}}{HNO_3} + \underset{\text{(Conc.)}}{3HCl} \longrightarrow NOCl + \underset{\text{Nascent Chlorine}}{2[Cl]} + 2H_2O$

(29) $SO_2 + 2HNO_3 \longrightarrow H_2SO_4 + 2NO_2\uparrow$

(30) $3Ag + \underset{\text{(Dil.)}}{4HNO_3} \longrightarrow 3AgNO_3 + NO + 2H_2O$

3. How are the following conversions carried out ? Give equations only :

(1) $NH_3 \xrightarrow{A} NO \xrightarrow{B} NO_2 \xrightarrow{C} HNO_3 \xrightarrow{D} Cu(NO_3)_2$

(2)

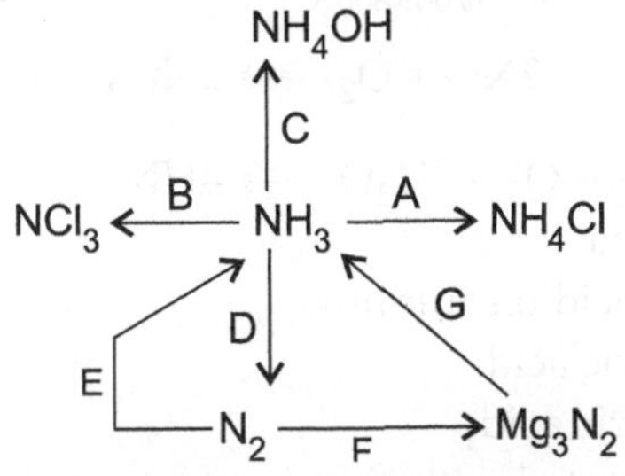

(3) $Na_2CO_3 \xrightarrow{A} NaNO_3 \xrightarrow{B} O_2$

(4) $NH_4Cl \underset{B}{\overset{A}{\rightleftharpoons}} NH_3 \xrightarrow{C} NH_4OH \xrightarrow{D} Cu(OH)_2 \xrightarrow{E} CuO \underset{F}{\overset{G}{\rightleftharpoons}} Cu(NO_3)_2$

Ans. (1) A. $$4NH_3 + 5O_2 \xrightarrow[800°C]{Pt} 4NO + 6H_2O$$

B. $$2NO + O_2 \longrightarrow 2NO_2$$

C. $$4NO_2 + 2H_2O + O_2 \longrightarrow 4HNO_3$$

D. $$CuO + \underset{(Dil.)}{2HNO_3} \longrightarrow Cu(NO_3)_2 + H_2O$$

(2) A. $$\underset{(Excess)}{8NH_3} + 3Cl_2 \longrightarrow 6NH_4Cl$$

OR

$$NH_3 + HCl \longrightarrow NH_4Cl$$

B. $$NH_3 + \underset{(Excess)}{3Cl_2} \longrightarrow NCl_3 + 3HCl$$

C. $$NH_3 + H_2O \longrightarrow NH_4OH$$

D. $$4NH_3 + 3O_2 \longrightarrow 2N_2 + 6H_2O$$

E. $$N_2 + 3H_2 \underset{400-500°C}{\overset{Fe/MO}{\rightleftharpoons}} 2NH_3 + \Delta$$

F. $$\underset{Burning}{3Mg} + N_2 \longrightarrow Mg_3N_2$$

G. $$Mg_3N_2 + 6H_2O \xrightarrow{Warm} 3Mg(OH)_2 + 2NH_3$$

(3) A. $$Na_2CO_3 + 2HNO_3 \longrightarrow 2NaNO_3 + CO_2 + H_2O$$

B. $$2NaNO_3 \xrightarrow{\Delta} 2NaNO_2 + O_2$$

(4) A. $$NH_4Cl + NaOH \xrightarrow{\Delta} NaCl + H_2O + NH_3$$

OR

$$2NH_4Cl + Ca(OH)_2 \xrightarrow{\Delta} CaCl_2 + 2H_2O + 2NH_3$$

B. $$8NH_3 + 3Cl_2 \longrightarrow 6NH_4Cl + N_2$$

OR

$$NH_3 + HCl \longrightarrow NH_4Cl$$

C. $$NH_3 + H_2O \longrightarrow NH_4OH$$

D. $$CuSO_4 + 2NH_4OH \longrightarrow Cu(OH)_2 + (NH_4)_2SO_4$$

E. $$Cu(OH)_2 \xrightarrow{\Delta} CuO + H_2O$$

F. $$CuO + \underset{(Dil.)}{2HNO_3} \longrightarrow Cu(NO_3)_2 + H_2O$$

G. $$2Cu(NO_3)_2 \xrightarrow{\Delta} 2CuO + 4NO_2 + O_2$$

4. Write the reactions involved in the preparation of nitric acid.

Ans. (1) $$4NH_3 + 5O_2 \xrightarrow[(700-800°C)]{Pt} 4NO + 6H_2O + 210 \text{ kcal}$$

(2) $$2NO + O_2 \longrightarrow 2NO_2$$

(3) $$4NO_2 + O_2 + 2H_2O \longrightarrow 4HNO_3$$

5. (1) Catalytic oxidation of ammonia.

(2) Action of concentrated nitric acid on sulphur.

(3) Laboratory preparation of nitric acid.

(4) Reaction of ammonia with nitric acid.

(5) When excess of ammonia is treated with chlorine.

(6) An equation to illustrate the reducing nature of ammonia.

Ans. (1) The equation for catalytic oxidation of ammonia is :

$$\underset{\text{Ammonia}}{4NH_3} + 5O_2 \longrightarrow \underset{\text{Nitric oxide}}{4NO} + 6H_2O$$

Catalyst is a wire mesh consisting of platinum and rhodium.

(2) $$\underset{\text{Sulphur}}{S(s)} + \underset{\text{Nitric acid}}{6HNO_3(aq)} \longrightarrow \underset{\text{Water}}{2H_2O(l)} + \underset{\text{Sulphuric acid}}{H_2SO_4(aq)} + \underset{\text{Nitrogen dioxide}}{6NO_2(g)}$$

(3) $$\underset{\text{Potassium nitrate}}{KNO_3(s)} + \underset{\text{Sulphuric acid}}{H_2SO_4(l)} \longrightarrow \underset{\text{Nitric acid}}{HNO_3(l)} + \underset{\text{Potassium hydrogen sulphate}}{KHSO_4(s)}$$

(4) Ammonia reacts with nitric acid to produce ammonium nitrate

$$\underset{\text{Ammonia}}{NH_3(g)} + \underset{\text{Nitric acid}}{HNO_3(aq)} \longrightarrow \underset{\text{Ammonium nitrate}}{NH_4NO_3(s)}$$

(5) $$8NH_3 + 3Cl_2 \longrightarrow 6NH_4Cl + N_2$$

(6) $$2NH_3 + 3CuO \longrightarrow 3Cu + 3H_2O + N_2$$

Chapter 8. (c) Study of Compounds : Sulphuric Acid

1. Write balanced chemical equation for the following :

(1) Action of concentrated sulphuric acid on carbon.
(2) Dilute sulphuric acid producing hydrogen.
(3) Dilute sulphuric acid is poured over sodium sulphite.
(4) Zinc reacts with conc. sulphuric acid.
(5) Sodium bicarbonate and dilute sulphuric acid.
(6) Sodium nitrate and conc. sulphuric acid.
(7) Iron reacts with dil. sulphuric acid.
(8) Sulphur is heated with concentrated sulphuric acid.
(9) Concentrated sulphuric acid is poured over sugar.
(10) Dilute sulphuric acid with copper carbonate.
(11) Dilute sulphuric acid with lead nitrate solution.
(12) Dilute sulphuric acid with zinc hydroxide.
(13) Concentrated sulphuric acid with barium chloride.
(14) Concentrated sulphuric acid with sulphur trioxide.
(15) Sodium sulphite with dilute sulphuric acid.
(16) Concentrated sulphuric acid with potassium chloride.
(17) Concentrated sulphuric acid with zinc nitrate.
(18) Concentrated sulphuric acid with potassium nitrate.
(19) Potassium hydrogen carbonate and sulphuric acid.
(20) Sodium nitrate and concentrated sulphuric acid.

Ans. (1) $$C + \underset{\text{(Conc.)}}{2H_2SO_4} \longrightarrow \underset{\text{Carbon dioxide}}{CO_2} + 2H_2O + \underset{\text{Sulphur dioxide}}{2SO_2}$$

(2) $$H_2SO_4 + Zn \longrightarrow ZnSO_4 + H_2$$

(3) $$Na_2SO_3 + H_2SO_4 \longrightarrow \underset{\text{Sodium sulphate}}{Na_2SO_4} + H_2O + SO_2\uparrow$$

(4) $$Zn + 2H_2SO_4 \longrightarrow ZnSO_4 + 2H_2O + SO_2$$

(5) $$NaHCO_3 + H_2SO_4 \longrightarrow NaHSO_4 + H_2O + CO_2$$

(6) $$NaNO_3 + H_2SO_4 \longrightarrow NaHSO_4 + HNO_3$$

(7) $$Fe + \underset{\text{(Dil.)}}{H_2SO_4} \longrightarrow FeSO_4 + H_2\uparrow$$

(8) $$S + 2H_2SO_4 \longrightarrow 2H_2O + \underset{\text{Sulphur dioxide}}{3SO_2}$$

(9) $$C_{12}H_{22}O_{11} + H_2SO_4 \longrightarrow 12C + 11H_2O + H_2SO_4$$

(10) $CuCO_3 + \underset{\text{(Dil.)}}{H_2SO_4} \longrightarrow CuSO_4 + H_2O + CO_2\uparrow$

(11) $Pb(NO_3)_2 + \underset{\text{(Dil.)}}{H_2SO_4} \longrightarrow PbSO_4\downarrow + 2HNO_3$

(12) $Zn(OH)_2 + \underset{\text{(Dil.)}}{H_2SO_4} \longrightarrow ZnSO_4 + 2H_2O$

(13) $BaCl_2 + \underset{\text{(Conc.)}}{H_2SO_4} \longrightarrow \underset{\text{White ppt.}}{BaSO_4}\downarrow + 2HCl$

(14) $SO_3 + \underset{\text{(Conc.)}}{H_2SO_4} \longrightarrow \underset{\text{Oleum}}{H_2S_2O_7}$

(15) $Na_2SO_3 + \underset{\text{(Dil.)}}{H_2SO_4} \longrightarrow Na_2SO_4 + H_2O + SO_2\uparrow$

(16) $KCl + \underset{\text{(Conc.)}}{H_2SO_4} \longrightarrow KHSO_4 + HCl$

(17) $Zn(NO_3)_2 + H_2SO_4 \longrightarrow ZnSO_4 + 2HNO_3$

(18) $KNO_3 + H_2SO_4 \xrightarrow{< 200°C} KHSO_4 + HNO_3$

(19) $2KHCO_3 + H_2SO_4 \longrightarrow K_2SO_4 + 2H_2O + CO_2\uparrow$

(20) $NaNO_3 + \underset{\text{(Conc.)}}{H_2SO_4} \longrightarrow NaHSO_4 + HNO_3$

2. Complete and balance the following chemical equations :

(1) $SO_3 + \underset{\text{(Conc.)}}{H_2SO_4} \longrightarrow$

(2) $Zn + \underset{\text{(Dil.)}}{H_2SO_4} \longrightarrow$

(3) $Na_2SO_3 + H_2SO_4 \longrightarrow$

(4) $NaCl + \underset{\text{(Conc.)}}{H_2SO_4} \xrightarrow{200°C}$

(5) $S + \underset{\text{(Conc.)}}{H_2SO_4} \longrightarrow$

(6) $C_6H_{12}O_6 \xrightarrow[\Delta]{\text{Conc. } H_2SO_4}$

(7) $HCOOH \xrightarrow[\Delta]{\text{Conc. } H_2SO_4}$

(8) $H_2C_2O_4 \xrightarrow[\Delta]{\text{Conc. } H_2SO_4}$

(9) $H_2S + \underset{\text{(Conc.)}}{H_2SO_4} \longrightarrow$

(10) $Cu + \underset{\text{(Conc.)}}{H_2SO_4} \longrightarrow$

Ans. (1) $SO_3 + \underset{\text{(Conc.)}}{H_2SO_4} \longrightarrow H_2S_2O_7$

(2) $Zn + \underset{\text{(Dil.)}}{H_2SO_4} \longrightarrow ZnSO_4 + H_2\uparrow$

(3) $Na_2SO_3 + H_2SO_4 \longrightarrow Na_2SO_4 + H_2O + SO_2\uparrow$

(4) $NaCl + \underset{\text{(Conc.)}}{H_2SO_4} \xrightarrow{200°C} NaHSO_4 + HCl$

(5) $S + \underset{\text{(Conc.)}}{H_2SO_4} \longrightarrow 3SO_2\uparrow + 2H_2O$

(6) $C_6H_{12}O_6 \xrightarrow[\Delta]{\text{Conc. } H_2SO_4} 6C + 6H_2O$

(7) $HCOOH \xrightarrow[\Delta]{\text{Conc. } H_2SO_4} CO\uparrow + H_2O$

(8) $$H_2C_2O_4 \xrightarrow[\Delta]{\text{Conc. } H_2SO_4} CO\uparrow + CO_2\uparrow + H_2O$$

(9) $$H_2S + \underset{\text{(Conc.)}}{H_2SO_4} \longrightarrow 2H_2O + SO_2\uparrow + S\downarrow$$

(10) $$Cu + \underset{\text{(Conc.)}}{H_2SO_4} \longrightarrow CuSO_4 + SO_2\uparrow + 2H_2O$$

3. Give balanced chemical equations for the action of sulphuric acid on each of the following :

(1) Potassium hydrogen carbonate

(2) Sulphur

Ans. (1) $$\underset{\text{Potassium hydrogen carbonate}}{2KHCO_3} + \underset{\text{Sulphuric acid (dil.)}}{H_2SO_4} \longrightarrow \underset{\text{Potassium sulphate}}{K_2SO_4} + \underset{\text{Water}}{2H_2O} + \underset{\text{Carbon dioxide}}{2CO_2\uparrow}$$

(2) $$\underset{\text{Sulphur}}{S} + \underset{\text{Sulphuric acid (conc.)}}{2H_2SO_4} \longrightarrow \underset{\text{Sulphuric dioxide}}{3SO_2\uparrow} + \underset{\text{Water}}{2H_2O}$$

Chapter 9. Organic Chemistry

1. (1) Monochloroethane is hydrolysed with aqueous KOH.

(2) A mixture of sodalime and sodium acetate is heated.

(3) Ethanol under high pressure and low temperature is treated with acidified potassium dichromate.

(4) Water is added to calcium carbide.

(5) Ethanol reacts with sodium at room temperature.

(6) Reaction between 1, 2-dibromoethane and alcoholic potassium hydroxide.

(7) Preparation of ethane from sodium propionate.

(8) Preparation of ethanol from monochloroethane and aq. sodium hydroxide.

(9) A saturated hydrocarbon from iodomethane.

(10) An unsaturated hydrocarbon from an alcohol.

(11) An unsaturated hydrocarbon from calcium carbide.

(12) An alcohol from ethyl bromide.

(13) Reaction between ethyl alcohol and acetic acid.

(14) Reaction of chlorine with excess of methane.

(15) Addition of chlorine to ethene at ordinary temperature.

(16) Burning of ethanol in air.

(17) Preparation of ethane from sodium propionate.

(18) Preparation of ethene from iodoethane.

(19) Preparation of ethyne from calcium carbide.

(20) Preparation of methanol from iodomethane.

(21) A mixture of methane and oxygen is heated in the presence of metallic oxide.

(22) A mixture of methane and air is heated at 400°C .

(23) Excess chlorine react with methane.

(24) Excess chlorine react with ethane.

(25) Ethene from bromoethane.*

(26) Ethyne using calcium carbide.*

(27) Methane from sodium acetate.*

(28) Ethene undergoes hydrogenation reaction.*

Ans. (1) $$C_2H_5Cl + \underset{\text{(aq)}}{KOH} \longrightarrow C_2H_5OH + KCl$$

(2) $$CH_3COONa + NaOH \xrightarrow[\text{(Soda lime)}]{CaO} \underset{\text{Methane}}{CH_4} + Na_2CO_3$$

(3) $$CH_3CH_2OH + \underset{\text{Nascent Oxygen}}{2[O]} \xrightarrow[K_2Cr_2O_7]{\text{Acidified}} \underset{\text{Acetic acid}}{CH_3COOH} + H_2O\uparrow$$

* Frequently asked previous years Board Exam Questions.

(4) $$CaC_2 + 2H_2O \longrightarrow Ca(OH)_2 + C_2H_2\uparrow$$
Calcium hydroxide

(5) $$2C_2H_5OH + 2Na \longrightarrow 2C_2H_5ONa + H_2\uparrow$$
Sodium ethoxide

(6) $$CH_2Br{\cdot}CH_2Br + 2KOH \longrightarrow CH \equiv CH + 2KBr + 2H_2O$$
(Alc.) Ethyne

(7) $$C_2H_5COONa + NaOH \xrightarrow{CaO} C_2H_6 + Na_2CO_3$$
Sodium propionate — Ethane — Sodium carbonate

(8) $$C_2H_5Cl + NaOH \longrightarrow C_2H_5OH + NaCl$$
Mono-chloro-ethane — (aq.) — Ethanol — Sodium chloride

(9) $$CH_3I + 2[H] \xrightarrow[\text{reducing agent}]{LiAlH_4} CH_4 + HI$$
Iodomethane — Methane

(10) $$C_2H_5OH + H_2SO_4 \xrightarrow{170°C} C_2H_4 + H_2O$$
Ethanol — Conc. — Ethene

(11) $$CaC_2 + 2H_2O \longrightarrow Ca(OH)_2 + C_2H_2$$
Calcium carbide — Ethyne

(12) $$C_2H_5Br + KOH \longrightarrow C_2H_5OH + KBr$$
Ethyl bromide — (aq) — Ethanol

(13) $$CH_3COOH + C_2H_5OH \longrightarrow CH_3COOC_2H_5 + H_2O$$
Acetic acid — Ethyl alcohol — Ethyl acetate

(14) $$CH_4 + Cl_2 \xrightarrow[\text{UV light}]{h\nu} CH_3Cl + HCl$$

(15) $$CH_2 = CH_2 + Cl_2 \xrightarrow[\text{(solvent)}]{CCl_4} CH_2Cl\text{—}CH_2Cl$$
Ethene — Ethylene chloride

(16) $$C_2H_5OH + O_2 \longrightarrow 2CO_2 + 2H_2O$$
Ethanol — (Air) — Carbon dioxide — Water

(17) $$CH_3CH_2COONa + NaOH \xrightarrow{CaO} C_2H_6 + Na_2CO_3$$

(18) $$CH_3CH_2I + KOH \xrightarrow{\Delta} C_2H_4 + KI + H_2O$$
(Alc.)

(19) $$CaC_2 + 2H_2O \longrightarrow Ca(OH)_2 + C_2H_2$$

(20) $$CH_3I + KOH \longrightarrow CH_3OH + KI$$
(aq.)

(21) $$CH_4 + O_2 \xrightarrow{\text{Molybdenum}} HCHO + H_2O$$
Formaldehyde

(22) $$CH_4 + O_2 \xrightarrow{400°C} HOOCH$$
Air — Formic acid

(23) $$CH_4 + 4Cl_2 \xrightarrow{U.V.} CCl_4 + 4HCl$$
Methane — Carbon tetrachloride

(24) $$CH_3\text{–}CH_3 \xrightarrow{Cl_2} CH_3CH_2Cl + HCl$$
Ethane — Ethyl chloride

$$CH_3CH_2\text{–}Cl \xrightarrow{Cl_2} C_2H_4Cl_2$$
Dichloroethane

(25) Ethene from bromoethane
$$H_2CBr\text{-}CH_3 + KOH \rightarrow H_2C{=}CH_2 + KBr + H_2O$$
Bromoethane — Ethene

(26) Ethyne using calcium carbide

$$\underset{\text{Calcium Carbide}}{CaC_2} + 2H_2O \rightarrow \underset{\text{Ethyne}}{HC \equiv CH} + Ca(OH)_2$$

(27) Methane from sodium acetate

$$\underset{\text{Sodium acetate}}{CH_3COONa} + \underset{\text{Sodium Hydroxide}}{NaOH} \xrightarrow[CaO]{\Delta} \underset{\text{Methane}}{CH_4} + \underset{\text{Sodium carbonate}}{Na_2CO_3}$$

(28) Ethane gas (C_2H_6) is evolved when ethene undergoes hydrogenation reaction.

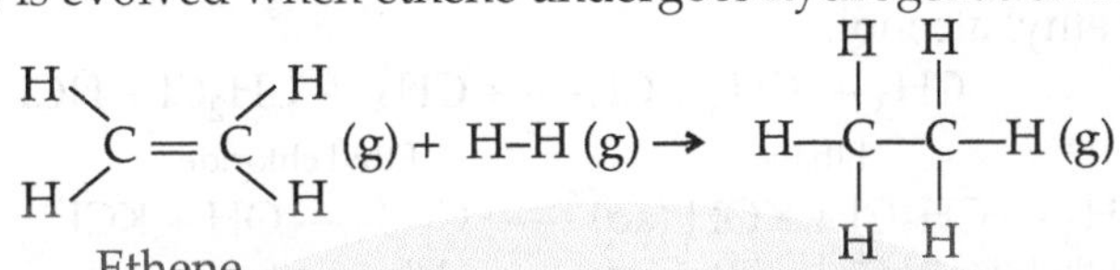

Ethene Ethane

2. How will you bring about the following conversion :

(1) Methane to methyl chloride.
(2) Methane to methyl alcohol.
(3) Methane to formaldehyde.
(4) Methane to formic acid.
(5) Ethane to ethyl chloride.
(6) Ethane to ethyl alcohol.
(7) Ethane to acetaldehyde.
(8) Ethane to acetic acid.
(9) Ethane to ethene.
(10) Ethyne to ethene.
(11) Ethene to ethane.
(12) Ethyne to ethane.
(13) Methane to formic acid.
(14) Ethane to acetic acid.
(15) Ethane to hexachloroethane

Ans. (1) When a mixture of methane and chlorine in the ratio of 1 : 1 by volume is exposed to diffused sunlight, methyl chloride is formed.

$$\underset{\text{Methane}}{CH_4} + Cl_2 \xrightarrow[\text{Diffused sunlight}]{\text{Chlorination}} \underset{\text{Methyl chloride}}{CH_3Cl} + HCl$$

(2) Methane is first converted into methyl chloride by chlorination, in the presence of diffused sunlight. Methyl chloride on hydrolysis forms methyl alcohol.

$$CH_4 + Cl_2 \xrightarrow[\text{Diffused sunlight}]{\text{Chlorination}} \underset{\text{Methyl chloride}}{CH_3Cl} + HCl$$

$$\underset{\text{Methyl chloride}}{CH_3Cl} + KOH\,(aq) \xrightarrow[\text{Diffused sunlight}]{\text{Chlorination}} \underset{\text{Methyl alcohol or Methanol}}{CH_3OH} + KCl + H_2O$$

(3) Methyl alcohol obtained from above is subjected to oxidation with sodium dichromate and dilute sulphuric acid to form formaldehyde.

$$\underset{\text{Methane}}{CH_4} + Cl_2 \xrightarrow{\text{Diffused sunlight}} \underset{\text{Methyl chloride}}{CH_3Cl} + HCl$$

$$\underset{\text{Methane}}{CH_3Cl + KOH\,(aq.)} \xrightarrow{\text{Hydrolysis}} \underset{\text{Methyl alcohol or Methanol}}{CH_3OH} + KCl + H_2O$$

$$\underset{\text{Methyl alcohol}}{CH_3OH + [O]} \xrightarrow[\text{Oxidation}]{Na_2Cr_2O_7/\text{Dil. } H_2SO_4} \underset{\text{Formaldehyde}}{HCHO} + H_2O$$

(4) Formaldehyde obtained from above, when subjected to oxidation with sodium dichromate and dilute sulphuric acid, forms formic acid.

$$\underset{\text{Methane}}{CH_4} + Cl_2 \xrightarrow{\text{Diffused sunlight}} \underset{\text{Methyl chloride}}{CH_3Cl} + HCl$$

$$\underset{\text{Methane chloride}}{CH_3Cl + KOH\,(aq)} \xrightarrow{\text{Hydrolysis}} \underset{\text{Methyl chloride or Methanol}}{CH_3OH} + KCl + H_2O$$

$$\underset{\text{Methyl alcohol}}{CH_3OH + [O]} \xrightarrow[\text{Oxidation}]{Na_2Cr_2O_7/\text{Dil. } H_2SO_4} \underset{\text{Formaldehyde}}{HCHO} + H_2O$$

$$HCHO + [O] \xrightarrow[\text{Oxidation}]{Na_2Cr_2O_7/\text{Dil. } H_2SO_4} \underset{\text{Formic acid}}{HCOOH}$$

(5) When equal volumes of ethane and chlorine are exposed to diffused sunlight, they react to form ethyl chloride (monochloro ethane).

$$\underset{\text{Ethane}}{CH_3 - CH_3} + Cl_2 \longrightarrow \underset{\text{Ethyl chloride}}{CH_3 - CH_2Cl} + HCl$$

(6) When equal volumes of ethane and chlorine are exposed to diffused sunlight, they form ethyl chloride, which when treated with aqueous solution of potassium hydroxide (KOH), undergoes hydrolysis to form ethyl alcohol.

$$\underset{\text{Ethane}}{CH_3 - CH_3} + Cl_2 \longrightarrow \underset{\text{Ethyl chloride}}{CH_3 - CH_2Cl} + HCl$$

$$\underset{\text{Ethyl chloride}}{CH_3 - CH_3Cl} + \underset{\text{Potassium hydroxide}}{KOH\ (aq)} \longrightarrow \underset{\text{Ethyl alcohol (Ethanol)}}{C_2H_5 - OH} + KCl$$

(7) Ethane is first converted to ethyl chloride by chlorination. Ethyl chloride undergoes hydrolysis, when treated with aqueous solution of potassium hydroxide, resulting in the formation of ethyl alcohol. Ethyl alcohol undergoes oxidation in the presence of sodium dichromate, acidified with dilute sulphuric acid, which provides nascent oxygen and oxidise ethyl alcohol to acetaldehyde (ethanol).

$$\underset{\text{Methane}}{C_2H_6} + Cl_2 \xrightarrow{\text{Chlorination}} \underset{\text{Ethyl chloride}}{C_2H_5Cl} + HCl$$

$$\underset{\text{Ethyl chloride}}{C_2H_5Cl} + KOH\ (aq.) \xrightarrow[\text{Ethyl alcohol}]{\text{Hydrolysis}} C_2H_5OH + HCl$$

$$\underset{\text{Ethyl chloride}}{C_2H_5OH} + [O] \xrightarrow{Na_2Cr_2O_7/\text{Dil. } H_2SO_4} \underset{\text{Acetaldehyde}}{CH_3 - CHO} + H_2O$$

(8) Acetaldehyde when subjected to oxidation with sodium or potassium dichromate and dilute sulphuric acid, gets oxidized to form acetic acid.

$$\underset{\text{Ethane}}{C_2H_6} + Cl_2 \xrightarrow{\text{Chlorination}} \underset{\text{Ethyl chloride}}{C_2H_5Cl} + HCl$$

$$\underset{\text{Ethyl chloride}}{C_2H_5Cl} + KOH\ (aq.) \xrightarrow[\text{Ethyl alcohol}]{\text{Hydrolysis}} C_2H_5OH + HCl$$

$$\underset{\text{Ethyl chloride}}{C_2H_5OH} + [O] \xrightarrow{Na_2Cr_2O_7/\text{Dil. } H_2SO_4} \underset{\text{Acetaldehyde}}{CH_3 - CHO} + H_2O$$

$$CH_3 - CHO + [O] \xrightarrow{Na_2Cr_2O_7/\text{Dil. } H_2SO_4} \underset{\text{Acetic acid}}{CH_3 - COOH}$$

(9) Ethane on chlorination forms ethyl chloride which when treated with alcoholic KOH, undergoes, dehydrochlorination to form ethene.

$$\underset{\text{Ethane}}{CH_3 - CH} + Cl_2 \xrightarrow{\text{Chlorination}} \underset{\text{Ethyl chloride}}{CH_2 - CH_2 - Cl} + HCl$$

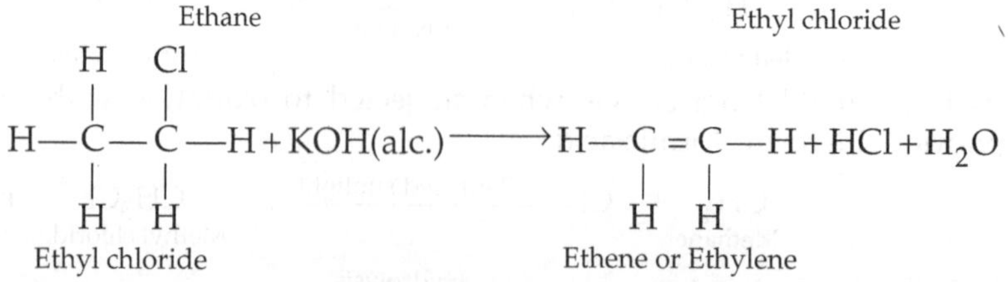

(10) 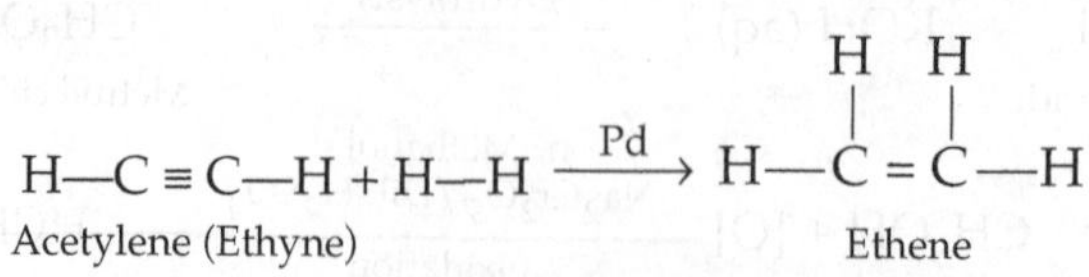

(11)

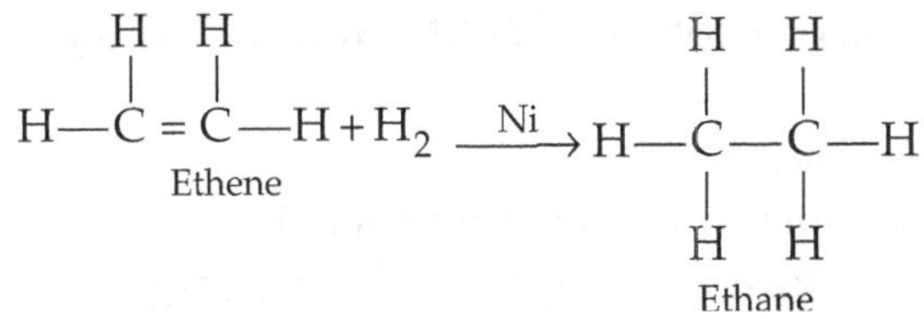

(12) $CH \equiv CH \xrightarrow[H_2]{Pd} C_2H_4 \xrightarrow[H_2]{Ni} C_2H_6$
Ethyne, Ethene, Ethane

(13) $CH_4 \xrightarrow{Cl_2} CH_3Cl \xrightarrow{NaOH} CH_4OH \xrightarrow{[O]} HCHO \xrightarrow{[O]} HCOOH$
Methane, Methyl chloride, Methanol, Formaldehyde, Formic acid

(14) $C_2H_6 \xrightarrow{Cl_2} CH_3CH_2Cl \xrightarrow{NaOH} C_2H_5OH$
Ethane, Ethyl chloride, Ethyl alcohol

$C_2H_5OH \xrightarrow{[O]} CH_3CHO \xrightarrow{[O]} CH_3COOH$
Acetaldehyde, Acetic acid

(15)
$$C_2H_6 + Cl_2 \longrightarrow C_2H_5Cl + HCl$$
Ethane
$$C_2H_5Cl_2 + Cl_2 \longrightarrow C_2H_4Cl_2 + HCl$$
$$C_2H_4Cl_2 + Cl_2 \longrightarrow C_2H_3Cl_3 + HCl$$
$$C_2H_3Cl_3 + Cl_2 \longrightarrow C_2H_2Cl_4 + HCl$$
$$C_2H_2Cl_4 + Cl_2 \longrightarrow C_2HCl_5 + HCl$$
$$C_2HCl_5 + Cl_2 \longrightarrow C_2Cl_2 + HCl$$
Hexachloroethane

2. Give balanced chemical equations for the following conversions :

(1) Ethanoic acid to ethyl ethanoate.
(2) Calcium carbide to ethyne.
(3) Sodium ethanoate to methane.
(4) Burning of ethane in plentiful supply of air.
(5) Action of water on calcium carbide.
(6) Heating of ethanol at 170°C in the presence of conc. sulphuric acid.
(7) Preparation of ethane from sodium propionate.

Ans. (1) $CH_3COOH(l) + C_2H_5OH(l) \xrightarrow{Conc.\ H_2SO_4} CH_3COOC_2H_5(l) + H_2O(l)$
Ethanoic acid, Ethanol, Ethyl ethanoate, Water

(2) $CaC_2(s) + 2H_2O(l) \longrightarrow Ca(OH)_2\ (s) + C_2H_2(g)$
Calcium carbide, Calcium hydroxide, Ethyne

(3) $CH_3COONa(s) + NaOH(s) \xrightarrow[\Delta]{CaO} Na_2CO_3\ (s) + CH_4(g)$
Sodium ethanoate, Sodium hydroxide (from soda lime), Sodium carbonate, Methane

(4) $2C_2H_6 + 7O_2 \longrightarrow 4CO_2 + 6H_2O$

(5) $CaC_2 + 2H_2O \longrightarrow Ca(OH)_2 + C_2H_2$

(6) $C_2H_5OH \xrightarrow[170°C]{Conc.\ H_2SO_4} C_2H_4 + H_2O$

(7) $C_2H_5COONa + NaOH \xrightarrow[\Delta]{CaO} C_2H_6 + Na_2CO_3$
Sodium propionate, Sodium hydroxide, Ethane, Sodium carbonate

Chapter 10. Practical Chemistry

1. Study the following conversion schemes :

(1) Give the equations for the following conversions A to E.

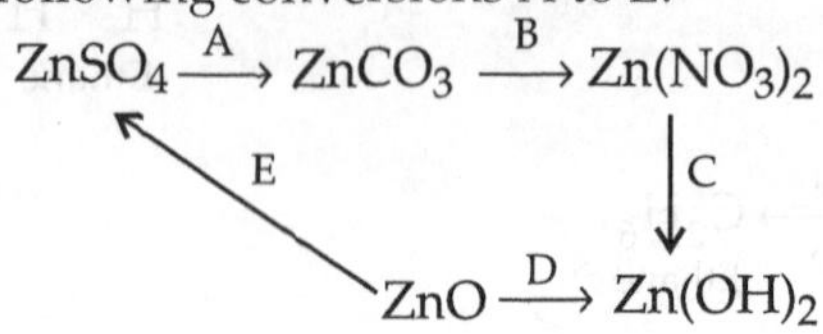

(2) Refer to the flow chart diagram below and give balanced equations with conditions, if any, for the following conversions A to D.

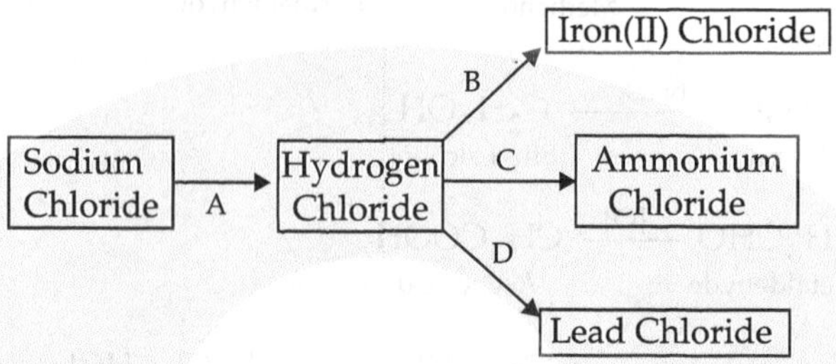

(3) Lead $\xleftarrow{A}$ Lead(II) oxide $\xrightarrow{B}$ Lead(II) nitrate $\xrightarrow{C}$ Lead(II) hydroxide

↓D

Lead(II) carbonate

(i) For each of the conversions A to D in the above state briefly how the conversions can be carried out.

(ii) Write equations for the conversions.

(4) For each of the conversion in the scheme given below, state briefly in words or by means of chemical equation, how the conversion is carried out ?

Copper $\xleftarrow{A}$ Copper(II) oxide $\xrightarrow{B}$ Copper sulphate

Copper(II) oxide ↓D Copper(II) carbonate; Copper sulphate ↓C Copper sulphide

(5) How are the following conversions carried out ? Give balanced chemical equations only.

(i) $PbO \underset{D}{\overset{A}{\rightleftharpoons}} Pb(NO_3)_2 \xrightarrow{B} Pb(OH)_2$ (C above the reverse arrow); $Pb(NO_3)_2 \xrightarrow{E} PbCl_2$

(ii) $Zn \xrightarrow{A} ZnS \xrightarrow{B} ZnSO_4 \xrightarrow{C} Zn(OH)_2$; $ZnS \xrightarrow{E} H_2S$; $ZnSO_4 \xrightarrow{F} ZnCO_3 \xrightarrow{G} ZnO$; $Zn(OH)_2 \xrightarrow{D} ZnO$

(iii) $Fe \xrightarrow{A} FeS \xrightarrow{B} FeCl_2 \xrightarrow{C} Fe(OH)_2$; $FeS \xrightarrow{E} H_2S$; $FeCl_2 \xrightarrow{F} FeCO_3 \xrightarrow{G} FeO$; $Fe(OH)_2 \xrightarrow{D} FeO$

(iv) $Fe \xrightarrow{A} FeCl_3 \xrightarrow{B} Fe(OH)_3 \xrightarrow{C} Fe_2O_3$

Ans. (1) [A] $ZnSO_4 + Na_2CO_3 \longrightarrow ZnCO_3 + Na_2SO_4$

Zinc sulphate, Sodium carbonate, Zinc carbonate, Sodium sulphate

[B] $ZnCO_3 + 2HNO_3 \longrightarrow Zn(NO_3)_2 + H_2O + CO_2$

Zinc carbonate, Nitric acid, Zinc nitrate, Water

[C] $Zn(NO_3)_2 + 2NaOH \longrightarrow Zn(OH)_2 + 2NaNO_3$

Zinc nitrate, Sodium hydoxide, Zinc hydroxide, Sodium nitrate

[D] $Zn(OH)_2 \xrightarrow{\text{Heated}} ZnO + H_2O$
Zinc hydroxide → Zinc oxide + Water

[E] $ZnO + H_2SO_4 \longrightarrow ZnSO_4 + H_2O$
Zinc oxide + Sulphuric acid → Zinc sulphate + Water

(2) [A] $NaCl + H_2SO_4 \xrightarrow{\Delta < 200°C} NaHSO_4 + HCl\uparrow$
conc. ... Hydrogen chloride

[B] $2HCl + Fe \longrightarrow FeCl_2 + H_2$
Iron(II) chloride

[C] $HCl(g) + NH_3(g) \longrightarrow NH_4Cl$
Ammonium chloride

[D] $2HCl + Pb(NO_3)_2(aq) \longrightarrow PbCl_2\downarrow + 2HNO_3$
(dil.) ... Lead chloride

(3) [A] Lead(II) oxide, when heated with powdered coke, reduced to metallic lead and carbon monoxide is formed.

$$PbO + C \longrightarrow Pb + CO$$

[B] Lead(II) oxide, when dissolved in the dilute nitric acid, lead nitrate is formed.

$$PbO + 2HNO_3 \longrightarrow Pb(NO_3)_2 + H_2O$$
Lead nitrate

[C] Lead(II) nitrate is dissolved in water and then sodium hydroxide solution is added, a white precipitate of lead hydroxide is obtained.

$$Pb(NO_3)_2 + 2NaOH \longrightarrow Pb(OH)_2 + 2NaNO_3$$

[D] Lead(II) nitrate is dissolved in water and then concentrated solution of sodium carbonate is added, a white precipitate of lead(II) carbonate is obtained.

$$Pb(NO_3)_2 + Na_2CO_3 \longrightarrow PbCO_3 + 2NaNO_3$$

(4) [A] Copper(II) oxide, when heated with coke powder is reduced to metallic copper.

$$CuO + C \longrightarrow Cu + CO$$
Metallic copper

[B] Copper(II) oxide is dissolved in dilute sulphuric acid to form copper sulphate.

$$CuO + H_2SO_4 \longrightarrow CuSO_4 + H_2O$$
Copper sulphate

[C] Copper sulphate is dissolved in water and hydrogen sulphide gas is passed, a black precipitate of copper sulphide is formed.

$$CuSO_4 + H_2S \longrightarrow CuS + H_2SO_4$$
Black ppt.

[D] Copper(II) oxide is dissolved in dilute hydrochloric acid to form copper chloride. To the solution of copper chloride, when a saturated solution of sodium carbonate is added, a light blue precipitate of copper(II) carbonate is formed.

$$CuCl_2 + Na_2CO_3 \longrightarrow CuCO_3 + 2NaCl$$

(5) (i) [A] $PbO + 2HNO_3 \longrightarrow Pb(NO_3)_2 + H_2O$
(dil.)

[B] $Pb(NO_3)_2 + 2NaOH \longrightarrow Pb(OH)_2 + 2NaNO_3$

[C] $Pb(OH)_2 \xrightarrow{\Delta} PbO + H_2O$

[D] $2Pb(NO_3)_2 \xrightarrow{\Delta} 2PbO + 4NO_2 + O_2$

[E] $Pb(NO_3)_2 + 2NaCl \longrightarrow PbCl_2 + 2NaNO_3$

(ii) [A] $Zn + S \xrightarrow{\Delta} ZnS$

[B] $ZnS + H_2SO_4 \longrightarrow ZnSO_4 + H_2S$
(dil.)

[C] $ZnSO_4 + 2NaOH \longrightarrow Zn(OH)_2 + Na_2SO_4$

[D] $Zn(OH)_2 \xrightarrow{\Delta} ZnO + H_2O$

[E] $ZnS + 2HCl \longrightarrow ZnCl_2 + H_2S$
(dil.)

[F] $ZnSO_4 + Na_2CO_3 \longrightarrow ZnCO_3 + Na_2SO_4$

[G] $ZnCO_3 \xrightarrow{\Delta} ZnO + CO_2$

(iii) [A] $Fe + S \xrightarrow{\Delta} FeS$

[B] $\underset{\text{(dil.)}}{FeS + 2HCl} \longrightarrow FeCl_2 + H_2S$

[C] $FeCl_2 + 2NaOH \longrightarrow Fe(OH)_2 + 2NaCl$

[D] $Fe(OH)_2 \xrightarrow{\Delta} FeO + H_2O$

[E] $FeS + 2HCl \longrightarrow FeCl_2 + H_2S$

[F] $FeCl_2 + Na_2CO_3 \longrightarrow FeCO_3 + 2NaCl$

[G] $FeCO_3 \xrightarrow{\Delta} FeO + CO_2$

(iv) [A] $\underset{\text{Dry}}{2Fe + 3Cl_2} \longrightarrow 2FeCl_3$

[B] $FeCl_3 + 3NaOH \longrightarrow Fe(OH)_3 + 3NaCl$

[C] $2Fe(OH)_3 \xrightarrow{\Delta} Fe_2O_3 + 3H_2O$

2. How can the following be converted to the respective chloride, sulphate and nitrate salts ? Give equations :

(1) Magnesium
(2) Zinc
(3) Iron
(4) Lead carbonate
(5) Copper oxide
(6) Potassium hydroxide
(7) Copper hydroxide
(8) Ammonium hydroxide.

Ans. (1) $\underset{\text{Magnesium}}{Mg(s)} + \underset{\text{Hydrochloric acid}}{2HCl(aq)} \longrightarrow \underset{\text{Magnesium chloride}}{MgCl_2\,(aq)} + \underset{\text{Hydrogen}}{H_2\,(g)}$

(2) $\underset{\text{Zinc}}{Zn(s)} + \underset{\text{Sulphuric acid}}{H_2SO_4(aq)} \longrightarrow \underset{\text{Zinc sulphate}}{ZnSO_4\,(aq)} + \underset{\text{Hydrogen}}{H_2\,(g)}$

(3) $\underset{\text{Iron}}{Fe(s)} + \underset{\text{Sulphur}}{S(s)} \longrightarrow \underset{\text{Iron(II) Sulphide}}{FeS(s)}$

(4) $\underset{\text{Lead carbonate}}{PbCO_3\,(s)} + \underset{\text{Nitric acid}}{2HNO_3(aq)} \longrightarrow \underset{\text{Lead nitrate}}{Pb(NO_3)_2(ag)} + H_2O(l) + CO_2(g)$

(5) $\underset{\text{Copper oxide}}{CuO(s)} + \underset{\text{Sulphuric acid}}{H_2SO_4(aq)} \longrightarrow \underset{\text{Copper sulphate}}{CuSO_4(aq)} + H_2O(l)$

(6) $\underset{\text{Potassium hydroxide}}{2KOH} + \underset{\text{Carbon dioxide}}{CO_2(g)} \longrightarrow \underset{\text{Potassium carbonate}}{K_2CO_3(aq)} + H_2O(l)$

(7) $\underset{\text{Copper hydroxide}}{Cu(OH)_2\,(s)} + \underset{\text{Sulphuric acid}}{H_2SO_4(aq)} \longrightarrow \underset{\text{Copper sulphate}}{CuSO_4(aq)} + 2H_2O$

(8) $\underset{\text{Ammonium hydroxide}}{3NH_4OH(aq.)} + \underset{\text{Phosphoric acid}}{H_3PO_4\,(aq)} \longrightarrow \underset{\text{Ammonium phosphate}}{(NH_4)_3PO_4\,(aq)} + 3H_2O(l)$

❑

IUPAC Nomenclature or Writing the Structural Formula

Set 10

Chapter 7. Metallurgy

1. Give the chemical formulae of the following naturally occuring ores :

(1) Cryolite
(2) Galena
(3) Corundum
(4) Dolomite
(5) Zincite
(6) Malachite
(7) Cinnabar
(8) Gypsum
(9) Horn silver
(10) Epsom salt
(11) Haematite
(12) Bauxite

Ans.

(1) Na_3AlF_6
(2) PbS
(3) Al_2O_3
(4) $CaCO_3.MgCO_3$
(5) ZnO
(6) $CuCO_3.Cu(OH)_2$
(7) HgS
(8) $CaSO_4.2H_2O$
(9) AgCl
(10) $MgSO_4.7H_2O$
(11) Fe_2O_3
(12) $Al_2O_3.2H_2O$

Chapter 9. Organic Chemistry

1. Give the structural formula of the following :

(1) Ethanol
(2) 1-propanal
(3) Ethanoic acid
(4) 1, 2, dichloroethane
(5) An isomer of *n*-butane
(6) 2-propanol
(7) Diethyl ether
(8) Methanoic acid
(9) Ethanal
(10) Ethyne
(11) Acetone
(12) 2-methyl propane
(13) Ethanoic acid
(14) But-2-yne
(15) Two isomers of Butane
(16) Ethane
(17) Vinegar
(18) Marsh gas
(19) An alcohol
(20) 2-Butyne
(21) 2, 3-dimethyl butane
(22) Propanoic acid

Ans. (1)

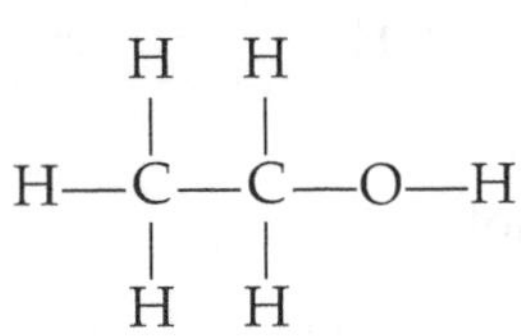

Ethanol

(3)

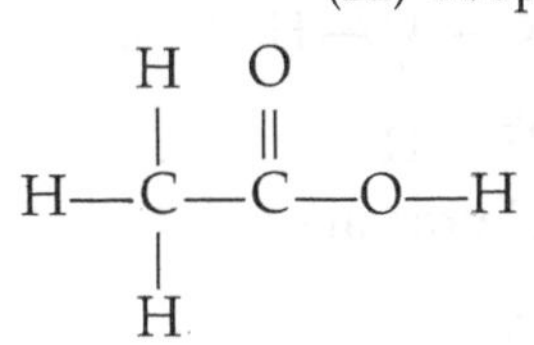

Ethanoic acid

(2)

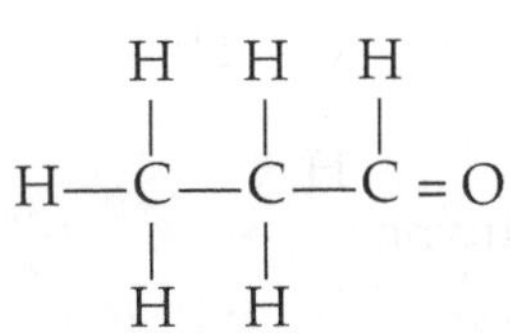

1- Propanal

(4)

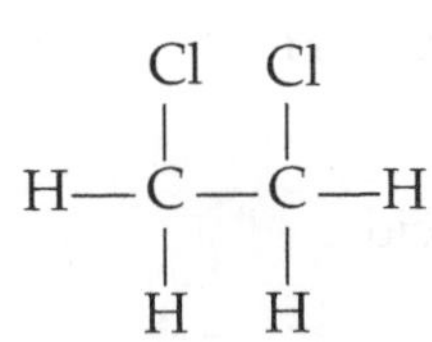

1, 2-dichloro ethane

(5) 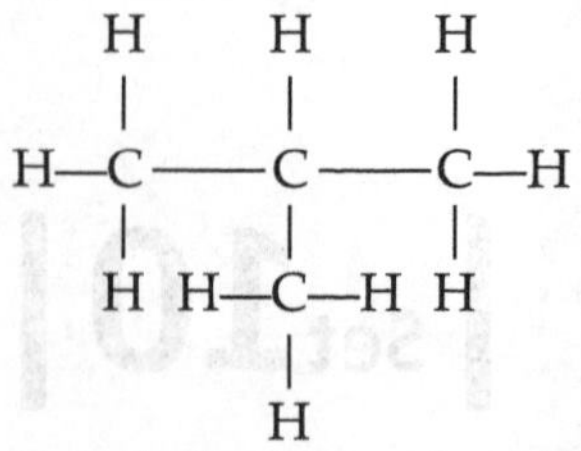

2-methyl propane (an isomer of *n*-butane)

(6) 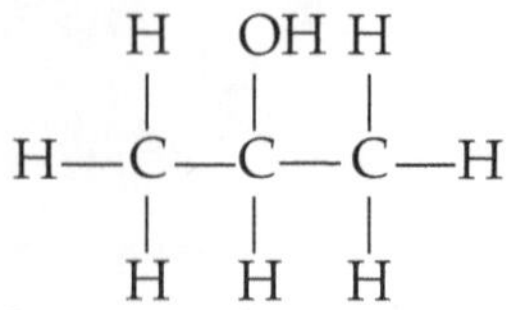

2-propanol

(7) 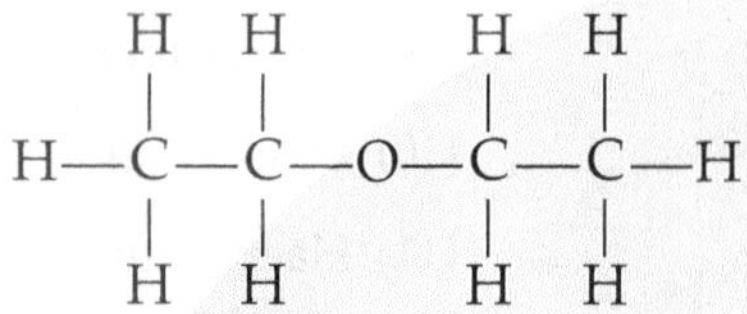

Diethyl ether

(8) 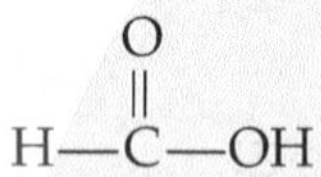

Methanoic acid

(9) 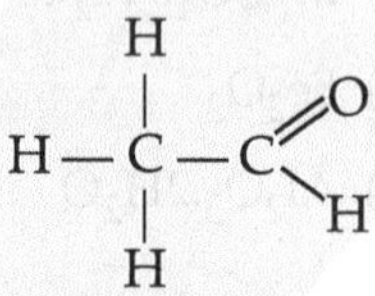

Ethanal

(10) $H-C\equiv C-H$

Ethyne

(11) 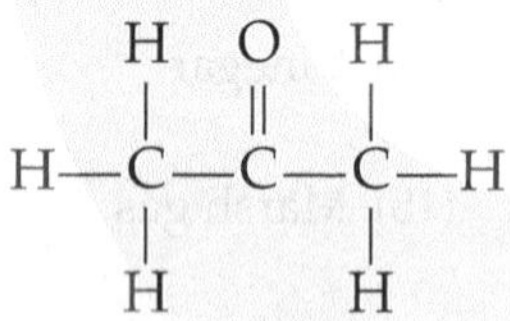

Acetone

(12) 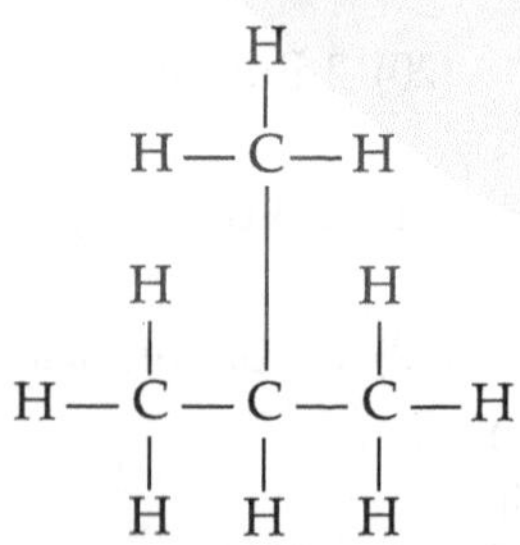

2-methyl propane

(13) 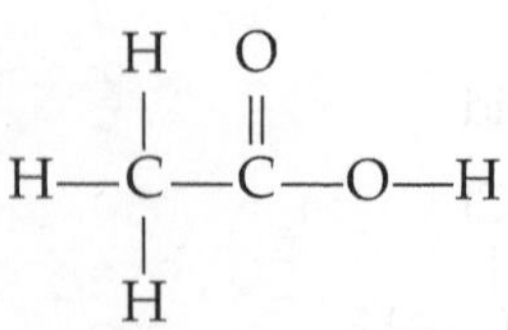

Ethanoic acid

(14) 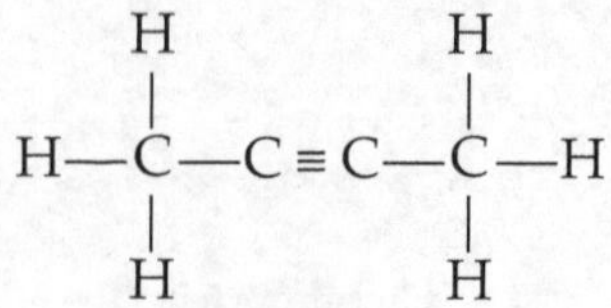

But-2-yne

(15) 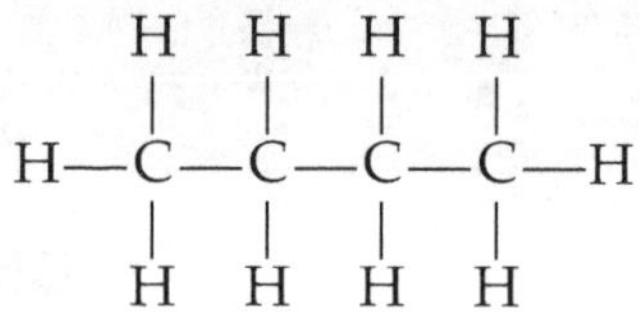

Normal butane

(16) 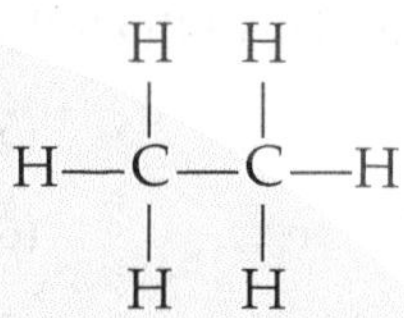

Ethane

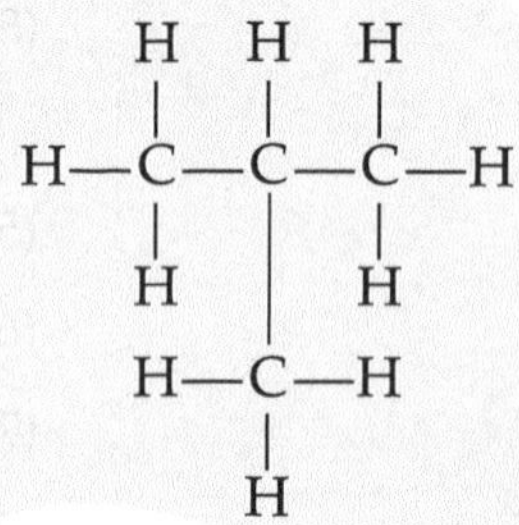

Iso-butane

(17) 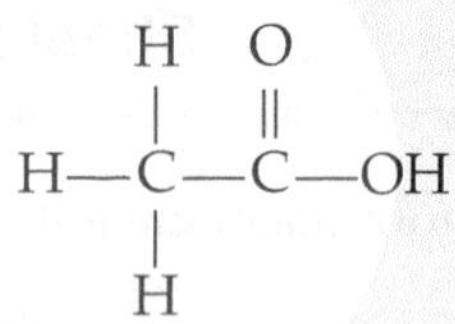

Vinegar

(18) 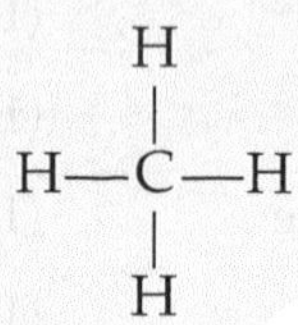

Marsh gas

(19) 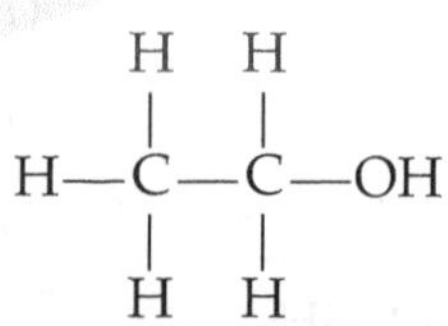

Alcohol

(20) 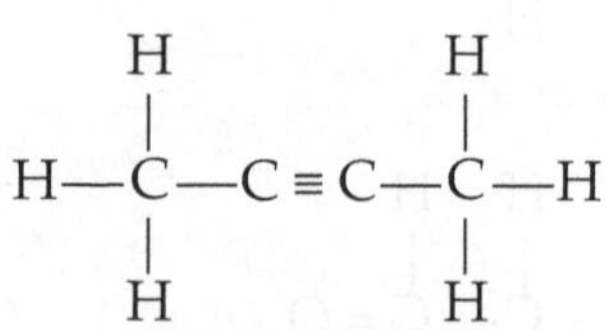

2-butyne

(21) 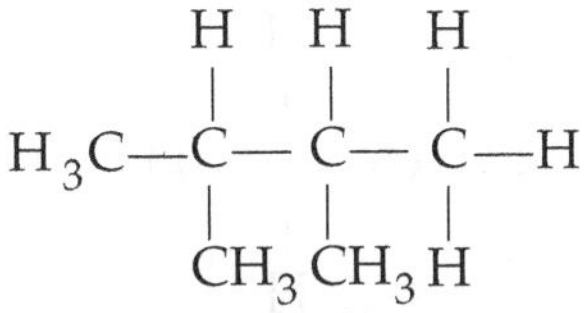

2, 3-dimethyl butane

(22) 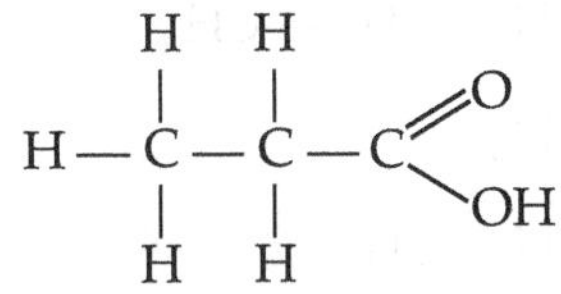

Propanoic acid

2. Give the correct IUPAC name structural formulae given below :

(1) 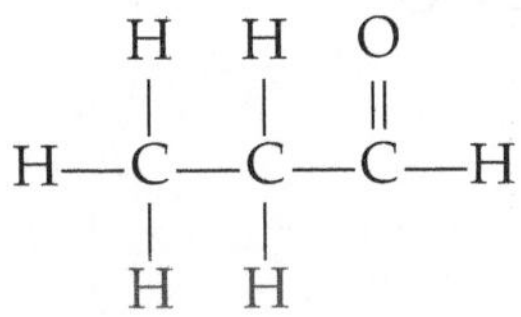

(2) 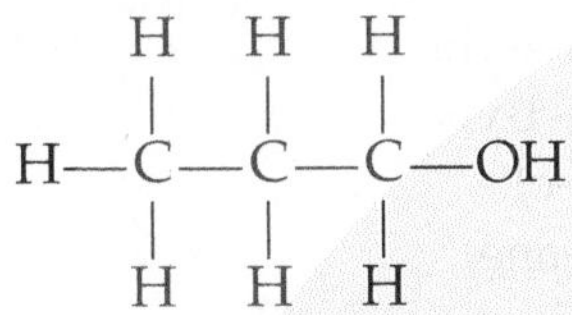

(3) 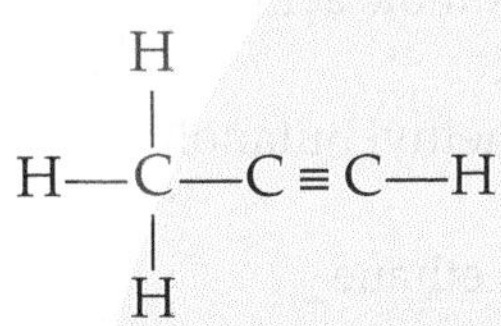

(4) 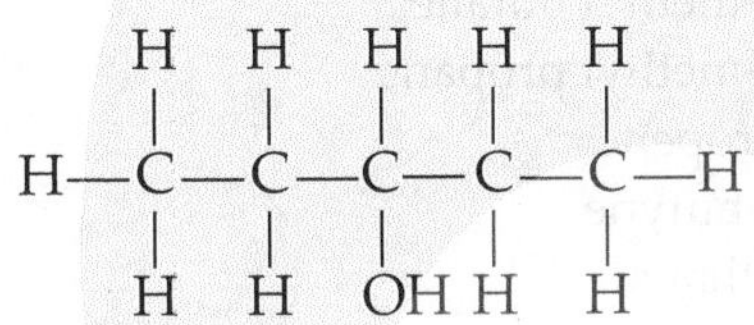

(5) 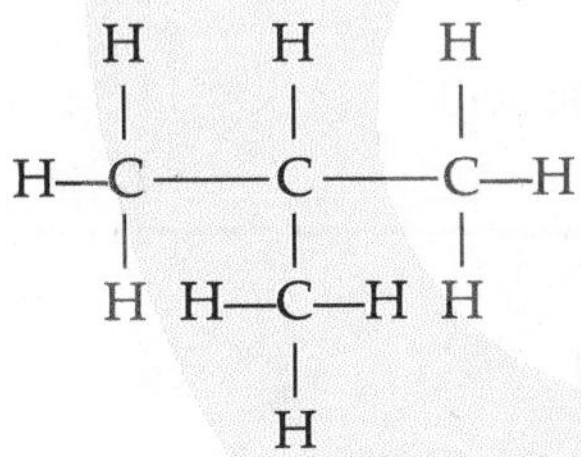

(6) 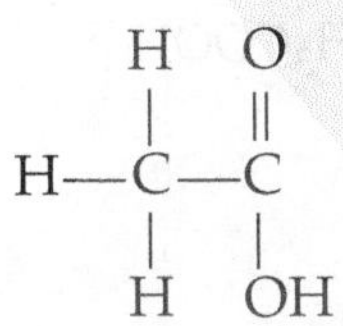

(7) 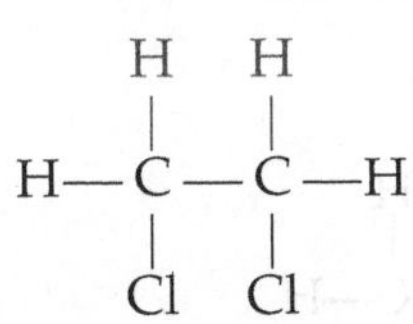

(8) 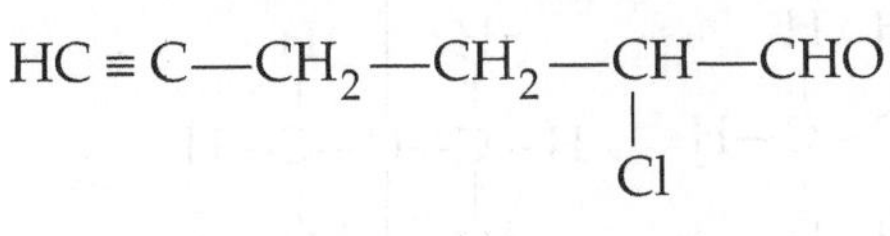

(9) 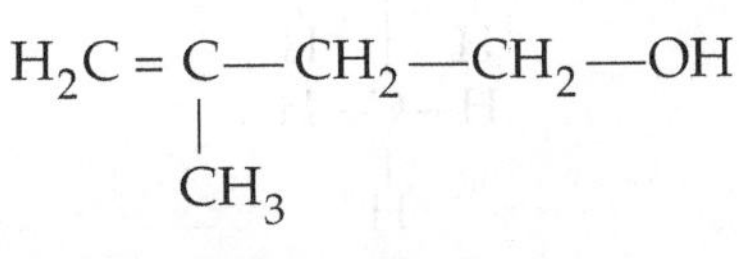

(10) 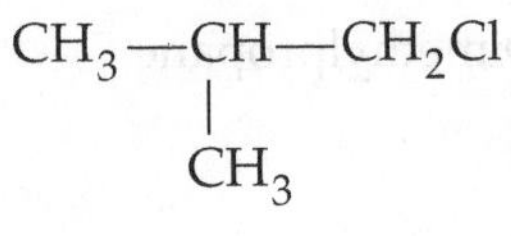

(11) 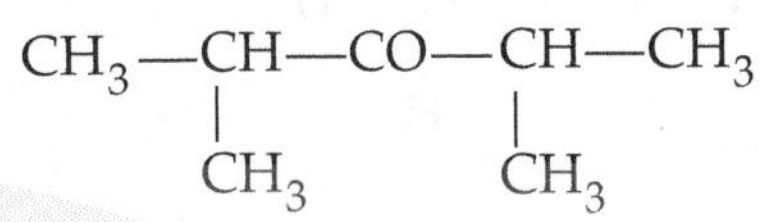

(12) 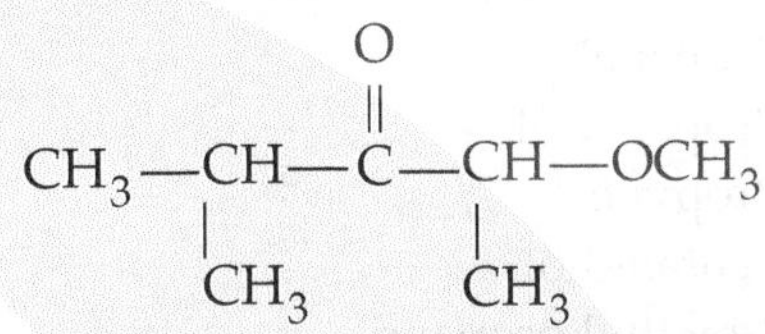

(13) 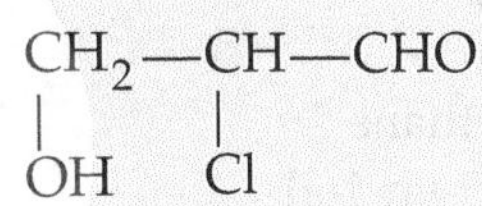

(14) 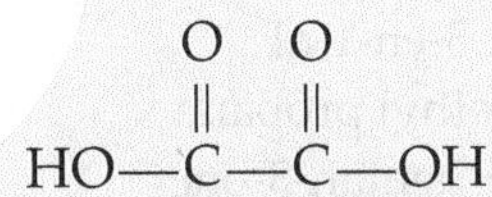

(15) $CH_2 = CH-CH_2-C \equiv CH$

(16) $CH_3-CH = CH-C \equiv CH$

(17) $CH_3CH_2CH_2OH$

(18) 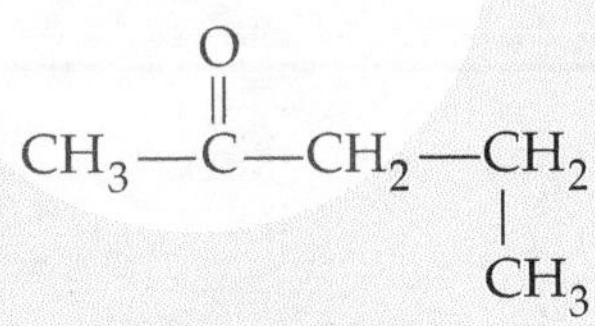

(19) 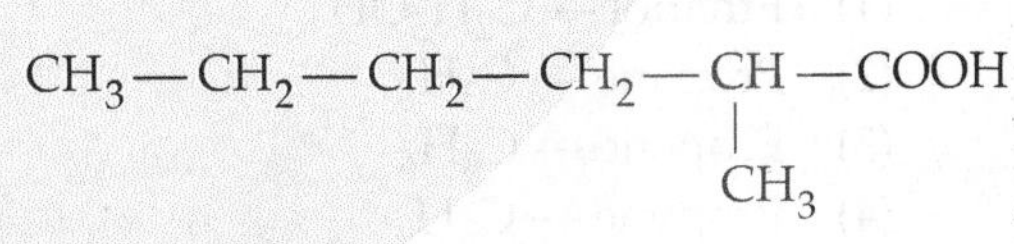

(20) $CH_3CH_2COCH_2CH_2CH_3$

(21) 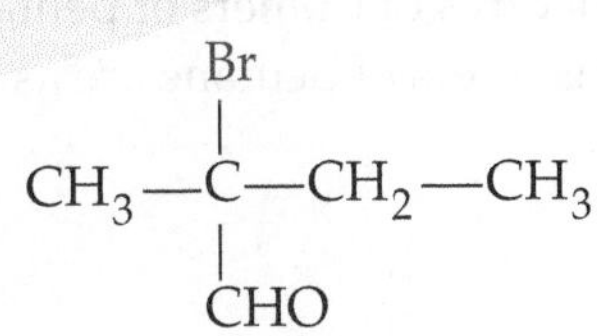

(22) 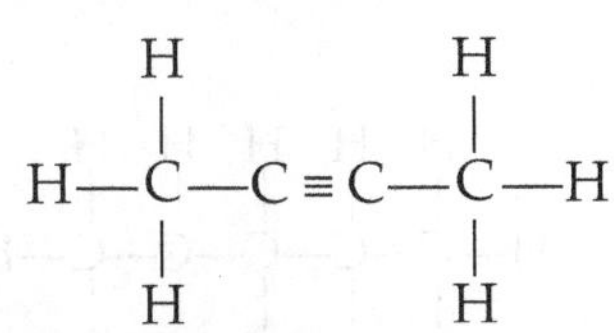

(23) 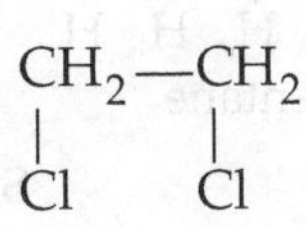

(24) 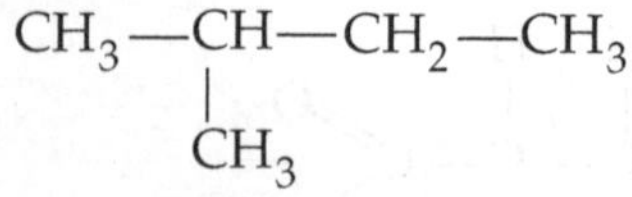

(25) 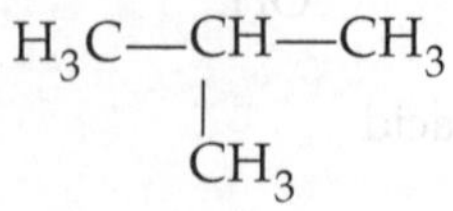

(26) 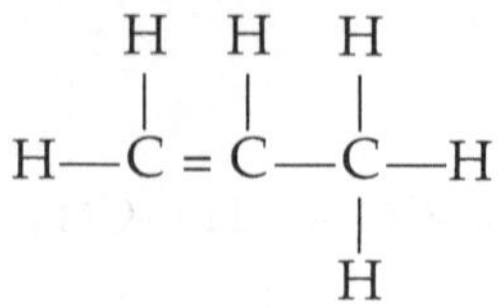

(27) 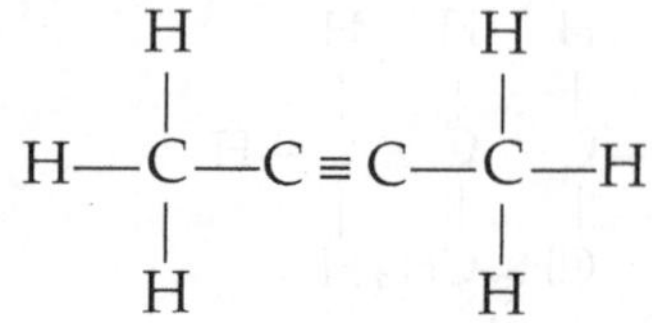

(28) 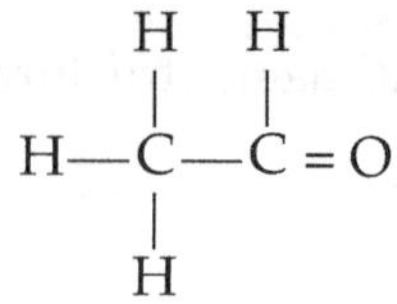

Ans.

(1) Propanal
(2) Propan-1-ol
(3) Propyne
(4) Pentan-3-ol
(5) 2-methyl propane
(6) Ethanoic acid
(7) 1, 2-dichloroethane
(8) 2-chloro hex-5-yn-1-al
(9) 3-methyl-but-3-en-1-ol
(10) 1-chloro-2-methyl propane
(11) 2, 4-dimethyl-pentan-3-one
(12) 2-methoxy-4-methyl pentan-3-one
(13) 2-chloro-3-hydroxy propanal
(14) Ethanedioic acid
(15) Pent-1-en-4-yne
(16) Pent-3-en-1-yne
(17) Propan-1-ol
(18) Pentan-2-one
(19) 2-methyl hexanoic acid
(20) Hexan-3-one
(21) 2-bromo-2-methyl butanol
(22) But-2-yne
(23) 1, 2-dichloro ethane
(24) 2-methyl butane
(25) 2-methyl propane
(26) Propene
(27) 2-butyne
(28) Ethanal

3. Give the formula of the next higher homologue of :

(1) Methanol
(2) Ethane
(3) Ethene
(4) Ethyne
(5) Propyl
(6) Methanoic acid
(7) Propane
(8) Butene
(9) Pentane
(10) Methanal

Ans.

(1) Ethanol — C_2H_5OH
(2) Propane — C_3H_8
(3) Propene — C_3H_6
(4) Propyne — C_3H_4
(5) Butyl — C_4H_9
(6) Ethanoic acid — CH_3COOH
(7) Butane — C_4H_{10}
(8) Pentene — C_5H_{10}
(9) Hexane — C_6H_{14}
(10) Ethanal — CH_3CHO

4. Draw the structures of isomers of pentane.*

Ans. Structures of isomers of pentane are as follows :

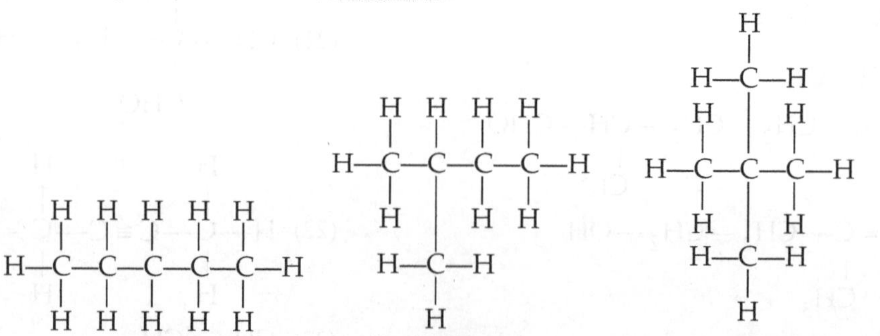

Pentane　　2-Methylbutane　　2, 2-Dimethylpropane

Structural isomers of C_5H_{12}

* Frequently asked previous years Board Exam Questions.

5. Name the following organic compounds :*

(1) The compound with 3 carbon atoms functional group is a carboxyl.
(2) The first homologue whose general formula is C_nH_{2n}.
(3) The compound that reacts with acetic acid to form ethyl ethanoate.
(4) The compound formed by complete chlorination of ethyne.

Ans. (1) The compound with three carbon atoms whose functional group is carboxyl-Propanoic acid, CH_3CH_2COOH

(2) The first homologue whose general formula is C_nH_{2n} is C_2H_4 that is Ethene.

(3) The compound that reacts with acetic acid to form ethyl ethanoate is ethanol-

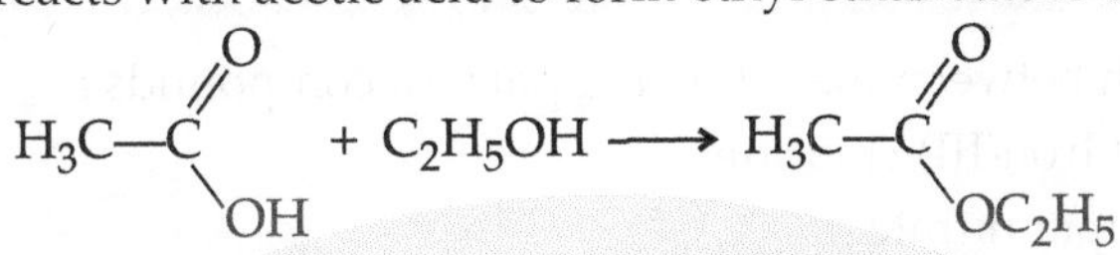

Ethanoic acid (acetic acid)   Ethyl ethanoate

(4) The compound formed by complete chlorination of ethyne is tetrachloroethane, $C_2H_2Cl_4$.

* Frequently asked previous years Board Exam Questions.

Chemical Tests | Set 11 |

Chapter 3. Study of Acids, Bases and Salts

1. How will you distinguish between the following pairs of compounds :

(1) Iron(II) chloride and Iron(III) chloride.

(2) Lead nitrate and copper nitrate.

(3) Zinc oxide and calcium oxide.

(4) Sodium carbonate and sodium nitrate.

(5) Sodium sulphate and sodium sulphite.

(6) How will you distinguish between ammonium hydroxide and sodium hydroxide using copper sulphate solution ?

Ans. (1) Iron(II) chloride is dissolved in water and then sodium hydroxide is added. A dirty green precipitate is obtained which confirms the presence of iron(II) chloride.

$$FeCl_2 + 2NaOH \longrightarrow Fe(OH)_2 + 2NaCl$$

Iron(III) chloride is also dissolved in water and then sodium hydroxide solution is added. A reddish brown precipitate is obtained, which confirms the presence of iron(III) chloride.

$$FeCl_3 + 3NaOH \longrightarrow Fe(OH)_3 + 3NaCl$$

(2) Lead nitrate is dissolved in water and then sodium hydroxide solution is added, a white precipitate is obtained, which is soluble in excess of sodium hydroxide.

$$Pb(NO_3)_2 + 2NaOH \longrightarrow Pb(OH)_2 + 2NaNO_3$$

Copper nitrate is dissolved in water and sodium hydroxide solution is added, a light blue precipitate of copper hydroxide is obtained.

$$Cu(NO_3)_2 + 2NaOH \longrightarrow Cu(OH)_2 + 2NaNO_3$$

(3) Zinc oxide is dissolved in dilute hydrochloric acid to form zinc chloride and then sodium hydroxide solution is added, a white precipitate is obtained, which is soluble in excess of sodium hydroxide to form a clear solution.

$$ZnO + 2HCl \longrightarrow ZnCl_2 + H_2O$$

$$ZnCl_2 + 2NaOH \longrightarrow Zn(OH)_2 + 2NaCl$$

$$Zn(OH)_2 + 2NaOH \longrightarrow \underset{\substack{\text{Sodium zincate} \\ \text{(A clear solution)}}}{Na_2ZnO_2} + 2H_2O$$

Calcium chloride is obtained by dissolving calcium oxide in dilute hydrochloric acid. To the solution of calcium chloride, sodium hydroxide solution is added, a white precipitate of calcium hydroxide is obtained, which is insoluble even in the excess of sodium hydroxide.

$$CaO + 2HCl \longrightarrow CaCl_2 + H_2O$$

$$CaCl_2 + 2NaOH \longrightarrow \underset{\text{Calcium hydroxide}}{Ca(OH)_2} + 2NaCl$$

(4) Sodium carbonate when treated with dilute hydrochloric acid, a vigorous reaction takes place and a colourless, odourless gas carbon dioxide is evolved which turns lime water milky.

$$Na_2CO_3 + 2HCl \longrightarrow 2NaCl + H_2O + CO_2$$

Sodium nitrate when mixed with copper filings and heated with concentrated sulphuric acid, reddish brown fumes of nitrogen dioxide are evolved.

$$2NaNO_3 + H_2SO_4 \longrightarrow Na_2SO_4 + 2HNO_3$$

$$Cu + 4HNO_3 \longrightarrow Cu(NO_3)_2 + 2NO_2 + 2H_2O$$

(5) Sodium sulphate is dissolved in water and barium chloride solution is added, an insoluble white precipitate of barium sulphate is obtained.

$$Na_2SO_4 + BaCl_2 \longrightarrow \underset{\text{White ppt.}}{BaSO_4} + 2NaCl$$

Sodium sulphite is warmed with dilute sulphuric acid, a colourless gas with a pungent and suffocating smell is evolved. When this gas is bubbled through acidified solution of potassium dichromate, it turns its colour from orange to green.

$$Na_2SO_3 + H_2SO_4 \longrightarrow Na_2SO_4 + H_2O + SO_2$$
$$K_2Cr_2O_7 + H_2SO_4 + 3SO_2 \longrightarrow K_2SO_4 + Cr_2(SO_4)_3 + H_2O$$

(6) When ammonium hydroxide solution is added drop by drop to copper sulphate solution, a pale blue or bluish white precipitate is formed which is soluble in excess of ammonium hydroxide and a deep blue or inky blue solution is formed with excess of ammonium hydroxide.

$$CuSO_4 + 2NH_4OH \longrightarrow Cu(OH)_2 + (NH_4)_2SO_4$$
$$Cu(OH)_2 + 4NH_4OH \longrightarrow [Cu(NH_3)_4](OH)_2 + 4H_2O$$

Copper solution forms a blue precipitate with sodium hydroxide solution. It is insoluble in excess of NaOH.

$$\underset{\text{Copper sulphate}}{CuSO_4} + \underset{\text{Sodium hydroxide}}{2NaOH} \longrightarrow \underset{\text{Copper hydroxide}}{Cu(OH)_2} + \underset{\text{Sodium sulphate}}{Na_2SO_4}$$

Chapter 4. Analytical Chemistry

1. Zinc nitrate solution and calcium nitrate solution.
2. Sodium nitrate solution and sodium chloride solution.
3. Iron(III) chloride solution and copper chloride solution.
4. Iron(II) sulphate solution and copper(II) sulphate solution.
5. Zinc nitrate solution and calcium nitrate solution.
6. Iron(II) chloride and Iron(III) chloride solution.

Ans. 1. Add NaOH solution in excess to the two solutions. The one in which white ppt. initially formed dissolves in excess of NaOH solution is $Zn(NO_3)_2$ solution and the other is $Ca(NO_3)_2$ solution.

2. Add freshly prepared ferrous sulphate solution to the two solutions. Then by the side of the test tube, pour conc. sulphuric acid to each slowly. The one in which brown ring appears is sodium nitrate solution while the other is sodium chloride solution.

3. Add NaOH solution to both the solutions. The one which gives a reddish brown ppt. is iron(III) chloride solution and the one which gives blue ppt. is copper chloride solution.

4. When sodium hydroxide solution is added to iron(II) sulphate solution, a dirty green precipitate is formed. When sodium hydroxide solution is added to copper(II) sulphate solution, light blue precipitate is formed.

5. When ammonium hydroxide is added to zinc nitrate solution, a white precipitate is formed. The white precipitate dissolves when excess of ammonium hydroxide is added.When ammonium hydroxide is added to calcium nitrate solution, no visible reaction occurs even with the addition of excess of NH_4OH.

6. When sodium hydroxide is added to iron(II) chloride, dirty green precipitate of $Fe(OH)_2$ is formed.

$$FeCl_2 + 2NaOH \longrightarrow Fe(OH)_2 \downarrow + 2NaCl$$

When sodium hydroxide is added to iron(III) chloride, reddish brown precipitate is formed.

$$FeCl_3 + 2NaOH \longrightarrow Fe(OH)_3 + 3NaCl$$

Chapter 8. (a) Study of Compounds : Hydrogen Chloride

1. Manganese dioxide and copper(II) oxide.
2. Hydrogen chloride gas and carbon dioxide gas.
3. Give three tests for HCl gas.

Ans. 1. When conc. hydrogen chloride is added to manganese dioxide, greenish yellow gas (Cl_2) is liberated.

$$\underset{\text{(Conc.)}}{MnO_2 + 4HCl} \longrightarrow MnCl_2 + 2H_2O + \underset{\text{Greenish yellow gas}}{Cl_2 \uparrow}$$

When conc. hydrogen chloride is added to copper(II) oxide, no gas is liberated but the solution turns bluish because of the formation of copper chloride.

$$CuO + \underset{\text{(Conc.)}}{HCl} \xrightarrow{\Delta} \underset{\text{Greenish blue solution}}{CuCl_2} + H_2O$$

2. When hydrogen chloride gas is passed into silver nitrate solution, it forms a curdy white precipitate of silver chloride.

$$HCl + AgNO_3 \longrightarrow \underset{\text{Curdy white ppt.}}{AgCl \downarrow} + HNO_3$$

When carbon dioxide gas is passed into lime water, it forms a milky white precipitate of calcium carbonate.

$$CO_2 + Ca(OH)_2 \longrightarrow \underset{\text{Milky white ppt.}}{CaCO_3} + H_2O$$

3. (i) When a glass rod dipped in ammonia, solution is held near the vapours of the acid, it form a dense white fumes of ammonium chloride.
   (ii) When hydrochloric acid is treated with silver nitrate solution, it forms curdy white precipitate which is soluble in excess of ammonium hydroxide solution.
   (iii) When hydrochloric acid is boiled with manganese dioxide, greenish yellow chlorine gas is evolved.

Chapter 8. (b) Study of Compounds : Ammonia and Nitric Acid

1. Dilute hydrochloric acid and dilute nitric acid.
2. Ammonium ion and hydroxide ion.
3. Nitric oxide gas and nitrous oxide gas.
4. Dilute hydrochloric acid and dilute sulphuric acid.
5. Oxygen gas and nitrous oxide gas.
6. Ammonium sulphate crystals and Sodium sulphate crystals.*

Ans. 1. Silver nitrate solution when added to dil. hydrochloric acid will give a white ppt. and when added to dil. nitric acid no change will be observed.
2. When iron(II) sulphate solution is added to ammonium ion, dirty green precipitate is obtained. No change will be observed with hydroxide ion.
3. If reddish brown fumes are produced in the atmosphere of oxygen then the given sample of gas is nitric oxide and if no brown fumes are produced and the mixture remains colourless, then the gas is nitrous oxide.
4. Dilute sulphuric acid will give a white precipitate of barium sulphate with barium chloride solution whereas no visible reaction occurs with dilute hydrochloric acid.
5. Carbon (red hot) burns in oxygen to form carbon dioxide gas only, but in nitrous oxide it forms carbon dioxide gas and nitrogen.
6. Ammonium sulphate crystals give pungent colourless gas Ammonia (NH_3) when heated. When NH_3 gas comes in contact with a glass rod dipped in HCl white fumes of NH_4Cl are produced. Sodium sulphate crystals do no undergo the above reaction sequence, hence can be differentiated from ammonium sulphate crystals.

Chapter 8. (c) Study of Compounds : Sulphuric Acid

1. How will you distinguish.
(1) Dilute sulphuric acid and dilute hydrochloric acid.

* **Frequently asked previous years Board Exam Questions.**

(2) Chlorine gas and sulphur dioxide gas.

(3) How will you distinguish between dilute hydrochloric acid and dilute sulphuric acid using lead nitrate solution ?

(4) Dilute hydrochloric acid and dilute sulphuric acid. (using lead nitrate solution)*

Ans. (1) When barium chloride solution is added to the dilute sulphuric acid, thick white precipitate of barium sulphate is formed which is insoluble in any mineral acid such as nitric acid or hydrochloric acid.

$$BaCl_2 + \underset{\text{(dil.)}}{H_2SO_4} \longrightarrow BaSO_4 \downarrow + 2HCl$$

With dilute hydrochloric acid, no effect is observe.

(2) Chlorine gas turns starch iodide paper blue and sulphur dioxide gas turns moist acidified potassium dichromate paper green.

(3) On adding lead nitrate to both acids, we will get a white precipitate. On heating the solution, the one whose precipitate will redissolve will be dil. HCl and the one with insoluble precipitate will be dil. H_2SO_4. Actually on adding lead nitrate to HCl, $PbCl_2$ precipitates out and on heating the solution it redissolves. But in case of H_2SO_4, $PbSO_4$ is formed which is insoluble even on heating it.

$$Pb(NO_3)_2 + \underset{\text{(dil.)}}{2HCl} \longrightarrow PbCl_2 + 2HNO_3$$

$$Pb(NO_3)_2 + \underset{\text{(dil.)}}{H_2SO_4} \longrightarrow PbSO_4 + 2HNO_3$$

(4) Lead nitrate reacts with dilute HCl to form the insoluble salt lead chloride, which appears as the white precipitate. The insoluble lead chloride reacts with excess Cl^- ions (of HCl) to form a soluble complex, the tetrachloroplumbate(II) ion,

$$Pb(NO_3)_2 + 2HCl \rightarrow PbCl_2\downarrow + 2HNO_3$$

$$PbCl_2 + 2HCl \rightarrow [PbCl_4]^{2-} + 2H^+(aq)$$

Soluble

Lead nitrate solution reacts with H_2SO_4 to give lead sulphate precipitate, which does not dissolve further in sulphuric acid solution.

$$Pb(NO_3)_2 + H_2SO_4 \rightarrow PbSO_4\downarrow + HNO_3$$

Chapter 9. Organic Chemistry

1. How will you distinguish.

(1) Ethene and ethane.

(2) Ethyne and ethane.

(3) Alkanes, alkenes and alkynes.

Ans. (1) Ethene gas decolourises bromine solution and potassium permanganate solution. But, ethane gas does not change the colour of these solutions.

(2) Ethyne gas forms a white precipitate with ammonical solution of silver nitrate and red ppt. with ammonical solution of copper(I) chloride. But, ethane does not respond to such tests.

(3)

S. No.	Test	Alkanes	Alkenes	Alkynes
(i)	Bromine Test. Add a few drops of sol. of bromine in carbon tetrachloride to the hydrocarbon.	No change takes place.	The red colour of bromine is decolourised.	The red colour of bromine is decolourised.
(ii)	Alkaline Potassium Permanganate Test : Add a few drops of alkaline pot. permanganate sol. To the hydrocarbon.	No change takes place.	The purple colour of Potassium permanganate is decolourised.	The purple colour of potassium permanganate is decolourised.

* Frequently asked previous years Board Exam Questions.

(iii)	Ammoniacal Cuprous Chloride Test : Add a few drops of ammonical cuprous chloride sol. to the hydrocarbon.	No change takes place.	No change takes place.	A red ppt. of copper acetylide is formed.

Chapter 10. Practical Chemistry

1. How will you distinguish.

(1) Zinc sulphate solution and zinc chloride solution.

(2) Iron(II) chloride solution and iron(III) chloride solution.

(3) Calcium nitrate solution and calcium chloride solution.

(4) Sodium carbonate and sodium sulphite.

(5) Ferrous nitrate and lead nitrate.

(6) Manganese dioxide and copper(II) oxide.

(7) Sodium chloride and sodium nitrate.

(8) Calcium nitrate and lead nitrate.

(9) Lead nitrate and zinc nitrate.

(10) Sodium sulphite and sodium sulphate.

(11) Ammonium chloride and ammonium nitrate.

(12) Zinc carbonate and lead carbonate.

(13) Potassium chloride and sodium chloride.

(14) Powdered coke and manganese dioxide.

(15) Copper oxide and manganese dioxide.

(16) Washing soda and baking soda.

(17) Potassium nitrate and potassium nitrite.

(18) Mercuric oxide and red lead.

(19) Ferrous ions and ferric ions.

OR

Ferrous sulphate and ferric sulphate solution.

(20) Magnesium chloride and Magnesium nitrate solution.*

Ans. (1) When $BaCl_2$ solution is added to the given solution, $ZnSO_4$ gives a white ppt. while no ppt. is obtained with $ZnCl_2$ solution.

(2) When NaOH solution is added to the given solution, iron(II) chloride gives dirty green ppt. while reddish brown ppt. is obtained with iron(III) chloride.

(3) When $AgNO_3$ solution is added to the given solution, $CaCl_2$ solution will give a white ppt. while no change is observed with calcium nitrate solution.

(4) Sodium carbonate when treated with dil. H_2SO_4 will liberate colourless odourless gas with brisk effervescence. The gas will turn lime water milky and will have no effect on acidified potassium permanganate solution.

Sodium sulphite when treated with dil. H_2SO_4 will liberate a colourless gas having suffocating smell of burning sulphur. The gas will turn acidified potassium permanganate solution from pink to colourless.

(5) Aqueous ferrous nitrate when treated with NaOH solution gives a dirty green ppt. Aqueous lead nitrate when treated with NaOH solution gives a white ppt. which dissolves in excess of NaOH.

(6) When manganese dioxide is heated with conc. HCl a pungent smelling greenish yellow chlorine gas is evolved.

* Frequently asked previous years Board Exam Questions.

When copper oxide is heated with conc. HCl no such gas is evolved.

(7) On heating with conc. H_2SO_4 sodium chloride gives pungent colourless gas which turns silver nitrate solution curdy white, whereas sodium nitrate gives a brown gas which shows no effect on silver nitrate solution.

(8) Aqueous solution of calcium nitrate gives a white precipitate with sodium hydroxide which is insoluble in excess of sodium hydroxide solution. The aqueous solution of lead nitrate gives a white precipitate with caustic soda solution which dissolves in excess of caustic soda solution.

(9) Aqueous solution of lead nitrate gives white precipitate with ammonium hydroxide which is insoluble in excess ammonium hydroxide while the solution of zinc nitrate gives a white precipitate which is soluble in excess ammonium hydroxide.

(10) Solution of both gives white precipitate with barium chloride solution. The precipitate of barium sulphite is soluble in dil. hydrochloric acid, white the precipitate of barium sulphite is insoluble in dil. hydrochloric acid.

(11) Heat each of the salt strongly in a hard glass test tube. In case of ammonium chloride, sublimation takes place and white powdery deposits settles near the mouth of test tube. In case of ammonium nitrate, an explosion takes place and nitrous oxide and stream are given off. No residue is left in the test tube.

(12) Heat each of the carbonates strongly in a hard glass test tube. In case of zinc carbonate, residue is yellow when hot and white then cold. In case of lead carbonate, residue is reddish-brown when hot and yellow when cold.

(13) Make a thick paste of each salt in HCl. Take a small amount of the paste on clean platinum wire and introduce it in non-luminous bunsen flame. In case of potassium chloride, a non-persistent lilac flame is formed. In case of sodium chloride, a persistent golden yellow flame is formed.

(14) Heat each of the substances with conc. hydrochloric acid. In case of coke, no visible reaction takes place. In case of manganese dioxide, greenish-yellow chlorine gas is evolved.

(15) Warm each of the oxides with dilute hydrochloric acid. In case of copper oxide, a greenish-blue solution of copper chloride is formed. In case of manganese dioxide, no reaction takes place.

(16) Heat each of the salt strongly in a hard glass test tube and pass the gas evolved through lime-water. In case of washing soda, lime-water does not turn milky. In case of baking soda, lime-water turns milky.

(17) Treat each of the salts with dilute sulphuric acid and warm gently. In case of potassium nitrate, no visible reaction takes place. In case of potassium nitrite, reddish-brown gas, nitrogen dioxide is given off.

(18) Heat each of the oxides strongly in a hard glass test tube. In case of mercuric oxide, tiny droplets of mercury are seen near the mouth of the test tube and residue is black in colour. In case of red lead, no tiny droplets are seen. The residue is reddish-brown when hot and yellow when cold.

(19) Treat the solution of each ion with sodium hydroxide solution. In case of ferrous ions, a dirty green ppt. of ferrous hydroxide is formed. In case of ferric ions, a reddish-brown ppt. of ferric hydroxide is formed.

OR

$$Fe_2(SO_4)_3\ (aq) + 6\ NaOH\ (aq) \longrightarrow Fe(OH)_3\ (s) + 3\ Na_2SO_4\ (aq)$$

(20) Magnesium chloride reacts with silver nitrate solution to give precipitate of silver chloride, whereas magnesium nitrate does not react with silver nitrate solution to give a precipitate.

$$MgCl_2\ (aq) + 2AgNO_3\ (aq) \longrightarrow Mg(NO_3)_2\ (aq) + 2AgCl\ (s)\downarrow$$

$$MgNO_3 + AgNO_3 \longrightarrow \text{No reaction}$$

❑

Reasoning Based Questions

Set 12

Chapter 1. Periodic Properties and Variations of Properties

1. Explain why the elements placed in the same group of the periodic table have the same chemical properties ?

Ans. Elements placed in the same group of the periodic table have the same chemical properties because they have the same number of valence electrons in the outermost shell of their atoms.

2. Why group IA elements are called alkali metals ?

Ans. Group IA elements are called alkali metals because their hydroxides are soluble in water and form strong bases.

3. Why sodium is a metal while sulphur is a non-metal ?

Ans. Sodium has a larger atomic radii and lower ionization potential than sulphur. Hence sodium is a metal while sulphur is a non-metal.

4. Alkali metals are good reducing agents.

Ans. Alkali metals are good reducing agents because alkali metals have one valence electron which they lose to attain stability. Hence, they themselves undergo oxidation causing reduction of others and are good reducing agents.

5. Why are the elements sodium and chlorine in the same period of the periodic table ?

Ans. Elements sodium and chlorine are in the same period of the periodic table because the atoms of both the elements have three shells containing the electrons.

6. Potassium atom is larger than sodium atom. Why ?

Ans. Potassium is placed below sodium in group 1. It, therefore, has one more electron shell. Na atom has three electron shells (2, 8, 1); K atom has four (2, 8, 8, 1). So, potassium atom is bigger than sodium atom.

7. Magnesium atom is smaller than calcium atom. Why ?

Ans. Magnesium atom precedes calcium atom in the same group, *i.e.*, group 2. Magnesium atom has got three electron shells (2, 8, 2) whereas calcium atom has four electron shells (2, 8, 8, 2). So, calcium atom is larger than the sodium atom.

8. Magnesium atom is smaller than sodium atom. Why ?

Ans. Magnesium come after sodium in the same period. Atoms of both elements have three electron shells (Na : 2, 8, 1 : Mg : 2, 8, 2).But the nuclear charge of sodium is + 11 and that of magnesium is + 12.Hence, the electron shells are pulled inward more strongly in Mg atom than in Na atom. Hence, Mg atom is smaller than Na atom.

9. Why the atomic size decreases in a period as we move from left to right ?

Ans. As we move from left to right across a period, the number of shells remain the same. As the atomic number increases, within the same electronic shell the nuclear charge increases and there is a greater attraction between the nucleus and the electrons. The atomic size, therefore, decreases across a period.

10. The reducing power of elements increases as one goes down a group ?

Ans. When an atom loses an electron then the element is said to be a reducing agent. The reducing power of an element depends upon how quickly it can lose electrons. As we move down the group, the effective nuclear charge decrease. Thus tendency to loose electron increases down the group and so the reducing power.

11. The reducing power of elements decreases as on one move from left to right in a period ?

Ans. The reducing power of an element depends upon, how quickly it can lose electrons in the valence shell. As one moves from left to right in periodic table, the electrons in valence shell are held more tightly because of increase of nuclear charge. Thus, the tendency of atoms to lose their electrons gradually decreases and so does the reducing power.

12. Why the oxidising power of elements increases on moving from left to right along a period in the periodic table ?

Ans. Oxidising power of elements increases on moving from left to right along a period in the periodic table because on moving from left to right along a period in the periodic table the electron affinity of elements increases.

13. Sodium atom, Na forms the positive ion Na^+, but chlorine atom Cl, does not form the positive Cl^+ ion.

Ans. Sodium is the first element of period 3 whereas chlorine is the last but one element of the same period. Since ionization potential increases across a period, the ionization potential of Na is much smaller than that of Cl. Hence, Na can lose an electron to form Na^+ ion whereas Cl cannot lose an electron to form Cl^+ ion.

14. Which is larger Na^+ or K^+ ? Why ?

Ans. K^+ is larger than Na^+ because the ionic radius increases in a particular group on moving from top to bottom due to increase in the principle energy shell though the number or electrons in the valence shell remain the same.

15. Mg^{2+} ions is smaller than O^{2-} ion although both are iso-electronic. Explain.

Ans. Mg^{2+} ion is smaller than O^{2-} ion though both are iso-electronic. The nuclear charge in Mg^{2+} is + 12 and O^{2-} is + 8, so with the increase in nuclear charge the size decreases and, hence $O^{2-} > Mg^{2+}$.

16. Why ionization potential of the element increases across a period ?

Ans. Across a period, the atomic radii decreases because of increase in nuclear charge due to addition of electrons which results in greater attraction of valence shell electrons. Hence, ionization potential increases.

17. Why the second ionization energy of an element is greater than its first ionization energy ?

Ans. More energy is required to remove an electron as it holds more firmly by the unipositive ion due to greater effective nuclear charge. Thus, the second ionization energy of an element is greater than its first ionization energy.

18. Why is ionisation energy of O less than that of N ?

Ans. Nitrogen (N) has more stable electronic configuration than O. Therefore, large amount of energy is required to remove valence electron from N-atom. Hence, ionization energy of O is less than that of N.

19. Why elements with low ionization potential exhibit metallic properties ?

Ans. Metallic character increases with decrease in ionization potential. Lower the value of ionization potential, the greater is the tendency of an atom to lose electrons. Hence, more is metallic character.

20. Noble gases have zero electron affinity values.

Ans. Noble gases have zero electron affinity values because they have stable electronic configuration and have no tendency to take an additional electron. Hence, no energy is released and their electron affinity is zero.

Chapter 2. Chemical Bonding

1. Why atoms combine with one another ?

Ans. The cause of chemical combination between atoms of the various elements is their tendency to acquire nearest stable noble gas configuration of octet of electrons and duplet of electrons in case of hydrogen atoms in their outermost shells.

2. Why do certain elements form ions ?

Ans. Every particle (molecule, atom or ion) has the tendency to attain the state of lowest energy.
As atoms of all elements except the noble gases, have one to seven electrons in their outermost shell, therefore, they are not in the state of minimum energy. They are reactive. In order to attain a minimum state of energy, *i.e.*, to acquire a duplet or octet structure in their valence shell, they either donate or accept electrons. In this process, they become charged particles or ions.

3. Why all atoms other then noble gas atoms are reactive ?

Ans. All atoms other then noble gas atoms are reactive because they are short of octet in their outermost shell. They tend to attain eight electrons in their outermost shell to acquire nearest inert gas electronic configuration.

4. Why a molecule of hydrogen is more stable than the uncombined atoms ?

Ans. When a molecule of hydrogen is formed from the atoms, energy is released (104 kcal/mol). The molecules possessing lower energy are more stable, than the atoms. Hence molecule of hydrogen is more stable than uncombined atoms.

5. Why is hydrogen ion called proton ?

Ans. An atom of hydrogen has one proton in its nucleus and one electron in its valence shell. It donates its valence electron, the residual ion consists of a single proton. It is on account of this fact, that hydrogen ion is called proton.

6. A cation is smaller than the atom from which it is formed. Why ?

Ans. In the formation of a cation, the atom loses the electrons of its outer shell. This result in increased effective nuclear charge. So, it is smaller than the atom.

7. An anion carries negative charge. Why ?

Ans. When an atom gains an extra electron in its valence shell, it acquires negative charge and hence, forms an anion.

8. Why electrovalent compounds form hard crystals ?

Ans. The crystals of electrovalent compounds are made up of crystal lattice containing oppositely charged ions. Each cation is surrounded by a definite number of anions and vice-versa. Their is a great electrostatic force of attraction among these oppositely charged ions and as a result, ionic compounds form hard crystals.

9. Why electrovalent compounds have high melting, boiling points and low volatility ?

Ans. There is a strong electrostatic force of attraction among the oppositely charged ions in the crystals of electrovalent compounds.Therefore, a large amount of energy is required to separate them. Due to these strong electrostatic forces of attraction, ionic compounds have high melting and boiling points and low volatility.

10. Why electrovalent compounds in crystalline state do not conduct electricity ?

Ans. Electrovalent compounds in crystalline state do not conduct electricity because the oppositely charged ions in them are held together by a strong electrostatic force of attraction. These ions occupy fixed position in the crystals and do not move when an electric field is applied.

11. Why do electrovalent compounds usually dissolve in water and molecular compounds dissolve in organic solvents ?

Ans. Water is polar molecule. It hydrates ions by its polar attraction and pulls the ions apart. In covalent compounds the polar charge on molecules is weak and does not have that much of attractive force on the molecules. Organic solvents exert a greater polar attraction on covalent molecules and hence dissolve them.

12. Why ionic compounds are generally soluble in water, but insoluble in organic solvents ?

Ans. Water molecules are polar in nature. It hydrates the ionic compounds by its polar attraction and pulls the ions apart (dissociation), resulting in free ions in the solution. Thus, dissolving the ionic compounds in water. However, organic solvents do not facilitate the formation of such ions.

13. Why molten NaCl conduct electricity but, CCl_4 does not ?

Ans. Molten NaCl contains Na^+ and Cl^- ions, which are free to move, hence it conducts electricity. Whereas Liquid CCl_4 does not contain any charged particles to conduct electricity.

14. Why is sodium ion (Na^+) not reactive, but sodium metal is very reactive ?

Ans. Sodium ion has eight electrons in its valency shell and it is in the minimum state of energy, while the sodium metal has one electron in valency shell, so the sodium metal is very reactive, whereas sodium ion (Na^+) is not reactive.

15. Why covalent compounds are generally liquids or gases ?

Ans. In covalent compounds, the molecules are held together by weak vander Waals forces. In liquids, the molecules are weakly attracted whereas in gases, these forces are almost non- existent. Hence, they are generally liquids or gases.

16. Why all covalent compounds are bad conductor of electricity ?

Ans. The covalent compounds do not have positive or negative ions in their fused state. Thus, when electric potential is supplied, no ions migrate to opposite poles and hence no conduction of electric current takes place. Thus all covalent compounds are bad conductor of electricity.

17. Why covalent compounds have low melting point and boiling point ?

Ans. The force of attraction between the molecules is very weak and so the amount of energy needed to separate them is small, consequently they have low melting points and boiling points.

18. Why most of the covalent compounds have density less than that of water ?

Ans. The covalent molecules are held very weakly by vander Waals forces. Thus, there are large inter molecular spaces between the molecules. In other words the number of molecules per unit volume is less. Thus mass per unit is also less and hence, covalent compound have low density.

19. Why do covalent compounds exist as gases, liquids or soft solids ?

Ans. Covalent compounds exists as gases, liquids or soft solids because they have weak forces of attraction between their molecules.

20. Why hydrogen chloride can be termed as a polar covalent compound ?

Ans. Pure covalent bond exists between two elements which have different electronegativities. In hydrogen chloride, chlorine being more electronegative attracts the shared pair of electrons towards itself. As a result hydrogen acquires partial positive charge and chlorine gets partial negative charge. Thus, hydrogen chloride can be termed as a polar covalent compound.

21. Why is methane molecule regarded as a non-polar covalent compound ?

Ans. Methane is said to be non-polar because the shared pair of electrons is equally distributed between all bonding atoms of methane.

22. Why the melting and boiling points of coordinate compounds are higher than covalent compounds and lower than ionic compounds ?

Ans. A coordinate bond is a union of one electrovalent and one covalent bond, the volatility of these compounds lies between that of covalent and ionic compounds. Thus, their melting and boiling points are higher than covalent compounds and lower than ionic compounds.

Chapter 3. Study of Acids, Bases and Salts

1. Hydrochloric acid is considered as a strong acid whereas acetic acid is a weak acid. Why ?

Ans. Hydrochloric acid is considered as a strong acid because it dissociates completely in water.

$$HCl\,(aq) \rightleftharpoons H^+\,(aq.) + Cl^-\,(aq)$$

Acetic acid is a weak acid as it dissociates partially when dissolved in water. Most of its molecules remain in molecular form in the solution.

$$CH_3COOH\,(aq) \rightleftharpoons CH_3COO^-\,(aq) + H^+\,(aq)$$

2. Why dilute sulphuric acid is stronger acid than concentrated sulphuric acid ?

Ans. The presence of water in dilute sulphuric acid increases the hydrogen ion concentration. Hence it is a stronger acid than concentrated sulphuric acid which contains comparatively less water.

3. Acetic acid is monobasic. Why ?

Ans. Acetic acid is monobasic because it has one ionisable hydrogen ion and combines with one hydroxyl ion of the base to give a single salt and water.

$$CH_3COOH \rightleftharpoons CH_3COO^- + H^+$$

$$NaOH + CH_3COOH \longrightarrow CH_3COONa + H_2O$$

4. Carbonic acid is a dibasic acid. Why ?

Ans. Carbonic acid (H_2CO_3) is a dibasic acid because it has two replaceable hydrogen atoms and hence it combines with two hydroxyl groups of the bases to form two kinds of salt and water.

$$H_2CO_3 \rightleftharpoons HCO_3^- + H^+$$

$$HCO_3^- \rightleftharpoons CO_3^{2-} + H^+$$

The displacement of two hydrogen atoms takes place in two steps.

5. Sodium hydroxide is a monoacidic base. Why ?

Ans. Sodium hydroxide is a mono-acidic base as it combines with only one hydrogen ion.

$$NaOH\,(aq) + HCl\,(aq) \longrightarrow NaCl\,(aq) + H_2O\,(l)$$

6. An aqueous solution of the salt ammonium chloride is acidic in nature while an aqueous solution of sodium chloride is neutral. Why ?

Ans. Ammonium chloride is a salt of weak base and strong acid, it undergoes salt hydrolysis to produce an acidic solution whereas sodium chloride is a salt of strong acid and strong base, it does not undergo salt hydrolysis hence its solution remains neutral.

7. An aqueous solution of the zinc sulphate is acidic in nature. Why ?

Ans. When zinc sulphate is dissolved in water, it is hydrolysed to form sulphuric acid and zinc hydroxide. Sulphuric acid is a strong acid, while zinc hydroxide is a weak base, hence, the solution is acidic in nature.

$$ZnSO_4 + 2H_2O \longrightarrow \underset{\text{Weak base}}{Zn(OH)_2} + \underset{\text{Strong acid}}{H_2SO_4}$$

8. An aqueous solution of ammonium acetate is neutral in nature. Why ?

Ans. Ammonium acetate, when dissolved in water is hydrolysed to form ammonium hydroxide and acetic acid. Ammonium hydroxide and acetic acid both are weak alkali and acid respectively, hence the solution is neutral in nature.

9. An aqueous solution of sodium carbonate is alkaline and that of ammonium chloride is acidic in behaviour. Why ?

Ans. Both of these salts react with water. Sodium carbonate reacts with water, producing a strong alkali, sodium hydroxide and a weak acid, carbonic acid. Hence, the solution is alkaline :

$$Na_2CO_3 + H_2O \longrightarrow 2NaOH + H_2CO_3$$

Ammonium chloride reacts with water to produce ammonium hydroxide, a weak alkali and hydrochloric acid, a strong acid. Hence the solution is acidic :

$$NH_4Cl + H_2O \longrightarrow NH_4OH + HCl$$

10. Lime juice has a sour taste while lime water is slightly bitter. Why ?

Ans. Lime juice contains citric acid which imparts a sour taste to it while lime water is alkaline and hence bitter to taste.

11. A universal indicator is more meaningful than an ordinary indicator.

Ans. A universal indicator is better than an ordinary indicator as it not only shows whether the solution is acidic or basic but also gives the pH values (approx.) by giving a wide range of colours corresponding to different pH values.

12. The heat of neutralization of a strong acid with strong base is always the same. Why ?

Ans. Strong acids, strong bases and their salts ionize completely in the solution. Consider the neutralization of hydrochloric acid by sodium hydroxide, which is as follows :

$$NaOH + HCl \longrightarrow NaCl + H_2O + Q \text{ cals.}$$

On ionization, we have

$$Na^+ + OH^- + H^+ + Cl^- \longrightarrow Na^+ + Cl^- + H_2O + Q \text{ cals.}$$

$$H^+ + OH^- \longrightarrow H_2O + Q \text{ cals.}$$

This shows that heat of neutralization of a strong acid by a strong base is nothing, but heat of formation of water molecule from hydrogen and hydroxyl ions, hence it is in the same and fixed quantity.

13. It is necessary to find out the ratio of reactants required in the preparation of sodium sulphate. Why ?

Ans. In the preparation of sodium sulphate, sodium hydroxide and sulphuric acid are both soluble and excess of any of them can not be removed by filtration. Thus it is necessary to find out the ratio of the solutions of the two reactants required for complete neutralisation.

14. While mixing strong sulphuric acid and water, why is the acid always poured slowly into water instead of water into the acid ?

Ans. When concentrated sulphuric acid is mixed with water, it evolves a considerable amount of heat. This heat is liberated due to the hydration of hydrogen ions derived from the acid.

$$H^+ + H_2O \longrightarrow H_3O^+ + \text{Heat}$$

Due to the production of heat, the water may be vaporised into steam and spill out corrosive drops of acid. To avoid this, the acid is always slowly added to water.

15. Lead chloride cannot be prepared by the action of hydrochloric acid on lead sulphate directly. Why ?

Ans. An insoluble salt cannot be prepared from another insoluble salt and since both lead chloride and lead sulphate are insoluble salt, we can not prepare lead chloride by the action of hydrochloric acid on lead sulphate directly.

16. Sodium sulphite (Na_2SO_3) and sodium hydrogen sulphite ($NaHSO_3$) are salts of sodium but sodium sulphite is called a normal salt and sodium hydrogen sulphite is called an acid salt. Why ?

Ans. Sodium hydrogen sulphite has replaceable hydrogen in its molecule so it is called an acid salt but sodium sulphite does not have, so it is called a normal salt.

$$NaHSO_3 \rightleftharpoons Na^+ + HSO_3^-$$
$$HSO_3 \rightleftharpoons H^+ + SO_3^{2-}$$

17. Fused calcium chloride is used in a desiccator. Why ?

Ans. Calcium chloride is a deliquescent substance *i.e.,* it absorbs water up to such an extent that it finally dissolves in it and more over, it does not react with water. When calcium chloride is put in a desiccator, it absorbs all water vapours present in it and thus dry atmosphere is created inside.

18. Why fused calcium chloride is used in the preparation of $FeCl_3$?

Ans. $FeCl_3$ is highly deliquescent, so it is kept dry with the help of fused calcium chloride and is used in the preparation of $FeCl_3$.

19. The solution of sodium carbonate is alkaline in nature. Why ?

Ans. Sodium carbonate when dissolved in water is hydrolysed to form a weak acid, carbonic acid and a strong base sodium hydroxide. Hence, an aqueous solution of sodium carbonate is alkaline in nature.

$$Na_2CO_3 + 2H_2O \longrightarrow \underset{\text{Weak acid}}{H_2CO_3} + \underset{\text{Strong base}}{2NaOH}$$

20. Why zinc chloride is stored in air-tight bottles ?

Ans. Zinc chloride absorbs moisture from the atmosphere and turns into solution because it is a deliquescent substance. In order to prevent this, it is stored in air tight bottles.

Chapter 5. Mole Concept and Stoichiometry

1. "When stating the volume of a gas, the pressure and temperatrue should be also given." Why ?

Ans. It is because, the volume of a gas changes, if the pressure or temperature or both change. Thus, while stating the volume of a gas, pressure and volume has to be specified alongwith volume.

2. "The number of atoms in one mole of hydrogen is twice the number of atoms in 1 mole of helium at the same temperature and pressure." Why ?

Ans. It is because, hydrogen is a diatomic gas, whereas helium is monoatomic gas. As number of atoms in one molecule of hydrogen are double, as compared to one molecule of helium, therefore, one mole of hydrogen has double the atoms, as compared to helium at the same temperature and pressure.

3. Why is the term relative atomic mass used for atomic mass of an element ?

Ans. Since the actual mass of an atom of element is extremely small, for comparing the masses of atoms of different elements, the mass of an atom of some light element is fixed as standard and the masses of other atoms are expressed relative to the standard mass, so the term relative atomic mass is used.

4. Why relative atomic mass is compared with 1/12 of carbon ?

Ans. Naturally occurring hydrogen has three isotopes (${}_1H^1$, ${}_1H^2$ and ${}_1H^3$) and its relative atomic mass works out 1·008 rather than 1. Atomic mass of carbon is 12. Thus, 1/12 mass of carbon work out as 1. It is on account of the above reason that 1/12 mass of carbon is used for comparing relative atomic masses of other elements.

5. Why atomic weight of many elements are fractional ?

Ans. The atomic weight of all the elements, is never whole number but the atomic weights of many elements are fractional.The atomic weight of an element is fractional because the atomic weight of an element is actually the average relative weight of all the naturally occurring isotopes of the element.

Chapter 6. Electrolysis

1. Why are acids, bases and salts classified as electrolytes ?

Ans. Acids, bases and salts are classified as electrolytes because these compounds dissociate into ions, conduct electricity and undergo chemical decomposition at the same time.

2. Metals like potassium, calcium, sodium, etc., can be extracted only by electrolysis. Why?

Ans. Metals like K, Ca, Na, etc., can be extracted only by electrolysis because conventional reducing agents such as coke, carbon monoxide, hydrogen do not supply sufficient energy to break ionic bonds between the active metals and their chlorides or oxides.

3. Dilute acids are strong electrolytes. Why ?

Ans. Dilute acids produce a large number of hydronium ions, so they behave like strong electrolytes.

4. Sea water is a strong electrolyte. Why ?

Ans. Sea water is a strong electrolyte because sodium chloride dissolved in it dissociates completely into free mobile ions.

$$Na^+Cl^- \rightleftharpoons Na^+ + Cl^-$$

5. Copper is a good conductor of electricity, but it is a non-electrolyte. Why ?

Ans. Copper being a metal, contains free electrons which conducts electricity. Thus, it is a good conductor of electricity. But copper does not dissociate into ions and hence, it is a non-electrolyte.

6. A solution of ionic compound is an electrolyte, while that of covalent compound is non- electrolyte ?

Ans. The solution of ionic compound has free ions, which can migrate to cathode and anode and discharge electrons. Thus, solution of ionic compound is good conductor of electricity and hence is an electrolyte. However, a solution of covalent compound, consists of only molecules and does not have any free ions, which could migrate to cathode or anode. Hence, it is a non-electrolyte.

7. Explain, why hydrochloric acid is a conductor of electricity ?

Ans. Hydrochloric acid dissociates into ions in aqueous solution as follows :

$$HCl \rightleftharpoons H^+ + Cl^-$$

When a current is passed through in aqueous solution of HCl, the ions move towards their respective electrodes. Thus, the hydrochloric acid is a conductor of electricity.

8. Does wax conduct electricity ? Give reason to justify your answer.

Ans. No, wax does not conduct electricity because wax, being a covalent compound, does not have positively or negatively charged ions which could not be weakened by heating or in aqueous solution. Therefore due to absence of free ions, wax does not conduct electricity.

9. Mercury is a liquid and allows the flow of electricity, though it is not an electrolyte.

Ans. An electrolyte is a substance which on dissolving in water breaks up into positively and negatively charged ions. But mercury is a metal, so on dissolving in water, it can not break up into cations and anions. When electric current passes through mercury, it does not undergo any decomposition and no new substance is formed. Electric current passes through mercury due to the presence of free electrons in its penultimate shell and not due to the formation of ions. Hence, mercury is a metallic conductor and not an electrolyte.

10. A solution of cane sugar does not conduct electricity, but a solution of sodium chloride is a good conductor.

Ans. The sugar cane solution is a covalent compound. When it is dissolved in water, it does not dissociate to give free ions which could migrate to cathode or anode. Hence, sugar solution is a bad conductor of electricity.

The sodium chloride solution mainly consists of free sodium and chloride ions which could migrate to positively charged electrodes. Hence, solution of sodium chloride is a good conductor of electricity.

11. Why does blue colour of $CuSO_4$ solution slowly disappear when iron rod is dipped in it ?

Ans. When an iron rod is dipped in $CuSO_4$ solution, iron displaces copper from $CuSO_4$ to form $FeSO_4$ which is light green in colour.

$$Fe(s) + \underset{\text{Blue}}{CuSO_4(aq)} \longrightarrow Cu(s) + \underset{\text{Light green}}{FeSO_4(aq)}$$

$$Fe + \underset{\text{Blue}}{Cu^{2+}} \longrightarrow \underset{\text{Light green}}{Fe^{2+}} + Cu$$

12. Hydroxyl (OH^-) ion is lower in the activity series, than chloride ion. Yet when a concentrated solution of hydrochloric acid is subjected to electrolysis, the hydroxyl ion does not get discharged. Why ?

Ans. If an electrolyte has a much higher concentration of ions that are higher in the electrochemical series than those that are lower, then the higher gets discharged in preference to the lower one.

Reasoning Based Questions

Concentrated hydrochloric acid being strong electrolyte gets fully dissociated furnishing H^+ and Cl^- ion. Water is less dissociated into H^+ and OH^- ion. Thus, concentration of chloride ion is more as compared to OH^- ion and hence, chloride ion (Cl^-) are discharged in preference to the OH^- ion.

13. While electrolysing concentrated sulphuric acid, the bulb glows very dimly but when diluted, the bulb glows brightly. Why ?

Ans. Concentrated (99%) sulphuric acid behaves like a weak electrolyte as it has very little hydronium ions in it. But when added to water, it becomes diluted and produces a large number of hydronium ions. Now, it behaves like a stronger electrolyte and hence the bulb glows brightly.

14. Why electrolysis of acidulated water is considered example of electrolysis ?

Ans. Electrolysis of acidulated water is considered as an example of electrolysis because the amount of sulphuric acid does not change when water is electrolysed. The sulphuric acid just helps in increasing the conductivity of water.

15. Why should water be acidified before proceeding with the electrolysis of water ?

Ans. Water should be acidified before proceeding with the electrolysis because the concentration of ions in pure water is very low due to slow rate of ionisation. Hence, it is a poor conductor of electricity. When dilute acid is added, the number of H+ ions increases, which further facilitates the dissociation of water molecules at much increased rate, catalysing the ionisation of water. This increase the number of free ions in the solution to conduct electricity efficiently.

$$H_2O\,(l) \rightleftharpoons H^+\,(aq) + OH^-\,(aq)$$
$$H_2SO_4\,(aq) \rightleftharpoons 2H^+\,(aq) + SO_4^{2-}\,(aq)$$

Chapter 7. Metallurgy

1. Why are metals called reducing agents ?

Ans. They tend to lose electrons and act as reducing agents.

2. Why are non-metals called oxidizing agents ?

Ans. They tend to gain electrons and act as oxidising agents.

3. Why iron is not found in free state in nature ?

Ans. Iron is quite reactive metal. It easily combines with other metals. Iron thus occurs in nature in the form of its compounds and not as a free element.

4. Iron liberates hydrogen from dilute sulphuric acid, while silver cannot. Why ?

Ans. In activity series of metal, iron occupies a higher position than hydrogen, while silver is placed below hydrogen, hence iron is more reactive than silver and is able to displace hydrogen from dilute sulphuric acid.

$$Fe + H_2SO_4 \longrightarrow FeSO_4 + H_2\uparrow$$

5. Zinc displaces lead from lead nitrate solution, while gold is unable to do so. Why ?

Ans. Zinc is above lead in the metal activity series. It is more reactive than lead while gold, a noble metal, lies far below lead in the activity series and it is less reactive or highly unreactive. Zinc reacts with lead nitrate solution to precipitate lead and zinc nitrate is formed. There is no reaction between gold and lead nitrate.

$$\underset{\text{Lead nitrate}}{Pb(NO_3)_2} + Zn \longrightarrow \underset{\text{Zinc nitrate}}{Zn(NO_3)_2} + \underset{\text{Lead}}{Pb\downarrow}$$

$$\underset{\text{Lead nitrate}}{Pb(NO_3)_2} + Au \longrightarrow \text{No reaction.}$$

6. Why is sodium metal always stored under kerosene oil ?

Ans. Sodium is a very reactive metal and on exposure to moist air, the surface of sodium metal is tarnished due to the formation of sodium carbonate.

$$4Na + O_2 \longrightarrow 2Na_2O$$
$$Na_2O + H_2O \longrightarrow 2NaOH$$
$$2NaOH + CO_2 \longrightarrow \underset{\text{Sodium Carbonate}}{Na_2CO_3} + H_2O$$

To avoid this sodium is always kept under kerosene oil.

7. Why carbon can reduce copper(II) oxide to copper but not calcium oxide to calcium ?

Ans. Carbon can reduce copper(II) oxide to copper but not calcium oxide to calcium because carbon has greater affinity for oxygen than copper and less affinity for oxygen than calcium.

8. Aluminium is highly electropositive metal, in spite of it aluminium does not oxidise rapidly in air. Why ?

Ans. In moist air, a thin layer of aluminium oxide is formed on it quickly which protects aluminium to oxidise.This is the reason why aluminium does not oxidise rapidly in air.

9. Why extraction of aluminium is difficult ?

Ans. Extraction of aluminium is difficult because :

(i) Pure aluminium oxide melts at 2050°C only. So, a large amount of energy is needed to maintain this high temperature.

(ii) A good amount of the aluminium vaporises at this temperature.

(iii) Fused alumina does not conduct electricity well.

10. During the extraction of aluminium, cryolite and fluorspar are added to alumina. Why ?

Ans. Cryolite and fluorspar are added to alumina :

(i) To lower the melting point of aluminium.

(ii) To make alumina a good conductor of electricity.

(iii) Cryolite acts as a solvent for alumina.

11. Aluminium transmission wires are preferred to copper transmission wires. Why ?

Ans. (i) Aluminium is lighter than copper.

(ii) It is a good conductor of electricity.

(iii) Aluminium is cheaper than copper.

12. Why in construction work alloy duralumin is used rather than aluminium ?

Ans. In construction work alloy duralumin is used rather than aluminium because duralumin is harder, stronger and more resistant to corrosion.

Chapter 8. (a) Study of Compounds : Hydrogen Chloride

1. Mixture of sodium chloride and concentrated sulphuric acid does not heated above the temperature of 170°C while preparing hydrogen chloride. Why ?

Ans. The mixture of sodium chloride and concentrated sulphuric acid is not heated above 170°C in preparing hydrogen chloride gas because at a higher temperature sodium sulphate is formed which is a hard substance and difficult to remove from the reaction flask.

$$2NaCl + H_2SO_4 \longrightarrow Na_2SO_4 + 2HCl$$

2. Hydrogen chloride gas cannot be dried over quick lime. Why ?

Ans. Hydrogen chloride gas cannot be dried over quick lime because quick lime is basic in nature and combines with moist hydrogen chloride gas forming calcium chloride.

3. Quick lime and phosphorus pentaoxide cannot be used for drying hydrochloric acid gas. Why ?

Ans. Quick lime and phosphorus pentaoxide cannot be used for drying HCl gas, because both reacts with HCl.

$$CaO + 2HCl \longrightarrow CaCl_2 + H_2O$$
$$2P_2O_5 + 3HCl \longrightarrow 3HPO_3 + POCl_3$$

4. Hydrogen chloride is not collected over water. Why ?

Ans. Hydrogen chloride is not collected over water because it is highly soluble in water.

5. When the stopper of a bottle full of hydrogen chloride gas is opened there are fumes in the air ?

Ans. When the stopper of a bottle full of hydrogen chloride gas is opened there are fumes in the air because hydrogen chloride gas has an affinity for water, hence, when the stopper is opened it immediately reacts with water vapour present in the atmosphere which leads to the formation of fumes.

6. Dilute hydrochloric acid cannot be concentrated by distilling (boiling) the dilute acid. Why ?

Ans. When dilute hydrochloric acid is distilled, a constant boiling mixture containing 20 - 24% of hydrochloric acid distills over unchanged at 760 mm Hg pressure. This constant boiling mixture cannot be separated into its constituents by simply distilling.

7. Anhydrous HCl is a poor conductor while aqueous HCl is an excellent conductor. Why ?

Ans. Anhydrous HCl is a poor conductor while aqueous HCl is an excellent conductor because anhydrous HCl does not contain any free ions. But when HCl is dissolved in water, it dissociates into hydronium ion (H_3O^+) and chloride ion (Cl^-). Due to the presence of free ions, aqueous solution of HCl conducts electricity.

8. Sodium is not used to prepare hydrogen from hydrochloric acid (or any other acid). Why ?

Ans. Sodium is not used to prepare hydrogen from acids because sodium metal is highly reactive. So, the reaction with acids is much exothermic and there are more chances of explosion. It is also very dangerous to handle sodium metal.

9. Silver nitrate crystals are dissolved in distilled water and not in tap water in order to prepare a solution of silver nitrate as a laboratory reagent.Why ?

Ans. Tap water always contains some amount of dissolved sodium chloride. Thus when the solution of silver nitrate is prepared in tap water, it reacts to form curdy white precipitate of silver chloride.

$$AgNO_3 + NaCl \longrightarrow AgCl + NaNO_3$$

To prevent the above chemical reaction, the solution of silver nitrate is prepared in distilled water.

10. Water for drinking purpose and in swimming pools, is treated with chlorine. Give reason.

Ans. Water for drinking purpose and in swimming pools is treated with chlorine because it sterilises the water. Due to its strong oxidizing action, it destroys bacteria, fungus and other micro-organisms.

11. Why is it inadvisable to use bleaching powder as a disinfectant when it is mixed with acids ?

Ans. When an acid is mixed into bleaching powder, chlorine gas is liberated which pollutes the atmosphere and makes it unsafe for breathing.

$$CaOCl_2 + H_2SO_4 \longrightarrow CaSO_4 + H_2O + Cl_2$$

12. Give reasons why :
    (i) Sodium chloride will conduct electricity only in fused or aqueous solution state.
    (ii) In the electroplating of an article with silver, the electrolyte sodium argento-cyanide solution is preferred over silver nitrate solution.
    (iii) Although copper is a good conductor of electricity, it is a non-electrolyte.

Ans. (i) Sodium chloride ions in the solid state are held by the electrostatic force of attraction, thus are not free to move and conduct electricity but in the fused state, the crystal lattice breaks down and the charged particles (ions) are free to move and thus are able to conduct electricity.

(ii) In aqueous solution of sodium argento-cyanide, silver ions migrate slowly as compared to in silver nitrate, thereby, ensuring even deposition of silver metal on the articles to be electroplated.

(iii) Copper is a good conductor of electricity but it is a non-electrolyte because it is a solid metal and has no mobile ions to conduct electricity.

Chapter 8. (b) Study of Compounds : Ammonia and Nitric Acid

1. Ammonia cannot be obtained in laboratory from ammonium nitrate and sodium hydroxide. Why ?

Ans. Ammonia cannot be obtained in laboratory from ammonium nitrate and sodium hydroxide because ammonium nitrate on heating decomposes explosively with the formation of nitrous oxide and water.

2. Why ammonia gas is not collected over water ?

Ans. Ammonia gas is not collected over water because it is highly soluble in water.

3. Ammonia cannot be dried by bubbling through concentrated sulphuric acid. Why ?

Ans. Ammonia gas cannot be dried by bubbling through concentrated sulphuric acids as it reacts with sulphuric acid to form ammonium sulphate.

$$2NH_3 + H_2SO_4 \longrightarrow (NH_4)_2SO_4$$

4. Quick lime is the only drying agent used for drying ammonia gas. Why ?

Ans. Quick lime is basic in nature and hence does not react with ammonia, which is also basic in nature, whereas all other drying agents are acidic in nature and hence react with ammonia.

$$\underset{\text{(Conc.)}}{2NH_3 + H_2SO_4} \longrightarrow (NH_4)_2SO_4$$

$$\underset{\text{(Anhydrous)}}{CaCl_2} + 8NH_3 \longrightarrow CaCl_2.8NH_3$$

$$6NH_3 + P_2O_5 + 3H_2O \longrightarrow 2(NH_4)_3.PO_4$$

5. Aqueous solution of ammonia gives a pungent smell. Why ?

Ans. Aqueous solution of ammonia gives a pungent smell because it is produced due to bacterial decomposition of urea (NH_2CONH_2), present in urine.

6. Ammonia solution is used as laboratory reagent to identify metal ions. Why ?

Ans. Ammonium hydroxide can precipitate insoluble hydroxide of metals from their salt solutions. Thus, it is used in analytical analysis of salts in laboratory.

7. Aqueous solution of ammonia turns red litmus blue. Give reason.

Ans. This is because ammonia on dissolving in water furnishes ammonium (NH_4^+) ions and hydroxyl (OH^-) ions. The presence of OH^- ions in the solution turns red litmus blue.

$$NH_3 + H_2O \rightleftharpoons NH_4^+ + OH^-$$

8. Ammonia solution is used to remove oil and grease stains from clothes. Give reason.

Ans. When ammonium chloride is heated and rubbed with the metal, the ammonia formed removes grease, oil, etc. It emulsifies fats, grease, etc.

$$\underset{\text{Ammonium chloride}}{NH_4Cl\,(s)} \longrightarrow \underset{\text{Ammonia}}{NH_3\,(g)} + \underset{\text{Hydrogen chloride}}{HCl(g)}$$

The hydrogen chloride formed removes the oxides of metals and thus cleans the metal surfaces before soldering, tinning, etc.

9. Liquid ammonia is used as refrigerant in ice plants. Give reason.

Ans. Liquid ammonia is used as a refrigerant as it has high latent heat of vaporisation equal to 5700 cal/mole and a low boiling point. When liquid ammonia evaporates, it takes large amount of heat from surroundings, resulting in the fall in temperature.

$$NH_3 \xrightarrow{\text{Vaporises}} NH_3\,(\text{gas}) - 5700 \text{ cals.}$$

10. In Haber's process before the gases are allowed to enter the oxidation chamber, the gases are sufficiently cooled. Why ?

Ans. The gases are cooled so that the complete oxidation of nitric oxide takes place or the rate of forward reaction increases or to minimize the decomposition of nitrogen dioxide.

11. The mixture of nitrogen and hydrogen gases entering the catalyst chamber must be pure. Why ?

Ans. The mixture of nitrogen and hydrogen gases entering the catalyst chamber must be pure, because the presence of carbon dioxide, carbon monoxide and traces of sulphur compound poison the catalyst. Therefore, the removal of these catalyst poison from nitrogen and hydrogen is very essential.

12. Nitrogen obtained from air is more dense as compared to nitrogen obtained from chemicals. Why ?

Ans. Nitrogen obtained from air contains traces of inert gases and therefore, it is more dense as compared to chemical nitrogen.

13. During the manufacture of nitric acid by Ostwald's process excess of oxygen is taken. Give reason.

Ans. Excess of oxygen is taken during the manufacture of nitric acid by Ostwald's process because each and every step requires oxygen.

14. In the laboratory preparation of nitric acid, it can be obtained below 200°C or above 200°C, but the lower temperature is preferred. Why ?

Ans. The lower temperature is favoured due to the following reasons :

(i) The glass apparatus may break at higher temperature.

(ii) A significant amount of nitric acid gets decomposed at higher temperature.

(iii) At higher temperature, potassium or sodium sulphate are formed, which cannot be easily removed from the apparatus.

15. Concentrated nitric acid fumes when kept open. Why ?

Ans. Concentrated nitric acid fumes when kept open because it contains considerable amount of nitrogen dioxide which escapes when exposed.

16. Nitric acid stains the skin yellow. Give reason.

Ans. Dilute nitric acid reacts with the proteins of the skin and forms a yellow compound called xanthoproteic acid. Hence, the skin becomes yellow. Conc. nitric acid causes blisters on the skin and is highly corrosive.

17. Iron is rendered passive with fuming nitric acid. Why ?

Ans. Iron is rendered passive with fuming nitric acid because of the formation of reddish brown layer of iron oxide which prevents its further reaction with the acid.

18. Conc. HCl is not used in place of conc. H_2SO_4 to prepare nitric acid from KNO_3. Why ?

Ans. Conc. HCl is not used in place of conc. H_2SO_4 for the preparation of conc. HNO_3 from KNO_3 because conc. HCl is volatile and hence, nitric acid vapours will carry HCl vapours. Lower volatility of conc. H_2SO_4 is used in the preparation of conc. HNO_3.

19. Only all-glass apparatus should be used for the preparation of nitric acid. Why ?

Ans. Nitric acid is highly corrosive and hence destroys rubber and cork of the apparatus. Therefore, only all glass apparatus should be used for the preparation of nitric acid.

20. Aluminium does not react with nitric acid of any concentration. Why ?

Ans. Aluminum metal is not attacked by nitric acid of any concentration because of the thin and unreactive protective layer of aluminum oxide formed on the metallic surface due to the reaction of aluminium metal with oxygen of air.

21. Pure nitric acid take on a yellowish brown colour when exposed to light. Why ?

Ans. When pure HNO_3 is exposed to light it decomposes to give a reddish brown NO_2 gas which dissolves in undecomposed nitric acid to give yellowish brown colour.

22. Burning magnesium continues to burn in nitric oxide, while burning sulphur is extinguished. Why ?

Ans. The heat produced by burning of magnesium is quite sufficient to decompose nitric oxide into nitrogen and oxygen. The oxygen so obtained supports combustion, but the burning of sulphur does not decompose nitric oxide hence it is extinguished.

23. During a thunderstorm, the rain water contains traces of nitric acid. Why ?

Ans. Nitrogen present in atmosphere combines with oxygen to form nitric oxide in the presence of electric spark due to lightening produced from a thunderstorm. Nitric oxide is further oxidized to nitrogen dioxide. Nitrogen dioxide dissolves in water in the presence of oxygen of air to form nitric acid.

$$N_2 + O_2 \longrightarrow 2NO$$

$$2NO + O_2 \longrightarrow 2NO_2$$

$$4NO_2 + O_2 + 2H_2O \longrightarrow \underset{\text{Nitric acid}}{4HNO_3}$$

24. Colourless nitric oxide forms brown fumes in air. Give reason.

Ans. Nitric oxide is oxidized by the oxygen of air to form brown coloured nitrogen dioxide :

$$2NO + O_2 \longrightarrow \underset{\substack{\text{Nitrogen dioxide} \\ \text{(Brown colour)}}}{2NO_2}$$

25. Nitric oxide gives a black colour with ferrous sulphate solution. Why ?

Ans. Nitric oxide gives a black colour solution, when it is passed through ferrous sulphate solution because nitric oxide combines with ferrous sulphate to form a black coloured complex compound known as "Nitroso ferrous sulphate".

$$FeSO_4 + NO \longrightarrow \underset{\text{Nitroso ferrous sulphate}}{[Fe(NO)]\,SO_4}$$

26. Nitrous oxide supports combustion more vigorously than air does. Give reason.

Ans. The percentage of oxygen present in nitrous oxide is more than air and it readily decomposes to form oxygen, which supports combustion.Nitrous oxide contains 36·37% of available oxygen, while air contains 21% of oxygen.

Chapter 8. (c) Study of Compounds : Sulphuric Acid

1. Why concentrated sulphuric acid is called the 'oil of vitriol' ?

Ans. Concentrated sulphuric acid is called 'Oil of vitriol' because of its oily appearance and the fact that it is present in vitrous or glassy substances like ferrous sulphate, alum, etc.

2. For the production of concentrated sulphuric acid, sulphur trioxide is not directly dissolved in water. Why ?

Ans. For the production of concentrated sulphuric acid, sulphur trioxide is not directly dissolved in water because with water, sulphur trioxide forms a mist of fine drops of sulphuric acid.

3. The impurity of arsenic oxide must be removed before passing the mixture of sulphur dioxide and air through the catalytic chamber in contact process. Why ?

Ans. This is because the impurity of arsenic oxide makes the catalyst poisonous.

4. Why concentrated sulphuric acid is kept in air tight bottles ?

Ans. Concentrated sulphuric acid readily absorbs moisture from atmosphere and gets diluted. Hence, it is kept in air tight bottles.

5. Why the level of concentrated sulphuric acid gets higher if it is left in an open vessel for a week ?

Ans. This is due to the hygroscopic nature of sulphuric acid. It absorbs water vapour from the atmosphere.

6. Why sulphuric acid behaves as an acid when diluted with water ?

Ans. When sulphuric acid is diluted with water, it ionizes almost completely into hydrogen ions (H^+) and sulphate ions (SO_4^{2-})

$$H_2SO_4 \rightleftharpoons H^+ + \underset{\text{Hydrogen sulphate ion}}{HSO_4^-}$$

$$HSO_4^- \rightleftharpoons H^+ + \underset{\text{Sulphate ion}}{SO_4^{2-}}$$

Since presence of H^+ ions imparts acidic character, therefore, solution of sulphuric acid in water behaves as an acid.

7. Why the wooden shelves on which conc. sulphuric acid bottles are kept, stained black ?

Ans. Concentrated sulphuric acid is a very powerful dehydrating agent. It removes atoms of hydrogen and oxygen in the form of water from the cellulose [$(C_6H_{12}O_5)_n$], leaving behind carbon. It is black carbon which appears in the form of black stains.

8. A black spongy mass is formed, when concentrated sulphuric acid is added to sugar. Why ?

Ans. Sulphuric acid has great affinity for water, hence when concentrated sulphuric acid is added to sugar, it absorbs water from sugar by removing hydrogen and oxygen atoms in the ratio of 2 : 1 from sugar molecules. The sugar is charred producing black spongy mass of carbon, which is known as sugar charcoal.

$$\underset{\text{Sugar}}{C_{12}H_{22}O_{11}} + H_2SO_4 \longrightarrow \underset{\text{Sugar charcoal}}{12C} + [H_2SO_4 + 11H_2O]$$

9. When blue crystals of copper(II) sulphate are added to concentrated sulphuric acid crystals turn white. Why ?

Ans. Hydrated copper(II) sulphate, when added to concentrated sulphuric acid, loses water of crystallization and thus white anhydrous copper(II) sulphate is formed. The blue coloured hydrated copper(II) sulphate turns white due to the loss of water of crystallization.

$$\underset{\substack{\text{Hydrated copper} \\ \text{sulphate} \\ \text{(Blue crystals)}}}{CuSO_4.5H_2O} + H_2SO_4 \longrightarrow \underset{\substack{\text{Anhydrous copper} \\ \text{sulphate} \\ \text{(White crystals)}}}{CuSO_4} + [H_2SO_4.5H_2O]$$

10. Why brisk effervescence is seen when H_2SO_4 is added to sodium carbonate ?

Ans. Brisk effervescence are seen due to the evolution of carbon dioxide gas.

Chapter 9. Organic Chemistry

1. Hydrocarbons are excellent fuels.Give reason.

Ans. Hydrocarbons are excellent fuels because they ignite easily at low temperature and liberate large amount of heat without producing harmful products.

2. Why alkanes are so inert ?

Ans. It is because in a molecule, a reactive site has one or more unshared pairs of electrons and a polar bond or an electron deficient atom. Alkanes have none of these, that is why they are so inert.

3. Why alkanes are insoluble in water ?

Ans. Alkanes are insoluble in water because they are nonpolar in nature (electronegativity difference between carbon and hydrogen is very low). As a result, they do not dissolve in polar solvents such as water.

4. Methane is called as marsh gas. Why ?

Ans. Methane is called as marsh gas because methane is formed by the decomposition of plant and animal matter lying under water in marshy areas.

5. Methane does not undergo addition reactions, but ethene does. Why ?

Ans. Methane does not undergo addition reactions, but ethene does because methane is saturated hydrocarbon while ethene is an unsaturated hydrocarbon. Addition reactions are characteristic properties of unsaturated hydrocarbons.

6. Why is it dangerous to burn methane in an insufficient supply of air ?

Ans. It is dangerous to burn methane in an insufficient supply of air because it will form carbon monoxide which is poisonous for human beings as it cuts off the oxygen supply by forming carboxy haemoglobin in the blood.

7. Why light or heat is necessary for chlorination of alkanes ?

Ans. The Cl–Cl bond must be broken to form Cl radicals, before the chlorination of alkanes can commence. The breaking of bond requires energy which is supplied either by heat or light.

8. Ethene undergoes addition reactions with halogens whereas ethane undergoes substitution reactions. Why ?

Ans. Ethene is an unsaturated hydrocarbon so, it adds up a molecule of halogen to give a saturated compound, whereas, ethane is a saturated hydrocarbon compound and hence, can only undergo substitution reaction with halogen.

$$\underset{\substack{\text{Ethene} \\ \text{(Unsaturated)}}}{CH_2 = CH_2} + Cl_2 \longrightarrow \underset{\substack{\text{Addition product} \\ \text{(Saturated)}}}{CH_2Cl—CH_2Cl}$$

9. Alkynes are unsaturated hydrocarbons.Give reason.

Ans. Alkynes have triple bonds, so they are unsaturated hydrocarbon.

10. Why ethyne is more reactive than ethane ?

Ans. Ethyne is an unsaturated hydrocarbon with a triple covalent bond. Ethane is a saturated hydrocarbon and hence is less reactive than ethyne.

11. Acetylene burns with sooty flame. Why ?

Ans. Acetylene has higher proportion of carbon and all carbon in it does not burn completely. Hence, unburnt carbon particles make the flame sooty.

12. Why pure acetic acid is known as glacial acetic acid ?

Ans. Pure acetic acid is known as glacial acetic acid because on cooling below its melting point (17°C) it solidifies and forms little ice-like crystals.

❑

Short Answers

Set 13

Chapter 1. Periodic Properties and Variations of Properties

1. State Newland's law of octaves. Why did the law of octaves fail? Write three important drawbacks of Newland's classification.

Ans. Newland's law of octaves : The elements with similar properties occur each time after every seven elements. Just the same way as the repetition of musical node in a octave.

The law failed because of the following reasons :

(i) The law was applicable only upto calcium. It could not include the other elements beyond calcium.

(ii) With the discovery of rare gases, it was the ninth element and not the eighth element having similar chemical properties.

Drawbacks of Newland's classification :

(i) Newland did not gave any place to transition elements.

(ii) Atomic weights were not correctly estimated.

(iii) Certain elements did not fit into the scheme.

2. Answer the following questions regarding Mendeleev's modified periodic table.

(i) How many periods are there in Mendeleev's periodic table?

(ii) How many groups are there in Mendeleev's periodic table?

(iii) How many groups consist of normal elements and transitional elements in Mendeleev's periodic table?

(iv) Which group of elements was missing from Mendeleev's original periodic table?

(v) What is the number of first and last vertical column?

(vi) In which vertical column hydrogen is placed.

(vii) What are horizontal rows known as? How many are they in number?

(viii) How many elements are there in the first horizontal row?

(ix) How many elements are there in the second and third horizontal rows each?

Ans. (i) There are seven periods in Mendeleev's periodic table.

(ii) There are nine groups, numbered from 0 to 8.

(iii) The groups IA, IIA, IIIA, IVA, VA, VIA and VIIA consist normal elements and groups, IIIB, IVB, VB, VIB, VIIB and IB, IIB consist of transition elements.

(iv) Zero group (Noble gases).

(v) The first column is numbered zero, whereas last group is numbered 8.

(vi) Hydrogen is placed in IA group.

(vii) The horizontal rows in periodic table are known as periods. There are seven periods in all.

(viii) There are only two elements in the first period.

(ix) The second and third periods have eight elements each.

3. What is modern periodic law? Name the elements of 'first transition series'.

Ans. The modern periodic law states that, "The properties of elements are the periodic functions of their atomic number". Scandium, titanium, vanadium, chromium, manganese, iron, cobalt, nickel, copper, zinc.

4. Who gave modern periodic law? What is meant by a group in the periodic table?

Ans. Henry Moseley gave the modern periodic law. Vertical column in a periodic table are called groups.

5. How many vertical columns are there in the modern periodic table? State whether the ionization potential increases or decreases on going down a group.

Ans. 18 vertical columns are there in the modern periodic table. Ionization potential decreases on going down a group.

6. How many elements are in the second period of modern periodic table? Write the names of first and last elements of second period.

Ans. There are eight elements in the second period of the periodic table. Lithium (Li) and neon (Ne) are the first and last element of the second period.

7. How many elements are in the third period of modern periodic table? Write the names of first and last elements of third period.

Ans. Eight elements in the third period of the periodic table. Sodium (Na) and Argon (Ar) are the first and last elements of the third period.

8. How many elements are in the fourth, fifth and sixth period of modern periodic table?

Ans. There are 18, 18 and 32 elements in the fourth, fifth and sixth period of the periodic table respectively.

9. Which period is the shortest one? In which period maximum number of elements are present?

Ans. The first period is the shortest period in the periodic table as it has only two elements *i.e.,* H and He. The period in which the maximum number of elements are present is 6th and 7th period. They are the longest period of the periodic table as they have 32 elements.

10. Give the number of the group and the period, of the element having three shells with three electrons in valence shell.

Ans. The element having three shells with three electrons in valence shell belong to 13th group and 3rd period.

11. What are the main characteristic of the last elements in the periodic table? What is the general name given to such elements?

Ans. The main characteristic of the last elements in the periods is the presence of 8 electrons in their valence shell (*i.e.,* octet is complete). The general name of such elements are noble gases or inert gases.

12. What are typical elements? Give the 'liquid elements' in periodic table.

Ans. Na, Mg, Al, Si, P, S, Cl, Ar are the elements present in third short period in periodic table. These elements are known as typical elements. Francium, mercury and bromine are known as liquid elements.

13. What do you know about metallic character of an element?

Ans. It is the property of an element by virtue of which an element possess utmost of 3 electrons in its outermost shell and because of this it is able to form positive ions by loosing there electrons.

14. What is shielding effect or screening effect? How does it govern the ionization energy of an atom?

Ans. When the number of inner electrons is greater, then they shield the outermost electron from the nucleus in such a way that the outermost electron becomes free from any nuclear attraction. This is called the shielding or screening effect.

The shielding effect makes the electron very less firmly held by the nucleus on descending a group, so the ionisation energy decreases down a group.

15. How does the chemical reactivity of alkali metals and halogens vary?

Ans. The reactivity of alkali metals increases as we move down the group. As the atomic number increases, the atomic radius also increases and the single electron in the valence shell is located further away from the nucleus. Hence, Li is least reactive and Francium is the most reactive. In case of halogens, the reactivity gradually decreases as we move down the group. All of them require one electron to complet, the outermost shell and the nuclear attraction is greater on the nearest shell (*i.e.,* IInd shell in F) than in the fathest shell (*i.e.,* VI shell in iodine). Hence, F is most reactive and I is least reactive.

16. What are iso-electronic ions? Account for the decrease in size of the following iso-electronic ions :

$$O^{2-} > F^{-} > Mg^{2+} > Al^{3+}$$

Ans. Iso-electronic ions are of different elements having the same number of electrons but differ from one another in magnitude of nuclear charge. The order of decrease in size is as follows :

Ions	O^{2-}	>	F^{-}	>	Na^{+}	>	Mg^{2+}	>	Al^{3+}
Nuclear charge	+8		+9		+11		+12		+13

On moving from left to right, the nuclear charge goes on increasing and the electrons are pulled more and more strongly decreasing the size of the ions.

17. Give the formula of one species positively charged and one negatively charged that will be iso- electronic with Ne.

Ans. Cation iso-electronic to neon = Mg^{2+}.

Anion iso-electronic to neon = F^-.

18. (i) State the number of elements in period 1, period 2 and period 3 of the periodic table.

(ii) Name the elements in Period 1.

(iii) What happens to atomic size of elements on moving from left to right in a period?

Ans. (i) Period 1 contains 2 elements; period 2 contains 8 elements; period 3 contains 8 elements.

(ii) Hydrogen and helium.

(iii) The atomic size of elements decreases on moving from left to right in a period.

19. (i) What is the common feature of the electronic configurations of the elements at the end of period 2 and period 3?

(ii) If an element is in group 7 (or group 7A) is it likely to be metallic or non-metallic character ?

(iii) Supply the missing word from those in brackets.

If an element has one electron in its outermost energy level (shell) then it is likely to be...... (metallic/non-metallic)

Ans. (i) The atoms of the elements at the end of period 2 and period 3 contains 8 electrons in their outermost shell.

(ii) Non-metallic.

(iii) Metallic.

20. An element Z has atomic number 16. Answer the following questions on Z :

(i) State the period and group to which Z belongs.

(ii) Is Z a metal or a non-metal?

(iii) State the formula between Z and hydrogen.

(iv) What kind of a compound is this?

Ans. Electronic configuration of metal Z = 2, 8, 6 :

(i) 3rd period, group 16th (ii) Z is a non-metal

(iii) H_2Z (iv) Covalent compound.

21. What is similar in the electronic structure of Li, Na and K?

Ans. The electronic configuration of Li = 2, 1.

The electronic configuration of Na = 2, 8, 1.

The electronic configuration of K = 2, 8, 8, 1.

All the three have one electron in the outermost shell which they lose to form positively charged species, *i.e.*, Li^+, Na^+ and K^+.

22. Answer the following questions regarding group 17 *i.e.*, halogens of the periodic table.

(i) What are halogens?

(ii) Which group and sub-group are they placed in?

(iii) What is their valency?

(iv) Why are they called halogens?

(v) Why do they not occur free in nature?

Ans. (i) Halogens are the members of VII A group which have seven electrons in their outermost shell.

(ii) They are placed in group VII and sub-group A.

(iii) Their valency is –1.

(iv) They are called halogens because chlorine which is a halogen, in the form of sodium chloride is the most abundant salt in nature. Halogen means salt producer.

(v) They do not occur free in nature because they all are very reactive as they need just one electron to complete their outermost shell and every member of the halogen group tries to attain stable electronic configuration.

23. State the factors which affect or influence the atomic size of the elements in a periodic table.

Ans. Factors which affect the atomic size of the elements in a periodic table are :

(i) Magnitude of nuclear charge (ii) Number of shells

(iii) Screening or shielding effect

24. In group 1 of the periodic table, three elements X, Y and Z have ionic radii 1.33Å, 0.95Å and 0.60Å respectively. Giving a reason, arrange them in the order of increasing atomic numbers in the group.

Ans. X – 1.33Å; Y – 0.95 Å; Z – 0.60 Å.

The order of increasing atomic numbers is : Z < Y < X.

This order is due to the fact, that greater is the ionic radii, greater is the atomic number of the element. As the ionic radii increases, the number of electrons increases, *i.e.,* the atomic number increases.

25. (i) A boy has reported the radii of Cu, Cu^+ and Cu^{2+} as 0.96Å, 1.22Å and 0.72Å respectively. However, it has been noticed that he interchanged the values by mistake. Assign correct values to different species.

(ii) Account for the difference in size of Na^+ (0.96Å) and Mg^{2+} (0.65Å) both of which have the same number of electrons.

Ans. (i) Cu = 1.22Å; Cu^+ = 0.96Å; Cu^{2+} = 0.72Å.

(ii) Mg^{2+} (0.65Å) is smaller than Na^+ (0.96Å) because as the effective nuclear charge per electron increases, the electrons are more strongly attracted and pulled towards the nucleus which causes a decrease in size of the Mg^{2+} ion.

26. Arrange the following elements as directed :

(i) Ar, He, Ne (in increasing order of electron shells)

(ii) Li, F, C, O (in increasing order of electron affinity)

(iii) Cl, Mg, P, Na (in increasing order of atomic size)

(iv) Cl, Li, F, N (in increasing order of electronegativity)

(v) Cl, S, Al, Na (in increasing order of ionisation potential)

(vi) I, Cl, Br, F (Increasing order of atomic size)

(vii) I, Cl, Br, F (Increasing non-metallic character)

(viii) I, Cl, Br, F (Increasing ionisation potential)

(ix) I, Cl, Br, F (Increasing electron affinity)

(x) I, Cl, Br, F (Decreasing electronegativity)

Ans. (i) He, Ne, Ar

(ii) Li, C, O, F

(iii) Cl, P, Mg, Na

(iv) Li, N, Cl, F

(v) Cl, S, Al, Na.

(vi) F > Cl > Br > I.

(vii) I < Br < Cl < F.

(viii) I < Br < Cl < F.

(ix) I < Br < F < Cl.

(x) F > Cl > Br > I.

27. (i) Arrange I, Cl and Br in an increasing order of their atomic size.

(ii) Arrange the following in the increasing order of their ionisation potential?

Li, Be, B, C, N, O, F.

(iii) What are the following groups known as :

(a) Group I (b) Group VII (c) Group VIIIA.

(iv) What was the significance of the Dobereiner triads or what is the achievement of Dobereiner's Triads?

(v) Arrange the following nine elements into three Dobereiner's triads. The atomic weights of some of these elements are to the left of their symbols. Supply the atomic weights of other elements by calculation.

^{137}Ba ^{80}Br ^{40}Ca ^{127}I K Cl ^{7}Li Sr ^{23}Na.

(vi) An element X belongs to 3rd period and group II of the periodic table. State :

(a) The number of valence electrons (b) Valency

(c) Metal or non-metal (d) Name of the element.

Ans. (i) Cl < Br < I.

(ii) Li, B, Be, C, O, N, F - Reason :

Be > B and N > O due to stable electronic configuration.

(iii) (a) Group I : Alkali metals (b) Group VII : Halogens

(c) Group VIII A : Noble gases.

(iv) This was the first attempt to relate the properties of the elements to their atomic weights.

(v) (a) ^{7}Li ^{23}Na ^{39}K (b) ^{40}Ca $^{88.5}Sr$ ^{137}Ba

(c) ^{33}Cl ^{80}Br ^{127}I

(vi) (a) 2 (b) +2
(c) Metal (d) Magnesium.

28. A group of elements in the periodic table are given below (Boron is first member of the group and Thallium is the last).

Boron
Aluminium
Gallium
Indium
Thallium

Answer the following questions in relation to the above group of elements :

(i) Which element has the most metallic character?
(ii) Which element would be expected to have the highest electro-negativity?
(iii) If the electronic configuration of Aluminium is 2, 8, 3, how many electrons are there in the outer shell of Thallium?
(iv) The atomic number of Boron is 5. Write the chemical formula of the compound formed when Boron reacts with Chlorine.
(v) Will the elements in the group to the right of this Boron group be more metallic or less metallic in character? Justify your answer.

Ans. (i) Thallium (ii) Boron
(iii) Three (iv) BCl_3
(v) Less metallic. Metallic character decreases from left to right in the periodic table.

29. How do the following change, on going from left to right in a period of the periodic table? Give example in support of your answer.
(i) Chemical reactivity of elements. (ii) Nature of oxides of the elements.

Ans. (i) On moving left to right in a period, the chemical reactivity of the elements first decreases and then increases.
For example, in period 3,
Order to chemical reactivity: Na > Mg > Al > Si < P < S < Cl
Most reactive metal — least reactive — most reactive non-metal

(ii) Oxides of elements in a particular period become less basic and finally becomes acidic in character, *e.g.*, oxides of third period :

Na_2O	MgO	Al_2O_3	SiO_2	P_2O_5	SO_3	Cl_2O_7
Strongly basic	Basic	Amphoteric	Feebly acidic	Acidic	More acidic	Most acidic.

30. The elements of one short period of the Periodic Table are given below in order from left to right :

Li Be B C O F Ne

(i) To which period do these elements belong?
(ii) One element of this period is missing. Which is the missing element and where should it be placed ?
(iii) Which one of the elements in this period shows the property of catenation?
(iv) Place the three elements fluorine, beryllium and nitrogen in the order of increasing electronegativity.
(v) Which one of the above elements belongs to the halogen series?

Ans. (i) Second period (ii) Nitrogen. It should be placed between carbon and oxygen.
(iii) Carbon (iv) Beryllium < Nitrogen < Fluorine
(v) Fluorine

31. (i) Electron affinities of two elements A and B are given as follows :
A = 3. 79 electron volts and B = 3.56 electron volts
Which of them will ionize more easily and why?
(ii) A, B and C are the elements or a Dobereiner's triad. If the atomic mass of A is 7 and that of C is 39, what should be the atomic mass of B?
(iii) Account for the difference in size of Fe^{3+} as Fe^{2+} = 0.76Å and Fe^{3+} = 0.64Å.

Ans. (i) A will ionize more easily than B because more is the electron affinity of an element, the more will be the tendency to accept electrons and hence get ionized easily.

(ii)

	A	B	C
	7	?	39

At. mass of B = $\frac{7+39}{2} = \frac{46}{2} = 23$.

(iii) The size of Fe^{3+} (0.64Å) is smaller than the size of Fe^{2+} (0.76Å) because the effective nuclear charge per electron increases. The electron are more strongly attracted and pulled towards the nucleus. This causes decrease in the size of the Fe^{3+} ion.

32. (i) For each of the following pairs, predict which one has greater ionization energy and greater electron affinity?

(a) I, I^- (b) B, C (c) Li, Li^+

(ii) Select the correct order of radii of three species : Ca, Ca^+ and Ca^{2+} :

(a) $Ca > Ca^+ > Ca^{2+}$ (b) $Ca^{2+} > Ca^+ > Ca$

(c) $Ca^+ > Ca > Ca^{2+}$ (d) $Ca^+ > Ca^{2+} > Ca$

Assign suitable reason.

Ans. (i)

Ionisation Energy	Electron Affinity
(a) $I^- > I$ (b) $C > B$ (c) $Li^+ > Li$	(i) $I > I^-$ (ii) $C > B$ (iii) $Li > Li^+$

(ii) The correct order of radii is (a), *i.e.*, $Ca > Ca^+ > Ca^{2+}$.

This is because due to the loss of electron, the effective nuclear charge per electron increases, the electrons are more strongly attracted and pulled towards the nucleus. This causes a decrease in the size of the positive ion.

33. (i) The elements calcium, strontium and barium were put in one group or family on the basis of their similar properties :

(a) What are those similar properties?

(b) What is the usual name of the group or family?

(ii) Chlorine, bromine and iodine elements were put in one group on the basis of their similar properties :

(a) What are those similar properties?

(b) What is the common name of this group or family?

Ans. (i) (a) All the three have the same number of valence electrons in their outermost shell, *i.e.*, 2. So they are reactive, metallic, have low ionisation energy and low electron affinity.

(b) The name of this group is alkaline earth metals.

(ii) (a) All are non-metals and bad conductors of heat and electricity; diatomic in the gaseous state and form ionic compounds with non-metals.

(b) The common name of this group or family is halogens.

34. (i) Two different elements A and B have atomic number 12 and 14 respectively, write down :

(a) The groups of periodic table to which they belong.

(b) The principal oxidation states.

(c) The formulae of their simplest chlorides.

(d) Give explanation :

The chloride of one is a good conductor of electricity in molten state while the other which is liquid, is not.

Ans. (i) The electronic configuration of A and B are :

$$A = 2, 8, 2$$

$$B = 2, 8, 4$$

(a) A belongs to IInd and B belongs to IVth group of the Periodic Table.

(b) The principal oxidation states of A and B are +2 and +4, respectively.

(c) The formulae of their chlorides are ACl_2 and BCl_4, respectively.

(d) The chloride of A is an electrovalent compound and conducts electricity in molten state. The chloride of B is a liquid covalent compound, hence is not a good conductor of electricity.

35. (i) An element has an atomic number 16. State

(a) the period to which it belongs

(b) the number of valence electrons

(c) whether it is a metal or non-metal

(ii) Within a group where would you expect to find the element with :

(a) The greatest metallic character

(b) The largest atomic size

Ans. (i) Electronic configuration – 2, 8, 6

(a) 3rd period (b) Six electrons (c) Non-metal

(ii) (a) At the bottom of the group (CS) (b) At the bottom of the group (CS)

36. With reference to the first three periods of the modern periodic table, answer the questions given below :

(i) Write the formula of the sulphate of the element with atomic number 13.

(ii) Name the element which has the highest ionisation potential.

(iii) What features of the atomic structure accounts for the similarities in the chemical properties of the elements in group VIIA of the periodic table?

(iv) How many electrons are present in the valence shell of the element with atomic number 18 ?

(v) What is the electronic configuration of the element in the third period which gains one electron to change into an anion?

(vi) What is the name given to the energy released when an atom in its isolated gaseous state accepts an electron to form an anion?

(vii) What type of bonding will be present in the oxide of the element with atomic number 1?

Ans. (i) $Al_2(SO_4)_3$

(ii) Helium

(iii) Seven electrons in their valence shells

(iv) Eight

(v) 2, 8, 7

(vi) Electron affinity

(vii) Covalent bonding

37. In Period 3 of the Periodic Table, element **B** is placed to the left of element **A.** On the basis of this information, choose the correct word from the brackets to complete the following statements :*

(i) The element **B** would have (lower/higher) metallic character than **A**.

(ii) The element **A** would probably have (lesser/higher) electron affinity than **B**.

(iii) The element **A** would have (greater/smaller) atomic size than **B**.

Ans. (i) The element **B** would have *higher* metallic character than element **A.**

(ii) The element **A** would have probably *higher* electron affinity than element **B.**

(iii) The element **A** would have *smaller* atomic size than element **B.**

Chapter 2. Chemical Bonding

1. The combination of atoms to form molecules is based on octet rule. Give two limitations of this rule.

Ans. (i) It cannot explain the formation of molecules like BeF_2, BF_3 in which the central atom has less than 8 electrons in its valence shell.

(ii) It cannot explain the formation of molecules such as PF_5, SF_6 etc., in which central atom has more than eight electrons in its valence shell.

2. Write the general characteristics of electrovalent compounds.

Ans. The general characteristics of electrovalent compounds are :

(i) Electrovalent compounds are mostly crystalline in nature.

(ii) Electrovalent compounds form hard crystals. These crystals are usually brittle.

(iii) Electrovalent compounds have high density with high melting and boiling points.

(iv) Electrovalent compounds are soluble in polar solvents.

(v) Electrovalent compounds exhibit isomorphism.

(vi) They react very fast.*

*** Frequently asked previous years Board Exam Questions.**

3. On what factors the formation of an electrovalent compound depends?

Ans. The formation of an ionic or electrovalent compound depends on the following factors :

(i) **Low ionisation energy :** Lesser is the ionisation energy of an atom, greater will be its tendency to form cation by losing the valence electron. Metals with lower ionisation energy values have a greater tendency to form ionic bonds.

(ii) **High electron affinity :** Higher the value of electron affinity, greater will be the tendency of the atom to gain electron and form an anion. Elements with high electron affinity values form ionic compounds.

(iii) **High Electronegativity difference :** Higher the difference in the electronegativities of the elements, easier the transfer of electrons. Hence, more ionic nature of the resulting compound.

4. What are the salient features of electrovalent bond?

Ans. Some salient features of electrovalent bond are as follows :

(i) An electrovalent bond is formed by loss or gain or transfer of electrons.

(ii) Ions are formed during the formation of an electrovalent bond.

(a) Positive ions $\longrightarrow$ Cations (Na^+, K^+, Ca^{2+}, etc.)

(b) Negative ions $\longrightarrow$ Anions (Cl^-, O^{2-}, N^{3-}, etc.)

(iii) An electrostatic force of attraction exist between the oppositely charged ions.

5. Write the important characteristics of covalent compounds.

Ans. The important characteristics of covalent compounds are as follows :

(i) Usually covalent compounds exist in gaseous, liquid or amorphous state.

(ii) Covalent compounds have low melting or boiling points.

(iii) Except for graphite, covalent compounds are bad conductors of electricity.

(iv) Covalent compounds are usually soluble in non-polar solvents (organic solvents). Polar covalent compounds are however soluble in polar solvents.

(v) Covalent compounds undergo reactions slowly.

6. (i) Name the charged particles which attract one another to form electrovalent compounds.

(ii) In the formation of electrovalent compounds, electrons are transferred from one element to another. How are electrons involved in the formation of a covalent compound?

(iii) The electronic configuration of nitrogen is 2, 5. How many electrons in the outer shell of a nitrogen atom are not involved in the formation of a nitrogen molecule?

(iv) In the formation of magnesium chloride (by direct combination between magnesium andchlorine), name the substance that is oxidised and the substance that is reduced.

Ans. (i) Cation and anion

(ii) There is mutual sharing of electrons to form a covalent compound.

(iii) 2

(iv) Magnesium is oxidised and chlorine is reduced.

7. In the formation of compound XY_2, atom X gives one electron to each Y atom. What is the nature of bond to XY_2? Give four properties of XY_2.

Ans. The bond in XY_2 is ionic.

Properties :

(i) It is hard and brittle.

(ii) It is soluble in water.

(iii) It has high melting and boiling point.

(iv) It does not conduct electric current in the solid state but conducts electric current in the molten or dissolved state.

8. There are three elements E, F, G with atomic numbers 19, 8 and 17 respectively.

(i) Classify the elements as metals and non-metals.

(ii) Give the molecular formula of the compound formed between E and G and state the type of chemical bond in this compound.

Ans. (i)

	Z	EC	
E	19	2, 8, 8, 1	Metal
F	8	2, 6	Non-metal
G	17	2, 8, 7	Non-metal

(ii) E – 2, 8, 8, 1

$E - 1e^- \rightarrow E^{+1}$

G – 2, 8, 7

$G + 1e^- \rightarrow G^{-1}$

$\overset{+1}{E} \times \overset{-1}{G} \Rightarrow$ EG or $E\cdot + \cdot\ddot{\underset{..}{G}}: \rightarrow E^+[:\ddot{\underset{..}{G}}:]^-$ or EG

Type of bond ⇒ Ionic bond.

9. (i) Give one property of hydrogen chloride which agrees with it being a covalent compound.

(ii) Give one property of magnesium chloride which agrees with it being an ionic compound.

(iii) Name one compound which is covalent, but on dissolving in water conducts electricity?

(iv) Which property of the above compound agrees with the being of a covalent compound?

Ans. (i) Hydrogen chloride is a gas at room temperature and in dry state it is a bad conductor of electricity.

(ii) Magnesium chloride in the molten state or in aqueous solution is a good conductor of electricity and therefore is an ionic compound.

(iii) Hydrogen chloride.

(iv) Hydrogen chloride is a gas. In dry state, it is bad conductor of electricity.

10. A compound has the formula H_2Y (Y = Non-metal). State the following :

(i) The outer electronic configuration of Y.

(ii) The valency of Y.

(iii) The bonding present in H_2Y.

(iv) The formula of the compound formed between calcium $[^{40}_{20}Ca]$ and Y.

Ans. (i) Y has six electrons in its valence shell

(iii) Covalent

(ii) 2

(iv) $\underset{2,8,8,2}{Ca:} + \ddot{\underset{..}{Y}}: \longrightarrow \underset{2,8,8,2}{Ca^{2+}[:\ddot{\underset{..}{Y}}:]^{2-}}$ or $Ca^{2+}Y^{2-}$ or CaY

11. Elements X, Y and Z have atomic number 6, 9 and 12 respectively. Which will :

(i) form an anion

(ii) form a cation

(iii) has four electrons in its valence shell.

Ans.

Element	Atomic No.	Eletronic configuration
X	6	2, 4
Y	9	2, 7
Z	12	2, 8, 2

(i) Y will form an anion

(iii) X has four electrons in its valence shell.

(ii) Z will form a cation

12. (i) Name two compounds that are covalent when taken pure but produce ions when dissolved in water.

(ii) For each compound, give the formulae of the ions formed in aqueous solutions.

(iii) Give the structure of hydroxyl ion.

Ans. (i) Hydrogen chloride and hydrogen bromide.

(ii) $HCl + H_2O \longrightarrow H_3O^+ + Cl^-$

$HBr + H_2O \longrightarrow H_3O^+ + Br^-$

(iii) Hydroxyl ion has the structure $[:\ddot{\underset{..}{O}}—H]^-$

13. How is a coordinate bond formed?

Ans. The formation of coordinate bond between two atoms occurs in the following steps as :

In the first step the donor atom (say A) transfers one electron of its lone pair to the acceptor atom (say B). Due to this the atom A develops a unit positive charge and atom B develops a unit negative change. This charge is known as formal charge and is similar to the formation of an ionic bond.

In the second step the two electrons, one each with A^+ and B^- are shared by both the ions. This is similar to the formation of covalent bond.

$$A: + B \longrightarrow A^{+\cdot} + \cdot B^{-}$$

$$A^{+\cdot} + \cdot B^{-} \longrightarrow A : B \text{ or } A \longrightarrow B$$

Thus, a coordinate bond is equivalent to a combination of an electrovalent bond and a covalent bond. Hence it is also called a semi polar bond or dative bond.

14. What are the characteristics of coordinate compounds?

Ans. The general characteristics of coordinate compounds are as follows :

(i) Coordinate compounds are identical to normal covalent compounds.
(ii) It is rigid and has directional properties.
(iii) These compounds are soluble in non-polar solvents and insoluble in polar solvent.
(iv) These compounds behave as non-conductors of electricity.
(v) The melting and boiling points are higher than covalent compounds and lower than ionic compounds.
(vi) These are stable compounds.

15. Which conditions are necessary for the formation of coordinate bond?

Ans. For the formation of a coordinate bond the atom acting as a donor must have one unshared pair of electrons which may be donated by it to the acceptor atom. The acceptor atom must have an empty orbital to accept the lone pair of electrons.

16. Explain the formation of H_3O^+ and NH_4^+ ion.

Ans. **Formation of Hydronium ion (H_3O^+) :** This ion is formed by the combination of H_2O molecule and H^+ ion. The water molecule has two O—H covalent bonds and central oxygen atom has two lone pairs of electrons. H^+ ion has one vacant (1s) orbital. During the formation of H_3O^+, one pair of lone pair from O-atom is donated to the vacant (1s) orbital of H^+ ion and $O \rightarrow H^+$ coordinate bond is formed. Thus H_3O^+ ion contains two O—H covalent bonds and one $O \rightarrow H^+$ coordinate bond. After the formation $O \rightarrow H_+$ coordinate bond becomes identical to the two O—H covalent bonds. Hence in H_3O^+ ion all the three bonds are identical.

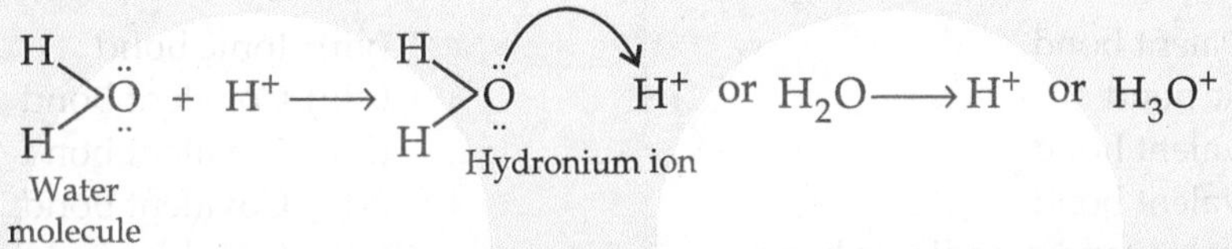

Formation of NH_4^+ Ammonium ion : This ion is formed by the combination of NH_3 molecule and H^+ ion. In NH_3 molecule each of three H-atoms is linked to N-atom by a covalent bond. Thus in this molecule N-atom is left with a lone pair of electrons after completing its octet by sharing three of its valence shell electrons with three H-atoms. The electrons of lone pair on N-atom are donated to H^+ ion and thus a N—H coordinate bond is established in NH_4^+ ion.

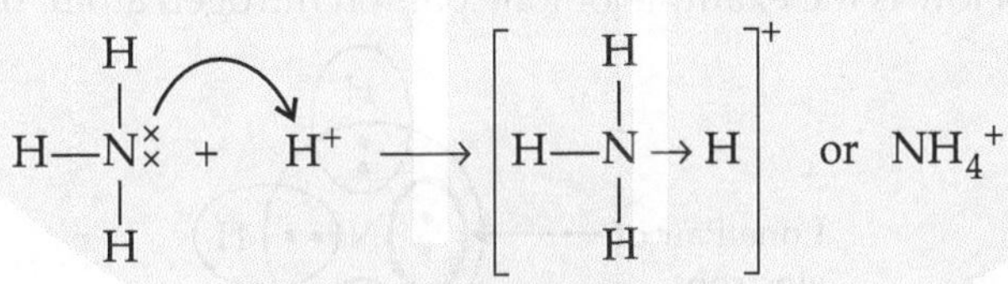

Ammonium ion

17. What is lone pair effect? In what kind of compound does this effect occur?

Ans. When the unshared pair of electrons around an atom in the middle of a molecule is completely shared by another atom or an ion, it is called lone pair effect. Lone pair effect is shown by polar covalent compounds such as HCl and NH_3.

18. (i) Which of the following is not a common characteristic of an electrovalent compound?

(a) High melting point
(b) Conducts electricity when molten
(c) Consists of oppsitely charged ions
(d) Ionises when disolved in water

(ii) What are the terms defined below :

(a) A bond formed by a shared pair of electrons with both electrons coming from the same atom.
(b) A bond formed by a shared pair of electrons, each bonding atom contributing one electron to the pair.

Ans. (i) All the characters are common for electrovalent compounds.

(ii) (a) Coordinate bond (b) Covalent bond

19. (i) Acids dissolve in water to produce positively charged ions. Draw the structure of these positive ions.
(ii) Explain why carbon tetrachloride does not dissolve in water.
(iii) Elements Q and S react together to form an ionic compound. Under normal conditions, which physical state will the compound QS exist in?
(iv) Can Q and S, both be metals? Justify your answer.

Ans. (i)

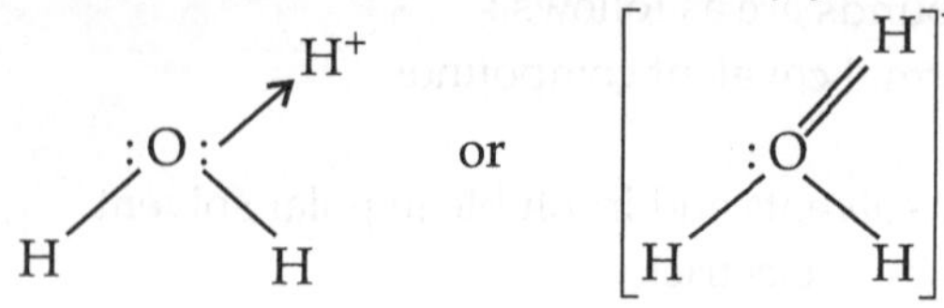

(ii) Carbon tetrachloride does not dissolves in water because carbon tetrachloride is a non-polar covalent compound whereas water is a polar covalent solvent.
(iii) Solid state.
(iv) No Q and S both can not be metals because to form an ionic compound if one element gives electrons, the other element should accept electrons. Metals can only lose electrons to provide positive ions.

20. Predict the type of bonding in the following molecules :
(i) Oxygen
(ii) Calcium oxide
(iii) Water
(iv) Methane
(v) Ammonium ion
(vi) Nitrogen
(vii) Magnesium chloride
(viii) Carbon dioxide
(ix) Carbon tetrachloride
(x) Hydrogen chloride
(xi) Hydrogen cyanide

Ans.
(i) Covalent bond
(ii) Ionic bond
(iii) Covalent bond
(iv) Covalent bond
(v) Covalent and coordinate bonds
(vi) Covalent bond
(vii) Ionic bond
(viii) Covalent bond
(ix) Covalent bond
(x) Covalent bond
(xi) Covalent and coordinate bonds.

21. (i) What do you understand by a lone pair of electrons ?*
(ii) Draw the electron dot diagram of hydronium ion. (H = 1; O = 8)

Ans. (i) A lone pair is an electron pair in the outermost shell of an atom that is not shared or bonded to another atom. Below is the example of lone pair on nitrogen atom of ammonia molecule (NH_3).

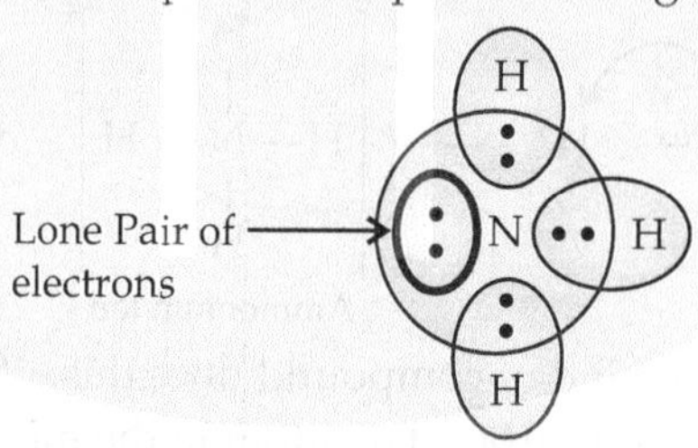

(ii) Electron dot diagram of hydronium ion :

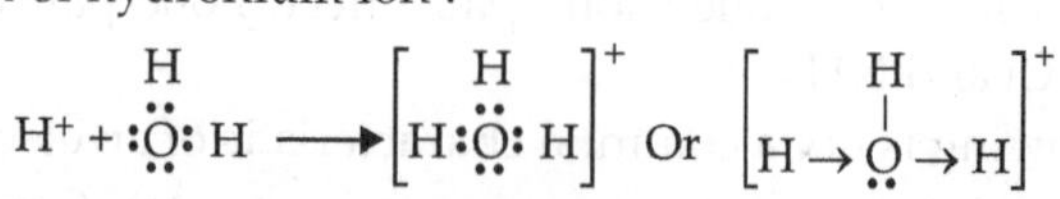

Chapter 3. Study of Acids, Bases and Salts

1. Solution A is a strong acid, solution B is a weak acid, solution C is a strong alkali.
(i) Which solution contains solute molecules in addition to water molecules.
(ii) Which solution could be a solution of glacial acetic acid?
(iii) Give an example of a solution which is a weak alkali.

* Frequently asked previous years Board Exam Questions.

Ans. (i) B (ii) B (iii) Ammonium hydroxide

2. (i) What is basicity of an acid. Give examples.
(ii) Give the basicity of nitric acid, acetic acid and phosphoric acid.
(iii) State two properties of an acid.
(iv) On what factors the strength of an acid depend?
(v) Name an acid used : (a) as a flavour and to preserve food, (b) in a drink, (c) to remove ink spots, (d) as an eye wash.

Ans. (i) The number of replaceable hydrogen atoms, present in one molecule of an acid is known as basicity of the acid. For example hydrochloric acid is a monobasic acid, sulphuric acid is a dibasic acid and phosphoric acid is a tribasic acid.

(ii)
$$\text{Nitric acid } (HNO_3) \longrightarrow \text{Monobasic}$$
$$\text{Acetic acid } (CH_3COOH) \longrightarrow \text{Monobasic}$$
$$\text{Phosphoric acid } (H_3PO_4) \longrightarrow \text{Tribasic.}$$

(iii) (a) An acid turns methyl orange into pink and blue litmus to red.
(b) An acid with a base forms salt and water only.
$$\underset{\text{(Alkali)}}{NaOH} + \underset{\text{(Acid)}}{HCl} \longrightarrow \underset{\text{(Salt)}}{NaCl} + \underset{\text{(Water)}}{H_2O}$$

(iv) The strength of an acid depends on the following factors :
(a) Reactivity of the acid. (b) Degree of ionisation to H^+ ion.

(v) (a) Citric acid. (b) Carbonic acid.
(c) Oxalic acid. (d) Boric acid.

3. (i) What is acidity of a base. Give examples.
(ii) State three properties of an alkali.

Ans. (i) The number of hydroxyl ions that one molecule of a base gives on dissociation is known as acidity of a base. *e.g.*, The acidity of sodium hydroxide (NaOH) is one, it is a monoacid, the acidity of $Ca(OH)_2$ is two, it is a diacid base, $Al(OH)_3$ is a triacid base.

(ii) An alkali turns :
(a) Red litmus blue (b) Phenolphthalein to pink
(c) Methyl orange to yellow.

4. Name the kind of particles present in :
(i) Sodium hydroxide solution. (iii) Sugar solution.
(ii) Carbonic acid

Ans. (i) Sodium ions and hydroxide ions.
(ii) Carbonic acid molecules, carbonate ions and hydronium ions.
(iii) Sugar molecules and water molecules.

5. Solution A is a sodium hydroxide solution. Solution B is a weak acid. Solution C is dilute sulphuric acid. Which solution will :
(i) Liberate sulphur dioxide from sodium sulphite.
(ii) Give a white precipitate with zinc sulphate solution.
(iii) Contain solute molecules and ions.

Ans. (i) C (ii) A (iii) B

6. (i) What are indicators?
(ii) Name three indicators and their colours in different medium.
(iii) What is reaction of neutralization?
(iv) Give two practical application of neutralization.

Ans. (i) Indicators are complex organic compounds, which are used to distinguish between acidic and alkaline solutions by the abrupt change in their colour, *e.g.* litmus, methyl orange and phenolpthalein.

(ii) The three indicators and their colours in different medium are given below :

Indicator	Colour in acidic solution	Colour in alkaline solution
(i) Litmus	Red	Blue
(ii) Methyl orange	Red (Pink)	Yellow
(iii) Phenolphthalein	Colourless	Pink

(iii) Neutralization is an exothermic reaction, in which an acid reacts with a base to form a salt and water only.

$$\underset{\text{Base}}{NaOH} + \underset{\text{Acid}}{HNO_3} \longrightarrow \underset{\text{Salt}}{NaNO_3} + \underset{\text{Water}}{H_2O}$$

In the reaction of neutralization, the hydrogen ions of an acid combines with the hydroxyl ions of an alkali to form water, at the same time a salt is formed.

(iv) (a) Acidity of soil is reduced by adding slaked lime.

(b) Antacid tablets are given to a person suffering from acidity.

7. (i) Explain pH value of a solution.

(ii) What is the pH of :

(a) Pure water (b) Milk (c) Human blood

(iii) The pH value of three solutions is given below. Which one of them is acidic, neutral and alkaline in nature?

(a) pH = 7 (b) pH = 10 (c) pH = 3

(iv) What is pH number assigned to a solution?

Ans. (i) pH value of a solution is the negative logarithm to the base 10 of hydrogen ion concentration expressed in gram-ions per litre.

$$pH = -\log_{10} [H^+]$$

(ii) The pH of :

(a) Pure water is 7 (b) Milk is 6·6 (c) Human blood is 7·3

(iii) (a) Neutral (b) Alkaline (c) Acidic

(iv) It is a number assigned to express the acidity or alkalinity of a solution which varies from 1 to 14. The pH number less than 7 indicates acidity, while pH number more than 7 indicates alkalinity. The pH of pure water is 7 and it is neutral.

pH value = 7 indicates a neutral solution.

pH value > 7 indicates an alkaline solution.

pH value < 7 indicates an acidic solution.

8. (i) (a) What is the purpose of the pH scale?

(b) What is the pH of pure water.

(c) A is a soluble acidic oxide, B is a soluble base compared to the pH of pure water, what will be the pH of : (i) a solution of A, (ii) a solution of B.

(ii) Taking sodium carbonate as an example, give the meaning of the following terms :

(a) Water of crystallization (b) Anhydrous (c) Efflorescence.

Ans. (i) (a) To measure the degree of acidity or alkalinity.

(b) The pH of pure water is equal to 7.

(c) (i) The pH is less than 7. (ii) The pH is more than 7.

(ii) (a) Water of crystallization is that definite number of water molecules which a crystal contain when it is crystallizing out from an aqueous solution. *e.g.*, Sodium carbonate crystal contains 10 molecules of water *i.e.*, $Na_2CO_3.10H_2O$.

(b) When $Na_2CO_3.10H_2O$ is heated, the water of crystallization is evaporated leaving anhydrous sodium carbonate Na_2CO_3.

(c) The loss of water by a hydrate salt on exposure to air is called efflorescence, *e.g.* $Na_2CO_3.10H_2O$ on exposure to air, looses 9 molecules of water of crystallization.

$$Na_2CO_3.10H_2O \xrightarrow{\text{Air}} Na_2CO_3.H_2O + 9H_2O.$$

9. A solution has a pH of 7 explain how you would :

(i) (a) Increase its pH, (b) Decrease its pH.

(ii) If a solution changes the colour of litmus red to blue what can you say about its pH?

(iii) What can you say about the pH of a solution that liberates carbon dioxide from sodium carbonate.

Ans. (i) (a) pH is increased by adding any caustic alkali to solution.

(b) pH is decreased by adding any mineral acid to solution.

(ii) pH of solution is more than 7.

(iii) pH of solution is less than 7.

10. (i) The pH value is utilised in medicine and agriculture. Explain.

(ii) The pH value of pure water is 7. Compare the pH values of sulphur dioxide solution and ammonia solution with that of pure water.

(iii) How does pH value of a solution and hydrogen ion concentration of solution vary?

Ans. (i) (a) **In medicine :** The pH values of urine and blood are used to diagnose various diseases.

(b) **In agriculture :** The pH of soil is checked to ensure fast growth of crops. For example, citrus fruits need slightly alkaline soil, rice crop needs an acidic soil, sugarcane needs a neutral soil.

(ii) Sulphur dioxide is a non-metal oxide, such oxides are acidic with pH less then 7. Ammonia solution is alkaline with pH more than 7.

(iii) Both varies inversely. It is seen that lower the pH value the greater the H^+ ion concentration and higher the pH value the lower the H^+ ion concentration.

11. Three solutions P, Q and R have pH value of 3.5, 5.2 and 12.2 respectively. Which one of these is a :*

(i) Weak acid ?

(ii) Strong alkali ?

Ans. (i) Solution Q is a weak acid as its pH is 5.2

(ii) Solution R is a strong alkali as its pH is 12.2.

12. State the action of a base :

(i) on acid (ii) on carbonates

Ans. (i) When base react with an acid, salt and water is formed. For example,

$NaOH + HCl \longrightarrow NaCl + H_2O$

(ii) When a base reacts with carbonate, no reaction occurs because carbonate is basic in nature. For example,

$NaOH + K_2CO_3 \longrightarrow$ no reaction

13. What is the action of dilute acids on the following :

(i) Sulphites (ii) Carbonates

(iii) Sulphides (iv) Bicarbonates

Ans. (i) $ZnSO_3 + 2HCl \longrightarrow ZnCl_2 + H_2O + SO_2$

(ii) $ZnS + 2HCl \longrightarrow ZnCl_2 + H_2S$

(iii) $CaCO_3 + 2HCl \longrightarrow CaCl_2 + H_2O + CO_2$

(iv) $NaHCO_3 + HNO_3 \longrightarrow NaNO_3 + H_2O + CO_2$

14. Which of the following salts give acidic solutions, alkaline solutions, and neutral solutions in water : potassium nitrate, KNO_3; ammonium sulphate, $(NH_4)_2SO_4$; potassium carbonate, K_2CO_3 ; sodium chloride, NaCl; sodium acetate, CH_3COONa; and copper sulphate, $CuSO_4$.

(i) Acidic solutions ....

(ii) Alkaline solutions ....

(iii) Neutral solutions ....

Ans. (i) **Acidic solutions :** ammonium sulphate $(NH_4)_2SO_4$, copper sulphate $(CuSO_4)$.

(ii) **Alkaline solutions :** potassium carbonate (K_2CO_3), sodium acetate (CH_3COONa).

(iii) **Neutral solutions :** potassium nitrate (KNO_3), sodium chloride (NaCl).

15. You are provided with the following chemicals :
Ammonium hydroxide, chlorine, copper oxide, iron, lead nitrate and dilute sulphuric acid; using only chemicals of this given list, write equation for the following salt preparations :

(i) A salt by direct combination.

(ii) A soluble salt by neutralisation of an alkali.

(iii) A soluble salt from an insoluble base.

(iv) A salt by double decomposition (Precipitation).

(v) A soluble salt from a metal.

Ans. (i) $$2Fe + 3Cl_2 \longrightarrow \underset{\text{Ferric chloride}}{2FeCl_3}$$

* Frequently asked previous years Board Exam Questions.

(ii) $2NH_4OH + H_2SO_4 \longrightarrow (NH_4)_2SO_4 + 2H_2O$

(iii) $\underset{\text{Insoluble base}}{CuO} + H_2SO_4 \longrightarrow \underset{\text{Soluble salt}}{CuSO_4} + H_2O$

(iv) $Pb(NO_3)_2 + H_2SO_4 \longrightarrow \underset{\text{(ppt.)}}{PbSO_4\downarrow} + 2HNO_3$

(v) $Fe + H_2SO_4 \longrightarrow \underset{\substack{\text{Soluble Salt} \\ \text{(Ferrous sulphate)}}}{FeSO_4} + H_2$

16. How many types of salts can be prepared from orthophosphoric acid? Is there any difference in the salts formed by the acid?

Ans. There are three types of salts prepared from orthophosphoric acid [H_3PO_4], two acidic salts and one normal salt. These salts are obtained by the stepwise replacement of hydrogen atom from orthophosphoric acid. For example :

$$NaOH + H_3PO_4 \longrightarrow \underset{\substack{\text{Sodium di-hydrogen} \\ \text{phosphate}}}{NaH_2PO_4} + H_2O$$

$$NaOH + NaH_2PO_4 \longrightarrow \underset{\substack{\text{Di-sodium hydrogen} \\ \text{phosphate}}}{Na_2HPO_4} + H_2O$$

$$NaOH + Na_2HPO_4 \longrightarrow \underset{\text{Sodium phosphate}}{Na_3PO_4} + H_2O$$

The acidic salts obtained from orthophosphoric acid have the capacity to ionize further to yield hydrogen ion and thus they can change blue litmus paper red, whereas normal salt does not.

17. Name, from the list of substances given below, the substance which you would use to prepare each of the following salts named in Part (i) to (iv).

The substances are :

Dilute sulphuric acid, copper, lead, dilute nitric acid, dilute hydrochloric acid, copper oxide, lead carbonate, sodium carbonate, sodium and zinc.

(i) Lead sulphate (ii) Copper sulphate

(iii) Sodium sulphate (iv) Zinc sulphate

(v) What are the two steps necessary to change lead carbonate into lead chloride?

(vi) Give the name of a soluble lead salt and write the equation for the action of heat on this salt.

Ans. (i) For lead sulphate, lead carbonate and dilute sulphuric acid are required.

(ii) For copper sulphate, copper oxide and dilute sulphuric acid are required.

(iii) For sodium sulphate, sodium carbonate and dilute sulphuric acid are required.

(iv) For zinc sulphate, zinc and dilute sulphuric acid are required.

(v) $PbCO_3 + 2HNO_3 \longrightarrow Pb(NO_3)_2 + H_2O + CO_2$

$Pb(NO_3)_2 + HCl \longrightarrow PbCl_2 + HNO_3$

(vi) Lead nitrate, $2Pb(NO_3)_2 \xrightarrow{\Delta} 2PbO + 4NO_2 + O_2\uparrow$

18. Name the method used for preparation of the following salts from the list given below :

(i) Sodium nitrate (ii) Iron(III) chloride

(iii) Lead chloride (iv) Zinc sulphate

(v) Sodium hydrogen sulphate

List :

(a) Simple displacement (b) Neutralization

(c) Decomposition by acid (d) Double decomposition

(e) Direct synthesis

Ans. (i) (b) Neutralisation (ii) (e) Direct synthesis

(iii) (d) Double decomposition (iv) (a) Simple displacement

(v) (c) Decomposition by acid

19. Name three classes of substances, which react with an acid to form salts. Write equations to describe their reactions with suitable acids.

Ans. Three different classes of substances are a metal, a base and a metallic carbonate, which react with an acid to form salts.

(i) Zinc, a metal reacts with dilute sulphuric acid to form zinc sulphate and hydrogen gas is liberated.

$$Zn + H_2SO_4 \longrightarrow ZnSO_4 + H_2$$

(ii) Sodium hydroxide, a base reacts with dilute nitric acid to form sodium nitrate and water.

$$NaOH + HNO_3 \longrightarrow NaNO_3 + H_2O$$

(iii) Magnesium carbonate a metallic carbonate reacts with dilute hydrochloric acid to form magnesium chloride, water and carbon dioxide gas is liberated.

$$MgCO_3 + 2HCl \longrightarrow MgCl_2 + H_2O + CO_2$$

20. Zinc sulphate is called a 'salt', sulphur dioxide an 'acidic oxide' and lead monoxide a 'basic oxide'. What is meant by these terms?

Ans. (i) Zinc sulphate is a salt, because it is formed by the complete replacement of the hydrogen of sulphuric acid by zinc.

$$Zn + H_2SO_4 \longrightarrow \underset{\text{Zinc sulphate}}{ZnSO_4} + H_2$$

(ii) Sulphur dioxide is an acidic oxide, because it is an oxide of non-metal, which dissolves in water to form an acid known as sulphurus acid. It reacts with a base to form salt and water.

$$SO_2 + H_2O \longrightarrow \underset{\text{Sulphurous acid}}{H_2SO_3}$$

$$SO_2 + 2NaOH \longrightarrow \underset{\text{Sodium sulphite}}{Na_2SO_3} + H_2O$$

(iii) Lead monoxide is a basic oxide, because it is a metallic oxide and it reacts with an acid to form salt and water only.

$$PbO + 2HCl \longrightarrow \underset{\text{Lead chloride}}{PbCl_2} + H_2O$$

21. (i) From the list of substances given below, choose the pair required to prepare the salt (a) to (c) in laboratory and write down the relevant equations.

The substances are :

Chlorine, iron, lead, lead nitrate solution, sodium nitrate solutions, iron(III) carbonate, lead carbonate, iron(III) chloride, sodium hydroxide solution and dilute hydrochloric acid.

The salts are :

(a) Sodium chloride. (b) Lead chloride.

(c) Anhydrous iron(III) chloride.

(ii) All ammonium salts are decomposed on heating. What other properties do ammonium salts have in common?

Ans. (i) (a) Hydrochloric acid and sodium hydroxide solution react to form sodium chloride.

$$HCl + NaOH \longrightarrow NaCl + H_2O$$

This reaction represent the formation of a salt by the acid base neutralization.

(b) Lead chloride prepared by adding dilute hydrochloric acid to a solution of lead nitrate.

$$Pb(NO_3)_2 + 2HCl \longrightarrow PbCl_2 + 2HNO_3$$

(c) Anhydrous iron(III) chloride is obtained from action of dry Cl_2 gas over heated powdered iron in a pyrex tube.

$$2Fe + 3Cl_2\,(dry) \longrightarrow 2FeCl_3$$

(ii) On heating, ammonium salt dissociates into ammonia and HCl. The other property of ammonium salt is that when heated with alkali, NH_3 is formed.

$$NH_4Cl + NaOH \longrightarrow NaCl + H_2O + NH_3$$

22. Give equation and indicate briefly, how will you prepare crystals of zinc sulphate from zinc filings?

Ans. Take about 50 cm^3 of dilute sulphuric acid in a beaker and add 5 g of zinc filings and heat it on a wire gauge with constant stirring. Zinc metal reacts with dilute sulphuric acid to liberate hydrogen gas and zinc sulphate is formed.

$$Zn + H_2SO_4 \longrightarrow \underset{\text{Zinc sulphate}}{ZnSO_4} + H_2$$

More zinc filings are added till dilute sulphuric acid is completely consumed and effervescence of hydrogen stops. Filter the solution to remove undissolved zinc and collect the filtrate in an evaporating dish. Heat the solution till a saturated solution is obtained. Cool the solution to obtain the crystals of zinc sulphate with seven molecules of water of crystallization.

$$ZnSO_4 + 7H_2O \longrightarrow \underset{\text{Heptahydrate zinc sulphate}}{ZnSO_4.7H_2O}$$

23. (i) Name four soluble salts.
(ii) Name four insoluble salts.

Ans. (i) (a) Zinc sulphate (b) Iron(II) sulphate
(c) Copper(II) sulphate (d) Sodium sulphate
(ii) (a) Calcium carbonate (b) Lead sulphate
(c) Silver chloride (d) Barium sulphate

24. (i) What happens to the crystals of washing soda when exposed to air? Name the phenomenon exhibited.
(ii) Classify the following compounds into : Deliquescent, Efflorescent, None of these
(a) Magnesium chloride crystals (b) Zinc chloride crystals
(c) Lead nitrate crystals (d) Ferrous sulphate crystals
(e) Lead chloride (f) Copper sulphate crystals
(g) Zinc hydroxide (h) Magnesium chloride

Ans. (i) When exposed to air washing soda crystals lose their water of crystallisation and become amorphous.
∴ The phenomenon is called **efflorescence**.
(ii) (a) Deliquescent (b) Deliquescent (c) None of these
(d) None of these (e) None of these (f) Efflorescent
(g) None of these (h) Highly deliquescent

25. $CuSO_4.5H_2O$, $Na_2CO_3.10H_2O$, CaO and anhydrous calcium chloride are chemicals commonly available in laboratory. Answer the following questions relating your answer to the list of chemicals given above.
(i) Which salt is blue in colour?
(ii) Which salt is efflorescent in nature?
(iii) Which salt is hygroscopic in nature?
(iv) Which salt is deliquescent in nature?
(v) State your observations when solution of calcium chloride is mixed with solution of sodium carbonate.

Ans. (i) $CuSO_4.5H_2O$ is blue in colour.
(ii) $Na_2CO_3.10H_2O$ is efflorescent in nature.
(iii) CaO is hygroscopic in nature.
(iv) Anhydrous calcium chloride is deliquescent in nature.
(v) A white ppt. of calcium carbonate appears when the solutions of $CaCl_2$ and Na_2CO_3 are mixed. The white ppt. gradually settles at the base of test tube.

26. Answer the questions given below, relating your answers only to salts given in the following list : Sodium chloride, calcium chloride, copper sulphate.5 H_2O.
(i) What name is given to the water in compound copper sulphate. 5 H_2O?
(ii) If copper sulphate. 5 H_2O is heated, the water is driven off leaving anhydrous copper sulphate.
(a) What is the colour of anhydrous copper sulphate?
(b) By what means, other than heating, could you dehydrate copper sulphate. 5 H_2O and obtain anhydrous copper sulphate?
(iii) What is deliquescence.
(iv) Which one of the salts in the given list is deliquescent?

Ans. (i) Water of crystallization
(ii) (a) White (b) By adding conc. H_2SO_4

(iii) Compounds that take up enough water from the air to dissolve in the water, are called deliquescent, *e.g.*, calcium chloride ($CaCl_2$) and sodium hydroxide are deliquescent.

(iv) Calcium chloride

27. (i) Give chemical names of the following :

(a) Green vitriol (b) Blue vitriol (c) White vitriol

(ii) State the colour of the following salts :

(a) Copper chloride (b) Ferric chloride

(c) Copper nitrate (d) Lead nitrate

(e) Magnesium carbonate (f) Zinc hydroxide

(iii) State the colours of the aqueous solution of the following salts :

(a) Calcium sulphate crystals (b) Ferrous chloride crystals

(c) Ferric chloride crystals (d) Ferrous sulphate crystals

(e) Ferric sulphate crystals (f) Copper sulphate crystals

Ans. (i) (a) Green vitriol—Ferrous sulphate (b) Blue vitriol—Copper sulphate

(c) White vitriol—Zinc sulphate

(ii) (a) Brown (anhydrous) (b) Black (anhydrous) (c) Blue

(d) White (e) White (f) White

(iii) (a) Colourless (b) Blue green (c) Yellow

(d) Green (e) Yellowish white (f) Blue

28. Making use only of substances chosen from those given below :

Dilute sulphuric acid	Sodium carbonate
Zinc	Sodium sulphite
Lead	Calcium carbonate

Give the equations for the reactions by which you could obtain :

(i) Hydrogen (ii) Sulphur dioxide

(iii) Carbon dioxide (iv) Zinc carbonate

Ans. (i) $Zn + H_2SO_4 \longrightarrow ZnSO_4 + H_2$

(ii) $Na_2SO_3 + H_2SO_4 \longrightarrow Na_2SO_4 + H_2O + SO_2$

(iii) $Na_2CO_3 + H_2SO_4 \xrightarrow{\Delta} Na_2SO_4 + H_2O + CO_2$

OR

$CaCO_3 \xrightarrow{\Delta} CaO + CO_2$

(iv) $Zn + H_2SO_4 \longrightarrow ZnSO_4 + H_2$

$ZnSO_4 + Na_2CO_3 \longrightarrow ZnCO_3 + Na_2SO_4$

29. Give equations and indicate briefly the procedure, you would adopt to prepare :

(i) Crystals of ferrous sulphate from iron filings.

(ii) Crystals of zinc sulphate from zinc carbonate.

(iii) Copper(II) carbonate from copper(II) sulphate.

Ans. (i) Take about 50 cm^3 of dilute sulphuric acid in a beaker and heat it on a wire gauge. Add iron fillings with constant stirring till effervescence of hydrogen stops. Filter to remove excess of iron fillings and evaporate the solution till a saturated solution is formed. Cool the solution, the crystals of ferrous sulphate are separated.

$$Fe + H_2SO_4 \longrightarrow FeSO_4 + H_2.$$

(ii) Take about 5 g of zinc carbonate in a beaker, add dilute sulphuric acid and heat till whole of zinc carbonate is dissolved and effervescence of carbon dioxide stops. Filter the solution in an evaporating dish and heat it till a saturated solution is formed. Allow to cool the solution to obtain the crystals of zinc sulphate.

$$ZnCO_3 + H_2SO_4 \longrightarrow ZnSO_4 + H_2O + CO_2$$

$$ZnSO_4 + 7H_2O \longrightarrow \underset{\text{Hydrated zinc(II) sulphate}}{ZnSO_4.7H_2O}$$

(iii) Take about 5 g of copper(II) sulphate in a beaker and dissolve it in water. Now add a concentrated solution of sodium carbonate to this solution. A light blue precipitate of copper(II) carbonate is obtained. Filter the solution, the sodium sulphate passes into the filtrate and the blue precipitate of copper carbonate is obtained as a residue on the filter paper.

$$CuSO_4 + Na_2CO_3 \longrightarrow \underset{\text{Light blue colour}}{CuCO_3} + Na_2SO_4$$

Chapter 4. Analytical Chemistry

1. What are the use of sodium and ammonium hydroxide in analytical chemistry?

Ans. Sodium hydroxide and ammonium hydroxide are used in analytical chemistry to precipitate insoluble metal hydroxide.

2. Out of the following metallic ions : Al^{3+}, Ca^{2+}, Cu^{2+}, Pb^{2+} and Zn^{2+}; which one forms a white hydroxide which dissolves in an excess of either aqueous sodium hydroxide or ammonium hydroxide?

Ans. Zn^{2+}.

3. Which reagent can be used to distinguish a solution containing a lead salt from a solution containing a zinc salt?

Ans. Ammonium hydroxide.

4. A metal, whose alloy is used in the construction of aircrafts, in the powdered form was added to sodium hydroxide solution, a colourless gas was evolved and after the reaction was over, the solution was colourless.
   (i) Name the powdered metal added to sodium hydroxide solution.
   (ii) Name the gas evolved.

Ans. (i) Aluminium (ii) Hydrogen

5. State the colour of the precipitate observed when caustic soda solution is added to the following solutions (prepared in water) :
   (i) Copper sulphate crystals
   (ii) Ferrous sulphate crystals
   (iii) Ferric chloride crystals
   (iv) Lead nitrate crystals
   (v) Zinc chloride crystals
   (vi) Calcium chloride
   (vii) Zinc sulphate.

Ans. (i) Blue
(ii) Dull green
(iii) Reddish brown
(iv) White
(v) White
(vi) White curdy
(vii) White gelatinous.

6. The questions (i) to (v) refer to the following salt solutions listed A to F :
   (A) Copper nitrate
   (B) Iron(II) sulphate.
   (C) Iron(III) chloride
   (D) Lead nitrate
   (E) Magnesium sulphate
   (F) Zinc chloride.
   (i) Which two solutions will give a white precipitate when treated with dilute hydrochloric acid followed by barium chloride solution?
   (ii) Which two solutions will give a white precipitate when treated with dilute nitric acid followed by silver nitrate solution?
   (iii) Which solution will give a white precipitate when either dilute hydrochloric acid or dilute sulphuric acid is added to it?
   (iv) Which solution becomes a deep or inky blue colour when excess of ammonium hydroxide is added to it?
   (v) Which solution gives a white precipitate with excess ammonium hydroxide solution?

Ans. (i) B and E (Iron (II) sulphate and magnesium sulphate).
(ii) C and F (Iron (III) chloride and zinc chloride)
(iii) D (lead nitrate)
(iv) A (copper nitrate)
(v) F (Zinc chloride)

7. What do you observe when caustic soda solution is added to the following solutions first a little and then in excess :

(i) $FeCl_3$ (ii) $Al_2(SO_4)_3$ (iii) $ZnSO_4$ (iv) Pb $(NO_3)_2$ (v) $CuSO_4$

Also give balanced chemical equations.

Ans. (i) A reddish brown ppt. of ferric hydroxide is obtained which is insoluble in excess of caustic soda solution.

$$FeCl_3 + 3NaOH \longrightarrow \underset{\substack{\text{Ferric hydroxide} \\ \text{(Reddish brown ppt.)}}}{Fe(OH)_3 \uparrow} + 3NaCl$$

(ii) A gelatinous white ppt. of aluminium hydroxide is obtained which is soluble in excess of caustic soda solution.

$$Al_2(SO_4)_3 + 6NaOH \longrightarrow \underset{\substack{\text{Aluminium hydroxide} \\ \text{(White, gelatinous ppt.)}}}{2Al(OH)_3 \downarrow} + 3Na_2SO_4$$

$$Al(OH)_3 + \underset{\text{(Excess)}}{NaOH} \longrightarrow \underset{\substack{\text{Sodium meta-aluminate} \\ \text{(Soluble)}}}{NaAlO_2} + 3Na_2SO_4$$

(iii) A gelatinous white ppt. of zinc hydroxide is obtained which is soluble in excess of caustic soda solution.

$$ZnSO_4 + 2NaOH \longrightarrow \underset{\substack{\text{Zinc hydroxide} \\ \text{(White gelatinous ppt.)}}}{Zn(OH)_2 \downarrow} + Na_2SO_4$$

$$Zn(OH)_2 + \underset{\text{(Excess)}}{2NaOH} \longrightarrow \underset{\substack{\text{Soluble} \\ \text{(Colourless)}}}{NaZnO_2} + 2H_2O$$

(iv) A white ppt. of lead(II) hydroxide is obtained which is soluble in excess of caustic soda solution.

$$Pb(NO_3)_2 + 2NaOH \longrightarrow \underset{\substack{\text{Lead hydroxide} \\ \text{(White ppt.)}}}{Pb(OH)_2 \downarrow} + 2NaNO_3$$

$$Pb(OH)_2 + \underset{\text{(Excess)}}{2NaOH} \longrightarrow \underset{\substack{\text{Sodium plumbite} \\ \text{(Colourless)} \\ \text{(Soluble)}}}{Na_2PbO_2} + 2H_2O$$

(v) A pale blue ppt. of copper(II) hydroxide is obtained which is insoluble in excess of caustic soda solution.

$$CuSO_4 + 2NaOH \longrightarrow \underset{\substack{\text{Copper(II)} \\ \text{hydroxide} \\ \text{(Pale blue ppt.)}}}{Cu(OH)_2 \downarrow} + Na_2SO_4$$

8. (i) What are amphoteric metals? Describe their reactions with hot caustic alkali.

(ii) What are amphoteric oxides? Why these oxides react with NaOH?

Ans. (i) Those metals which react with both alkalies as well as acids are called amphoteric metals *e.g.*, Zn, Sn, Al etc. They react with caustic alkalies like NaOH, KOH on heating and liberates H_2 gas. *e.g.*,

$$\underset{\text{Zinc}}{Zn} + \underset{\text{(Conc.)}}{2NaOH\,(aq)} \longrightarrow \underset{\text{Sodium zincate}}{Na_2ZnO_2} + H_2\uparrow$$

$$\underset{\text{Zinc}}{Zn} + \underset{\text{(Conc.)}}{2KOH\,(aq)} \longrightarrow \underset{\text{Potassium zincate}}{K_2ZnO_2} + H_2\uparrow$$

$$\underset{\text{Tin}}{Sn} + \underset{\text{(Conc.)}}{2NaOH\,(aq)} + H_2O \longrightarrow \underset{\text{Sodium stannate}}{Na_2SnO_3} + 2H_2\uparrow$$

Short Answers

$$Sn + 2KOH (aq) + H_2O \longrightarrow K_2SnO_3 + 2H_2\uparrow$$

Tin (Conc.) Potassium stannate

$$2Al + 2NaOH + 2H_2O \longrightarrow 2NaAlO_2 + 3H_2\uparrow$$

Aluminium Sodium aluminate

$$2Al + 2KOH + 2H_2O \longrightarrow 2KAlO_2 + 3H_2\uparrow$$

Aluminium Potassium aluminate

(ii) Those oxides which react with both acids as well as bases are called amphoteric oxides. Oxides of amphoteric metals like Zn, Sn, Al, etc., react with strong alkalies like NaOH to form complex salt and water, *e.g.*

$$ZnO + 2NaOH (aq) \longrightarrow Na_2ZnO_2 + H_2O$$

Zinc oxide (Conc.) Sodium zincate

$$SnO_2 + 2NaOH (aq) \longrightarrow Na_2SnO_3 + H_2O$$

Tin oxide (Conc.) Sodium stannate

$$Al_2O_3 + 2NaOH (aq) \longrightarrow 2NaAlO_2 + H_2O$$

Aluminium oxide (Conc.) Sodium aluminate

Chapter 5. Mole Concept and Stoichiometry

1. State Gay-Lussac's law of combining volumes.

Ans. The law states That : Under same conditions of temperature and pressure, the volume of gases taking part in a chemical reaction show simple whole number ratio to one another and to those of products if gaseous.

2. (i) When gases react together, their reaction volume bears a simple ratio to each other, under the same conditions of temperature and pressure. Who proposed this gas law?

(ii) What is the volume (measured in dm^3 or litres occupied by one mole of gas at S.T.P.?

Ans. (i) This law was proposed by 'Gay Lussac'.

(ii) One mole of gas occupies 22·4 litre at S.T.P.

3. Under the same conditions of temperature and pressure, one collects 2·2 litre of CO_2, 3·3 litres of Cl_2, 5·5 litres of hydrogen, 4·4 litres of nitrogen and 1·1 litres of SO_2. In which gas sample will there be :

(i) Greatest number of molecules. (ii) The least number of molecules.

Ans. We know that 22·4 litres of any gas at S.T.P. has $6{\cdot}023 \times 10^{23}$ molecules. If the volume of gas at S.T.P. is more than 22·4 litres, then the number of molecules will be greater and vice-versa.

(i) 5·5 litres of hydrogen will contain greatest number of molecules.

(ii) 1·1 litres of sulphur dioxide will contain least number of molecules.

4. Under the same conditions of temperature and pressure, you collect 2 litres of carbon dioxide, 3 litres of chlorine, 5 litres of hydrogen, 4 litres of nitrogen and 1 litre of sulphur dioxide. In which gas will there be?

(i) The greatest number of molecules (ii) The least number of molecules. Justify your answer.

Ans. The greatest number of molecules will be in 5 litres of hydrogen and the least number of molecules in 1 litre of sulphur dioxide. The justification is based on Avogadro's law, which states that equal volumes of all gases, under conditions of same temperature and pressure, contain same number of molecules. So greater the volume, greater will be the number of molecules.

5. How does Avogadro's law explain Gay-Lussac's law of gaseous volumes?

Ans. Avogadro's law states that equal volumes of all gases contain equal number of molecules under similar conditions of temperature and pressure. Since, when gases react chemically, they do so in volumes which bear a simple whole number ratio to each other and to the volume of products, provided the products are also in gasesous state under similar conditions of temperature and pressure. This is what Gay-Lussac's law states. For example, in the reaction of carbon monoxide with oxygen, two volumes of carbon monoxide react with one volume of oxygen to give two volumes of carbon dioxide under similar conditions of temperature and pressure.

$$\underset{\text{2 vol.}}{2CO} + \underset{\text{1 vol.}}{O_2} \longrightarrow \underset{\text{2 vol.}}{2CO_2}$$

The volume ratio of carbon monoxide, oxygen and carbon dioxide is 2 : 1 : 2.

6. What are the applications of Avogadro's law?

Ans. (i) It determines the molecular formula of a gas.

(ii) It determines atomicity of gases.

(iii) It explains Gay-Lussac's law of combining volumes.

(iv) It establishes the relation between molecular weight and vapour density of a gas.

(v) It establishes relation between gram molecular weight and gram molecular volume of a gas.

(vi) Avogadro's hypothesis modifies Dalton's atomic theory.

7. What are the limitations of a chemical equation?

Ans. Following are the limitations of a chemical equation :

(i) A chemical equation does not tell us the physical state of the reactants and the products in the reaction.

(ii) It does not tell us the actual concentration or dilution of the reactants used in the reaction.

(iii) It does not tell whether the reaction starts at its own or some heat is required to start the reaction.

(iv) It does not tell whether the reaction is violent in nature or not.

(v) The time taken by the reaction to complete itself is also not known.

8. What is the important information given by a balanced chemical equation of decomposition of hydrogen peroxide?

Ans. The decomposition of hydrogen peroxide is given by the equation :

$$2H_2O_2 \longrightarrow 2H_2O + O_2$$

The equation gives the following important informations :

(i) Hydrogen peroxide decomposes to form water and oxygen.

(ii) Two molecules of hydrogen peroxide when decomposed, form two molecules of water and one molecule of oxygen.

(iii) 68 g of hydrogen peroxide when decomposed, yield 36 g of water and 32 g of oxygen.

9. (i) What is 'mole scale' of a compound?

(ii) What is the relation between atomic mass and equivalent mass?

Ans. (i) The molecular weight of a compound expressed in gram is known as 'a mole' of a compound. The multiples of the fractions of a mole give different mole values of a compound. The molecular weight of carbon dioxide is 44, hence 44 g is one mole of carbon dioxide. 88 g and 176 g of carbon dioxide represent 2 moles and 4 moles respectively.

In general,

$$\text{Number of moles of a compound} = \frac{\text{Mass of the compound in g}}{\text{Molecular weight of the compound}}$$

(ii) Atomic mass = Equivalent mass × Valency.

10. How will you differentiate between atomic weight and actual weight of an element?

Ans.

	Atomic weight	Actual weight
(i)	It is the number of times an atom of an element is heavier than a carbon atom (C^{12}).	It is the weight of an atom as compared to a standard weight of 1g.
(ii)	Its numerical value is quite large and is conveniently used for comparing the weights of various atoms and making calculations.	Its numerical value is quite small and is inconvenient to use in comparing the weights of various atoms and making calculations.
(iii)	It represents the mass of $6{\cdot}023 \times 10^{23}$ atoms.	It represents the mass of one atom.
(iv)	It is simply a number as it expresses the ratio between two weights.	It is the absolute weight of an atom and is generally expressed in grams.

11. Explain the following :

(i) Is it possible to change the temperature and pressure of a fixed mass of gas without changing its volume?

(ii) One mole of hydrogen contains $2 \times 6{\cdot}023 \times 10^{23}$ atoms of hydrogen where as one mole of helium contains $6{\cdot}023 \times 10^{23}$ atoms of helium.

Ans. (i) Yes, by keeping the volume constant, the pressure of the gas increases with temperature or vice versa.

$$P \propto T \qquad \text{(Volume of gas is constant)}$$

Or

$$\frac{P_1}{T_1} = \frac{P_2}{T_2}$$

(ii) Hydrogen is a diatomic gas.

So one molecule of hydrogen = 2 atoms

$\therefore$ 1 mole or $6{\cdot}023 \times 10^{23}$ molecules of H_2 = $2 \times 6{\cdot}023 \times 10^{23}$ atoms

On the other hand, helium is mono-atomic gas,

$\therefore$ One molecule of helium = 1 atom of He

or $6{\cdot}023 \times 10^{23}$ molecules of He = $6{\cdot}023 \times 10^{23}$ atoms of He.

12. What is the relationship between gram molecular weight and gram molecular volume at S.T.P. ?

Ans. Density of gas is defined as mass per unit volume. Volume is usually taken as 1 dm^3 at S.T.P.

$$\therefore \quad \text{Density of gas} = \frac{\text{Mass of gas}}{1\ dm^3 \text{ of gas at S.T.P.}}$$

Molar volume of hydrogen :

Density of hydrogen = $0{\cdot}09\ g/dm^3$ at S.T.P.

Gram molecular weight of hydrogen = $2{\cdot}016$ g

$0{\cdot}09$ g of hydrogen occupies volume = 1 dm^3 at S.T.P.

$$\therefore \quad 2{\cdot}016 \text{ g of hydrogen occupies volume} = \frac{2.016}{0.09}\ dm^3 \text{ at S.T.P.}$$

$$= 22{\cdot}4\ dm^3 \text{ at S.T.P.}$$

As $2{\cdot}016$ g of hydrogen = 1 gram molecular weight

$\therefore$ 1 gram molecular weight of hydrogen occupies $22{\cdot}4\ dm^3$ at S.T.P.

13. What do you understand by the statement, "The vapour density of CO_2 is 22"?

Ans. The molecular weight of the gas is twice its vapour density.

Molecular weight = 2 × Molecular weight of carbon dioxide

$CO_2 = 12 + 2\,(16) = 12 + 32 = 44$ gram.

The vapour density of carbon dioxide

$$\text{V.D. }(CO_2) = \frac{\text{Molecular weight of } CO_2}{2}$$

$$\text{V.D. }(CO_2) = \frac{44}{2}$$

$$\text{V.D. }(CO_2) = 22$$

The vapour density of CO_2 is 22 indicates that the mass of *x* litres of hydrogen gas or the molecular mass of CO_2 is 44 g.

Thus, the vapour density of carbon dioxide (CO_2) is 22.

14. A compound is formed by 24 grams of X and 64 grams of oxygen, if atomic weight of X = 12 and O = 16. Find the simplest formula of the compound.

Ans.

Relative number of atoms	Simplest ratio
$X = \frac{24}{12} = 2$	$X = \frac{2}{2} = 1$
$O = \frac{64}{16} = 4$	$O = \frac{4}{2} = 2$

Therefore, simplest ratio between X and O is X : O = 1 : 2

Thus, empirical formula of the compound is XO_2.

Chapter 6. Electrolysis

1. Classify following substances under three headings : Strong electrolytes, Weak electrolytes, Non-electrolytes.

   Acetic acid, ammonium chloride, ammonium hydroxide, carbon tetrachloride, dilute hydrochloric acid, sodium acetate, dilute sulphuric acid.

Ans. Strong electrolytes —Ammonium chloride, dilute hydrochloric acid, dilute sulphuric acid.

Weak electrolytes — Ammonium hydroxide, acetic acid, sodium acetates.

Non-electrolyte — Carbon tetra chloride.

2. How will you distinguish between metallic conduction and electrical conduction?

Ans.

Metallic conduction	Electrical conduction
(i) By the movement of electrons.	By the movement of ions.
(ii) Does not involve the transfer of matter.	It involves the transfer of matter as ions.
(iii) No change in chemical properties of the conductor.	It involve the decomposition of electrolyte as a result of chemical reactions.
(iv) Increase in resistance with the rise of temperature.	Decrease in resistance with the rise of temperature.

3. Differentiate between an electrolytic cell and electrochemical cell.

Ans.

Galvanic cell	Electrolytic cell
It converts electrical energy into chemical energy.	It converts chemical energy into electrical energy.
Electrical energy brings out a chemical reaction with the help of an external force.	Chemical energy generates electrical energy with the help of redox reaction.
Used in purifying copper and electroplating.	Used in batteries for example, batteries present in TV remotes.

4. Differentiate between electrical conductivity of copper sulphate solution and copper metal.

Ans. Differences between electrical conductivity of copper sulphate solution and copper metal :

Copper sulphate solution	Copper metal
(i) Electric current is by flow of ions.	Electric current is by flow of electrons.
(ii) It is aqueous solution of ionic compound.	It is a metal in solid state.
(iii) Copper sulphate undergoes a chemical change.	Copper metal remains unchanged chemically.

5. Choose A, B, C or D to match the descriptions (i) to (v) below. Some alphabets may be repeated.

   (A) Non-electrolyte (B) Strong electrolyte

   (C) Weak electrolyte (D) Metallic conductor

   (i) Molten ionic compound

   (ii) Carbon tetrachloride

   (iii) An aluminium wire

   (iv) A solution containing solvent molecules, solute molecules and ions formed by the dissociation of solute molecules.

   (v) A sugar solution with sugar molecules and water molecules.

Ans. (i) B (ii) A (iii) D (iv) C (v) A

6. Give three differences between sodium atom and sodium ion.

Ans. (i) Sodium atom is neutral in nature, while sodium ion is a positively charged particle.

(ii) Sodium atom vigorously reacts with water to liberate hydrogen gas, while sodium ion does not react with water.

(iii) Sodium atom tends to lose an electron to form sodium ion with a complete octet in the outermost shell.

7. Explain how electrolysis is an example of redox reaction.

Ans. Redox reactions are called simultaneous oxidation-reduction reactions. In electrolytic reactions, the positively charged ions (cations) accept electrons from cathode to form neutral atoms, *i.e.*, at cathode reduction takes place. At anode, the negatively charged ions (anions) lose electrons to form neutral atoms, *i.e.*, at anode oxidation takes place. So, electrode reactions also signify oxidation-reduction reactions. Hence, they are also called redox reactions.

8. (i) Write equations to show the electrolytic dissociation of :
   (a) Two acids (b) Two bases

   (ii) When fused sodium chloride is electrolysed, explain exactly what happens at the electrodes and explain how the electricity is conducted?

Ans. (i) (a) Acids :

$$H_2SO_4 \rightleftharpoons 2H^+ + SO_4^{2-}$$
$$HNO_3 \rightleftharpoons H^+ + NO_3^-$$

(b) Bases :

$$NaOH \rightleftharpoons Na^+ + OH^-$$
$$KOH \rightleftharpoons K^+ + OH^-$$

(ii) When electricity is passed through fused sodium chloride, the electrolysis starts as follows :

$$NaCl \rightleftharpoons Na^+ + Cl^-$$

At cathode : $Na^+ + e^- \longrightarrow Na$

At Anode : $Cl^- - e^- \longrightarrow Cl$

$$Cl + Cl \longrightarrow Cl_2$$

Sodium metal is deposited at cathode, while chlorine gas is liberated at anode. Electricity is conducted with the help of free sodium and chloride ions, which are present in fused sodium chloride.

9. How is it possible to discharge Na^+ ions in preference to H^+ ions in electrolysis of NaCl solution?

Ans. By using cathode made of moving mercury, Na^+ ions are discharged in preference to H^+ ions because of the nature of electrode. Mercury has strong tendency to form an amalgam with sodium :

$$Na^+ + e^- \longrightarrow Na$$
$$Na + Hg \longrightarrow Na/Hg$$

When the sodium amalgam dissolves in water, the reaction is :

$$2Na/Hg + 2H_2O \longrightarrow 2NaOH + H_2 + Hg$$

10. A certain metal, say M, does not liberate hydrogen from dilute sulphuric acid, but displaces copper from aqueous copper(II) sulphate. State the most likely place for the metal in electro- chemical series.

Ans. The activity series is obtained, when we examine replacement of one metal ion from its solution by another metal. The metal (M) which displaces copper from aqueous copper(II) sulphate is placed at higher position as compared to copper in activity series.

11. M is a metal above hydrogen in the activity series and its oxide has the formula M_2O. This oxide when dissolved in water forms the corresponding hydroxide which is a good conductor of electricity. In the above context answer the following :
   (i) What kind of combination exists between M and O?
   (ii) How many electrons are there in the outermost shell of M?
   (iii) Name the group to which M belongs.
   (iv) State the reaction taking place at the cathode.
   (v) Name the product at the anode.

Ans. (i) Electrovalent bond exists between M and O.
(ii) One electron is there in the outermost shell.
(iii) M belongs to First group.
(iv) $M^+ + e^- \longrightarrow M$ (at cathode).
(v) Oxygen gas is liberated at anode.

12. How will you electrolyse the molten solution of lead bromide?

Ans. Molten lead bromide ($PbBr_2$) forms Pb^{2+} and Br^- ions. The positive lead ions (Pb^{2+}) move to the cathode, and gain two electrons and change into lead atom.

At cathode : $Pb^{2+} + 2e^- \longrightarrow \underset{\text{(Lead)}}{Pb}$

Short Answers

Lead is deposited at the cathode. The bromide ion migrates to the anode. It loses an electron and becomes a bromine atom. The two bromine atoms combine to form bromine molecule and liberated as amber colored vapours at anode :

At anode : $Br^- - e^- \longrightarrow Br$

$$Br + Br \longrightarrow Br_2$$

13. (i) What are the particles present in a non-electrolyte?

(ii) What is conductivity of metals due to?

(iii) What should be the physical state of lead bromide, if it is to conduct electricity?

(iv) What particles are present in pure lead bromide?

Ans. (i) Molecules are present in a non-electrolyte.

(ii) The conductivity of metals is due to movement of electrons.

(iii) Lead bromide should be in molten state, if it is to conduct electricity.

(iv) Lead ions and bromide ions are present in pure lead bromide in molten state.

14. (i) When the electrolysis of acidified water is carried out :

(a) What is the ratio of the volume of hydrogen produced to the volume of oxygen?

(b) Give the equation for the discharge of ions at the cathode.

(ii) To carry out the so-called 'electrolysis of water', sulphuric acid is added to water. How does the addition of sulphuric acid produce a conducting solution?

Ans. (i) (a) The ratio is 2 : 1

(b) $H^+ + e^- \rightarrow H$; $2H + 2H \longrightarrow 2H_2$

(ii) Addition of sulphuric acid causes dissociation of water molecules into $[H^+]$ and $[OH^-]$ ions.

15. What would happen if in the electrolysis of acidified water, copper electrodes were used instead of platinum ones?

Ans. **At anode :** OH^- and SO_3^{2-} would migrate to anode but neither would be discharged, instead copper atoms would get oxidised to Cu^{2+} and enter solution

$$Cu \longrightarrow Cu^{2+} + 2e^-$$

The electrolytic solution would become blue in colour. Anode would dissolve.

At cathode : H^+ ions would migrate to the cathode and get reduced. Thus, H_2 gas would be discharged at cathode

$$2H^+ + 2e^- \longrightarrow 2H \longrightarrow H_2$$

Later, as the solution turns blue due to formation of Cu^{2+} ions, and will get discharged to $2H^+$ ions as they are less electropositive

$$Cu^{2+} + 2e^- \longrightarrow Cu \downarrow$$

16. During the electrolysis of aqueous copper sulphate, between copper electrodes, the sulphate and hydroxyl ions remain as spectator ions. Why?

Ans. During electrolysis of aqueous copper sulphate using copper electrodes, the two anions OH^- and SO_4^{2-} migrate to the anode, but none of them get discharged because the copper of the anode dissolves in the solution producing copper ions and electrons. Hence, OH^- and SO_4^{2-} ions remain as spectator ions.

17. The following questions refer to the electrolysis of copper sulphate solution with copper electrodes.

(i) Compare the change in mass of the cathode with the change in mass of the anode.

(ii) What happens, when electrolysis of aqueous copper sulphate between platinum electrode, occurs.

(iii) What is the practical application of the electrolysis of copper sulphate solution? Briefly describe one such application.

Ans. (i) Mass of cathode increases where as that of anode decreases due to deposition of pure copper on cathode.

(ii) The blue colour of copper sulphate is due to the presence of cupric ions (Cu^{2+}). Cu^{2+} ions are discharged at the cathode and deposited as pinkish copper metal, but OH^- ions are discharged at anode. The electrolyte consists of hydrogen and sulphate ions which associate to form colourless sulphuric acid.

(iii) The electrolysis of copper sulphate solution is used in the purification of copper using pure copper plate as cathode and impure copper plate as anode.

18. During the electrolysis of copper(II) sulphate solution using platinum as cathode and carbon as anode :

(i) What do you observe at the cathode and at the anode?

(ii) What change is noticed in the electrolyte?

(iii) Write the reactions at the cathode and at the anode.

Ans. (i) At cathode red shiny metal deposits.

At anode bubbles of a colourless odourless gas are seen coming out.

(ii) Colour of electrolyte gradually fades from blue to colourless.

(iii) Reaction at cathode

$$Cu^{2+} + 2e^- \longrightarrow Cu$$

Reaction at anode

$$OH^- - e^- \longrightarrow OH$$
$$4OH \longrightarrow 2H_2O + O_2 \uparrow$$

19. Explain, how the blue colour of electrolyte fades during electrolysis of $CuSO_4$ solution?

Ans. The blue colour of electrolyte is due to the presence of copper ions in it. As the electrolysis is carried out, the copper ions discharge at the cathode.

$$Cu^{2+} + 2e^- \longrightarrow Cu$$

However, no copper ions enters in the electrolyte from anode. Thus concentration of copper ions goes on decreasing. This result in fading of blue colour. When copper ions completely finish the electrolyte becomes colourless.

Reaction of cathode :

$$H^+ + 1e\_ \longrightarrow H$$
$$H + H \longrightarrow H_{2(g)}$$

Reaction of anode :

$$OH^- - 1e^- \longrightarrow OH$$
$$4OH \longrightarrow 2H_2O + O_{2(g)}$$

20. Explain, how blue colour of electrolyte remains unchanged for time being during electrolysis of aqueous copper sulphate?

Ans. The decrease in the weight of copper anode is equal to the increase in weight of copper cathode.

$$CuSO_4 \rightleftharpoons Cu^{2+} + SO_4^{2-} \quad \text{(Dissociation of } CuSO_4)$$
$$H_2SO_4 \rightleftharpoons 2H^+ + SO_4^{2-} \quad \text{(Dissociation of } H_2SO_4)$$
$$H_2O \rightleftharpoons H^+ + OH^- \quad \text{(Dissociation of } H_2O)$$
$$Cu + 2e^- \longrightarrow Cu \quad \text{(at Cathode)}$$
$$Cu - 2e^- \longrightarrow Cu^{2+} \quad \text{(at Anode)}$$

From the above equations it can be deduced that for every Cu^{2+} discharging at cathode a copper atom ionises from anode to form Cu^{2+} ion.

Thus concentration of copper ions in electrolyte remain unchanged. As the blue colour of electrolyte is due to the presence of Cu^{2+} ions, therefore, it does not change. However, it fades continously with decreasing number of Cu^{+2} ions.

21. Mention the type of ions present, the products obtained and the electrode reactions that occur, when the following are electrolysed :

(i) Molten lead bromide between steel cathode and graphite anode.

(ii) Water acidified with sulphuric acid between platinum electrodes.

(iii) Aqueous copper sulphate between copper electrodes.

(iv) Aqueous copper sulphate between copper cathode and platinum anode.

Ans.

		Cathode	Anode
(i)	Electrode used	Steel	Graphite
	Ions present	Pb^{2+}	Br^-
	Products	Lead	Bromine vapours
	Reaction	$Pb^{2+} + 2e^- \longrightarrow Pb$	$2Br^- - 2e^- \longrightarrow Br_2$
(ii)	Electrode used	Platinum	Platinum
	Ions present	H^+	OH^- and SO_4^{2-}
	Products	Hydrogen	Oxygen
	Reaction	$4H^+ + 4e^- \longrightarrow 2H_2$	$4OH^- - 4e^- \longrightarrow 2H_2O + O_2$

(iii) Electrode used Ions present Products Reaction	Copper Cu^{2+} and H^+ Copper atom $Cu^{2+}(aq) + 2e^- \longrightarrow Cu(s)$	Copper OH^- and SO_4^{2-} Copper ions $Cu(s) - 2e^- \longrightarrow Cu^{2+}(aq)$
(iv) Electrode used Ions present Products Reaction	Copper Cu^{2+} and H^+ Copper atom $Cu^{2+} + 2e^- \longrightarrow Cu(s)$	Platinum OH^- and SO_4^{2-} Oxygen $4OH^- - 4e^- \longrightarrow 2H_2O + O_2$

22. Three different electrolytic cells A, B and C are connected in separate circuits. Electrolytic cell A contains sodium chloride solution. When the circuit is completed a bulb in the circuit glows brightly. Electrolytic cell B contains acetic acid solution and in this case the bulb in the circuit glows dimly. The electrolytic cell C contains sugar solution and the bulb does not glow. Give a reason for each of these observations.

Ans. **In cell A :** Sodium chloride being strong electrolyte dissociates completely and therefore, current flows better.

In cell B : Acetic acid being weak electrolyte ionises only partially and therefore, only a weak current flows.

In cell C : Sugar being a covalent compound does not ionise at all and therefore, no current flows.

23. With reference to electroplating answer the following :

(i) Why are articles electroplated?

(ii) Why a small current passed for a longer period is preferred over high current for a shorter period ?

(iii) Why the article to be electroplated is made a cathode?

(iv) Why a direct current is used?

Ans. (i) Articles are electroplated for the following reasons :

(a) To prevent it from corrosion.

(b) To improve the appearance of the metal articles.

(ii) To get a uniform and smooth coating of superior metal, a small current should be used for a longer time.

(iii) The article to be electroplated is always placed at cathode because the metal is always deposited at cathode.

(iv) Direct current is used to get smooth coating and the phase of the current is same at all instance of time.

24. How a spoon is electroplated with silver?

Ans. Silver nitrate solution is taken in a beaker. Silver wire and a spoon, which is to be plated with silver, are dipped in the solution. Silver wire is connected with the positive terminal of the battery which acts as anode and spoon is connected with the negative terminal of the battery which acts as cathode.

When electric current is passed through silver nitrate solution, electrolysis takes place and silver is deposited as a fine thin film at the surface of spoon. The spoon is coated with silver and looks as a silver spoon.

25. Element X is a metal with a valency 2.

Element Y is a non-metal with a valency 3.

(i) Write equations to show how X and Y form ions.

(ii) If Y is a diatomic gas, write the equation for the direct combination of X and Y to form a compound.

(iii) Write two applications of electrolysis in which the anode diminishes in mass.

(iv) If the compound formed between X and Y is melted and an electric current passed through the molten compound, the element X will be obtained at the ... and Y at the ... of the electrolytic cell. (Provide the missing words.)

Ans. (i)

$$X - 2e^- \longrightarrow X^{2+}$$

$$Y + 3e^- \longrightarrow Y^{3-}$$

(ii)

$$3X^{+2} + 2Y^{-3} \longrightarrow X_3Y_2$$

(iii) (a) Electroplating of metals.

(b) Electrorefining of metals.

(iv) If the compound formed between X and Y is melted and an electric current passed through the molten compound, the element X will be obtained at the cathode and Y at the anode of the electrolytic cell.

26. Mr. Ramu wants to electroplate his key chain with nickel to prevent rusting. For this electroplating :

(i) Name the electrolyte (ii) Name the cathode
(iii) Name the anode (iv) Give the reaction at the cathode
(v) Give the reaction at the anode.

Ans. (i) Nickel sulphate (ii) Key chain
(iii) Pure nickel plate (iv) $Ni^{2+} + 2e^- \longrightarrow Ni$
(v) $Ni - 2e^- \longrightarrow Ni^{2+}$

27. Choose the correct word which refers to the process of electrolysis from A to E, to match the description (i) to (iv) :*

A : Oxidation B : Cathode C : Anode
D : An electrolyte E : Reduction

(i) Conducts electricity in aqueous or in molten state.
(ii) Loss of electron takes place at anode.
(iii) A reducing electrode.
(iv) Electrode connected to the positive end or terminal of the battery.

Ans. (i) Conducts electricity in aqueous or in molten state – D: An electrolyte
(ii) Loss of electron takes place at anode – A: Oxidation
(iii) A reducing electrode – B: Cathode
(iv) Electrode connected to the positive end or terminal of the battery – C: Anode

28. Differentiate between the following pairs based on the information given in the brackets :*

(i) Conductor and electrolyte (conducting particles)
(ii) Cations and anions (formation from an atom)
(iii) Acid and Alkali (formation of type of ions)

Ans. (i) Conductor has electrons as the conduction particles whereas ions are the conducting species in electrolyte.
(ii) Cations are formed when a neutral atom loses electrons whereas anions are formed when a neutral atom gains electron.
(iii) An acid dissociates to furnish H^+ ions and the conjugate base which is negatively charged and a base dissociates to give OH^- ions and a conjugate acid which is positively charged.

29. Name the particles present in :*

(i) Strong electrolyte
(ii) Non-electrolyte
(iii) Weak electrolyte

Ans. (i) The particles present in strong electrolyte are molecules which easily and completely dissociate into ions. Example: Strong electrolyte such as NaCl which dissociates strongly into Na^+ and Cl^- ions.

(ii) The particles present in non-electrolyte are molecules which do not dissociate into ions. Example: Non-electrolyte such as urea. NH_2CONH_2 which do not dissociate.

(iii) The particles present in weak electrolyte are both molecules and ions which dissociate into ions to a very less extent. Example: CH_3COOH which dissociates feebly into CH_3COO^- and H^+ ions.

Chapter 7. Metallurgy

1. (i) Arrange Cu, Ca, Al, Fe, Mg, Pb, Na and Zn in the decreasing order, in which they appear in the activity series; putting down the most reactive metal first and least reactive in the last.

(ii) (a) Among the above metals, write the names of metals which will displace hydrogen from water or steam.

(b) Give two evidences to show that magnesium is more reactive than iron.

* Frequently asked previous years Board Exam Questions.

Ans. (i) The given metals are arranged in the activity series of metals as follows : Na, Ca, Mg, Al, Zn, Fe, Pb and Cu.

(ii) (a) (1) Sodium and calcium displace hydrogen from cold water.

$$2Na + 2H_2O \longrightarrow 2NaOH + H_2\uparrow$$
$$Ca + 2H_2O \longrightarrow Ca(OH)_2 + H_2\uparrow$$

(2) Magnesium and zinc metals are less reactive as they react with boiling water to liberate hydrogen gas.

$$Mg + 2H_2O \longrightarrow Mg(OH)_2 + H_2\uparrow$$
$$Zn + 2H_2O \longrightarrow Zn(OH)_2 + H_2\uparrow$$

(3) Iron which is less reactive, reacts in red hot conditions with steam to liberate hydrogen gas.

$$3Fe + 4H_2O \longrightarrow Fe_3O_4 + 4H_2\uparrow$$

(4) Lead and copper almost fail to liberate hydrogen gas in any conditions, because they are not so reactive. They lie just above and below hydrogen inactivity series of metals.

(b) (1) Magnesium reacts with boiling water to liberate hydrogen gas, while iron can do so with steam in red hot condition only.

(2) Magnesium can displace hydrogen from acids vigorously in cold but iron displaces hydrogen slowly.

2. (i) Na, Ca, Mg, Al, Zn, Fe, Pb and Cu, are well known metals.

(a) X, Y and Z are coded letters for three of the metals in the activity series of metals as given above.

Metal X, reacts violently with cold water and its hydroxide is not decomposed by heat.

Metal Y, has no reaction with water but its hydroxide decomposes, with slight warming, giving a black powder.

Metal Z, reacts vigorously with dilute hydrochloric acid but hardly at all with cold water. If it is heated in steam, a white solid A is formed and a colourless gas B is set free.

(1) which of the metals in the list is X?

(2) which of the metals in the list is Y?

(3) which of the metals in the list is Z?

(4) write the name of the solid A and gas B.

(b) State whether the following are soluble or insoluble in water.

(1) The carbonate of X.

(2) The carbonate of Y.

(3) The hydroxide of Z.

(ii) A certain metal does not liberate hydrogen from dilute sulphuric acid but it displaces silver from aqueous silver nitrate solution. State the most likely place for the metal in the activity series.

(iii) What would you expect to happen, if aluminium metal is heated with iron(III) oxide? Also write the equation.

Ans. (i) (a) (1) The metal X is sodium.

(2) The metal Y is copper.

(3) The metal Z is magnesium.

(4) The name of the solid A is magnesium hydroxide, while the gas B is hydrogen.

(b) (1) Soluble as sodium carbonate is soluble in water.

(2) Insoluble, as copper carbonate is insoluble in water.

(3) Soluble, as magnesium hydroxide is soluble in water.

(ii) The metal lies below hydrogen and above silver in the activity series of metals.

(iii) When aluminium metal is heated with iron(III) oxide with metallic iron, an enormous amount of heat is produced due to the exothermic nature of the reaction. Molten iron is thus produced, which can be used in welding.

$$Fe_2O_3 + 2Al \longrightarrow Al_2O_3 + 2Fe + Q.$$

3. (i) Arrange Ca, Pb, Fe, Na, Zn, Cu, and Al in the decreasing order of their reactivity.
   (ii) Answer the following question related to above (i) sequence :
   (a) Which of these is most likely to tarnish readily when exposed to the air?
   (b) Which of these is most likely to be found in free state in nature?
   (c) Which of these is most likely to react with cold water?

Ans. (i) The decreasing order of the given metals is as follows :

[Most reactive] Na, Ca, Al, Zn, Fe, Pb, and Cu [Least reactive]

(ii) (a) Sodium [Na] (b) Copper [Cu] (c) Sodium [Na] and calcium [Ca]

4. (i) From the metals copper, zinc, magnesium, sodium and iron, select the metal in each case which :
   (a) Does not react with dil. hydrochloric acid.
   (b) Has a hydroxide that reacts with both acids and alkalies.
   (c) Does not react with cold water but reacts with steam when heated.
   (d) Can form +2 and +3 ions.
   (ii) Arrange the metals in decreasing order of reactivity.

Ans. (i) (a) Copper (b) Zinc
(c) Magnesium (d) Iron

(ii) Sodium > Magnesium > Iron > Zinc > Copper.

5. Explain why :
   (i) In the electrolysis of alumina using the Hall Heroult's process the electrolyte is covered with powdered coke.
   (ii) Iron sheets are coated with zinc during galvanization.

Ans. (i) To reduce the heat loss by radiation and to prevent burning of anode.

(ii) Iron sheets are coated with zinc during galvanization to prevent them from rusting.

6. (i) (a) How will you show that sodium is a metal?
   (b) How will you show that sulphur is a non-metal?
   (ii) (a) Which gas is liberated when aluminium metal reacts with a solution of sodium hydroxide?
   (b) Which gas is generally liberated when metals react with dilute acid?

Ans. (i) (a) Sodium metal can form positive ions by the loss of one electron, this metal is electropositive.

$$Na - e^- \longrightarrow Na^+$$

or

$$Na \longrightarrow Na^+ + e^-$$

Sodium has high density and is less dense then water.

(b) Sulphur is non-metal because, it gives negative sulphur ions by gaining of two electrons. It dissolves in many liquid solvents, but it is non-conductor of electricity and heat.

$$S + 2e^- \longrightarrow S^{2-}$$

(ii) (a) When aluminium metal reacts with sodium hydroxide solution, hydrogen gas is liberated.

(b) Hydrogen gas is generally liberated when metals react with dilute acid.

7. (i) (a) With reference to the reduction of copper oxide, iron(II) oxide, lead(II) oxide and magnesium oxide by hydrogen; place the oxides in increasing order of reduction, *i.e.*, first the oxide that is most difficult to reduce; and at last, the oxide that is most easy to reduced.
   (b) (1) What is the type of bonding expected in metallic chloride?
   (2) If fused metallic chloride is electrolysed, at which electrode the metal will be obtained.
   (3) What metallic property is shown by the non-metal graphite?
   (c) (1) Cast iron contains about 4% of carbon. By which chemical process is the amount of carbon decreased to make steel?
   (2) Which metal is added to steel to make stainless steel?
   (ii) (a) For each substance listed below, explain its significance in the extraction of Aluminium :
   (1) Bauxite (2) Cryolite
   (3) Graphite (4) Sodium hydroxide

(b) The following questions relate to the extraction of aluminium by electrolysis :

(1) Give the equation for the reaction that takes place at the cathode.

(2) Explain why it is necessary to renew the anode from time to time.

Ans. (i) (a) The increasing order of oxides is :

Magnesium oxide > iron(II) oxide > lead(II) oxide > copper oxide.

The magnesium oxide is highly stable. It cannot be reduced by hydrogen, while the last three members are reduced by hydrogen to their metals according to reactivity series.

(b) (1) Electrovalent or ionic bond.

(2) Cathode.

(3) Non-metal graphite is good conductor of heat and electricity.

(c) (1) In the Bessemer converter it takes only a few minutes to convert cast iron into steel.

(2) Stainless steel is an alloy which contains about 18% of Cr, 8% Ni and 1% C.

(ii) (a) (1) **Bauxite :** It is the main ore of aluminium from which aluminium metal can easily be extracted.

(2) **Cryolite :** It is added to lower the fusion temperature of the electrolytic bath. The mixture melts at 950°C instead of 2050°C thereby saving electrical energy. It also increases conductivity alongwith fluorspar.

(3) Graphite acts as an anode. Here the anode gets oxidised to carbon dioxide, *i.e.*

$$C + O_2 \longrightarrow CO_2$$

or

$$2O^{2-} - 4e^- \longrightarrow O_2$$

Thus, electrodes are made of graphite.

(4) Sodium hydroxide when added to powdered bauxite and the mixture when heated under pressure for 2-3 hours, bauxite is converted to soluble sodium aluminate ($NaAlO_2$).

$$Al_2O_3.2H_2O + 2NaOH \xrightarrow{150 \text{ to } 200°C} 2NaAlO_2 + 3H_2O$$

This solution is used to obtain pure aluminium.

(b) (1) The following reaction (reduction) takes place at the cathode during the extraction of aluminium.

$$Al^{3+} + 3e^- \longrightarrow Al$$

(2) Oxygen gas is produced at the graphite anode, which combines with carbon to form carbon dioxide gas at high temperature and thus anode destroys away. Thus, it is necessary to renew the anode to continue the process of extraction of aluminium.

8. (i) Differentiate between :

(a) Slag and Flux. (b) Calcination and Roasting.

(ii) Compare the properties of a typical metal and a non-metal on the basis of the following :

(a) Electronic configuration (b) Nature of the oxides

(c) Oxidising or reducing action (d) Conductivity of heat and electricity.

(iii) What are the differences between a mineral and an ore?

Ans. (i) (a)

Slag	Flux
It is the product obtained by the combination of the flux with gangue in metallurgy.	It is a substance which is added along with charge to separate the gangue in metallurgy.

(b)

	Calcination	Roasting
1.	It is the process of heating concentrated ore in a limited supply of air to a temperature insufficient to melt the ore.	It is the process of heating concentrated ore in a free supply of air to a temperature insufficient to melt the ore.
2.	During calcination, no other chemical change occurs except decomposition.	During roasting, chemical changes like oxidation or reduction take place.

(ii) (a) Metals complete their octet by the loss of electrons whereas non-metals complete their octet by the gain of electrons.

Metals generally contain 1 to 3 valence electrons in their outermost shell whereas non-metals contain 4 to 7 valence electrons in their outermost shell.

(b) Metals form basic oxides whereas non-metals form acidic oxides.

(c) Metals are reducing agents whereas non-metals act as oxidising agents.

(d) Metals are generally good conductors of heat and electricity whereas non-metals are bad conductors of heat and electricity.

(iii) (a) The minerals contain a low percentage of metal, while the ores contain a large percentage of the metal.

(b) The metal cannot be extracted from mineral, on the other hand ores can be used for the extraction of metal.

(c) All minerals are not ores, but all ores are minerals.

9. (i) The ore zinc blende, is an important source of the metal zinc. What is the name of the zinc compound in zinc blende?

(ii) What is the zinc compound obtained by roasting zinc blende?

(iii) What is the type of chemical reaction carried out in order to obtain zinc?

(iv) Are liquid zinc and liquid lead miscible or immiscible?

(v) What is the name of the alloy formed between zinc and copper?

Ans. (i) Zinc sulphide (ZnS).

(ii) Zinc blende is oxidized to zinc oxide by roasting in presence of excess air.

(iii) Reduction of zinc oxide.

(iv) Immiscible.

(v) Brass [70% of Cu, 30% of Zn].

10. The following questions refer to the extraction of aluminium and iron from their ores :

(i) Name the principal ore from which; (a) iron and (b) aluminium are extracted.

(ii) What is the most important chemical process in the extraction of any metal? State how this essential step is carried out in the extraction of; (a) iron, (b) aluminium.

(iii) Iron and aluminium ores both, contain impurities. Explain briefly how these impurities are removed in each case.

(iv) What is the major impurity present in iron when it is removed from the blast furnace?

Ans. (i) (a) Haematite (Fe_2O_3). (b) Bauxite (Al_2O_3).

(ii) Reduction of the oxide is an important step in extraction of metal.

In case of iron, $Fe_2O_3 + 3CO \longrightarrow 2Fe + 3CO_2$

Al_2O_3 cannot be easily reduced, hence it is subjected to electrolysis. Aluminium is collected at the cathode.

(iii) Iron ore contains impurities of silica and sand. These are removed by magnetic separation. Bauxite and aluminium ore contains impurities of FeO and SiO_2.

Bauxite containing FeO is calcinated at high temperature when FeO is oxidised to Fe_2O_3. Calcinated ore is then treated with NaOH when Al_2O_3 is converted into soluble $NaAlO_2$. Fe_2O_3 can thus be filtered off. Bauxite containing SiO_2 is mixed with coke and heated to 1000°C in an atmosphere of N_2. Silica is reduced to Si which volatilises at the temperature of reaction. Aluminium oxide is converted into AlN which is hydrolysed with water to obtain $Al(OH)_3$.

(iv) Carbon is major impurity present in iron.

11. (i) What is bauxite? Which metal is extracted from it?

(ii) In the electrolysis of molten alumina, the carbon anode is gradually burnt away. Why?

(iii) Describe modern method of aluminium extraction.

Ans. (i) Bauxite is hydrated aluminium oxide [$Al_2O_3 .2H_2O$] and aluminium metal is extracted from bauxite.

Short Answers

(ii) In the electrolysis of molten aluminium oxide, oxygen gas is liberated which gradually burns away carbon anode at a higher temperature to form carbon dioxide.

$$C + O_2 \longrightarrow CO_2$$

(iii) In the modern method, pure alumina is dissolved in cryolite [Na_3AlF_6], which makes it good conductor of electricity.

When an electric current is passed through electrolyte, the heat is also produced which keeps the mass in molten state and alumina gets reduced to free aluminium metal according to the following reactions :

$$Na_3AlF_6 \longrightarrow 3NaF + AlF_3$$

$$2AlF_3 \rightleftharpoons 2Al^{3+} + 6F^-$$

$$6F^- + Al_2O_3 \longrightarrow 2AlF_3 + 3O^{2-}$$

At cathode : $2Al^{3+} + 6e^- \longrightarrow 2Al$

At anode : $3O^{2-} - 6e^- \longrightarrow 3O$

$$3O + 3O \longrightarrow 3O_2$$

12. The following questions are relevant to the extraction of aluminium :

(i) State the reason for addition of caustic alkali to bauxite ore during purification of bauxite.

(ii) Give a balanced chemical equation for the above reaction.

(iii) Alongwith cryolite and alumina, another substance is added to the electrolyte mixture. Name the substance and give one reason for the addition.

Ans. (i) Caustic alkali dissolves aluminium oxide forming soluble sodium aluminate while impurities remains insoluble and ppt. as red mud.

(ii) $Al_2O_3{\cdot}2H_2O + NaOH \longrightarrow 2NaAlO_2 + 3H_2O$

(iii) The name of substance is Fluorspar (CaF_2) and it increases conductivity of the electrolyte.

13. 'Alumina (aluminium oxide) has a very high melting point of over 2,000°C so that it cannot readily be liquiefied. However, conversion of alumina to aluminium and oxygen, by electrolysis, can occur when it is dissolved in some other substance.'

(i) Which solution is used to react with bauxite as a first step in obtaining pure aluminium oxide?

(ii) The aluminium oxide for the electrolytic extraction of aluminium is obtained by heating aluminium hydroxide. Write the equation for this reaction.

(iii) Name the element which serves both as the anode and the cathode in the extraction of aluminium.

(iv) Write the equation for the reaction that occurs at the cathode during the extraction of aluminium by electrolysis.

(v) Give the equation for the reaction which occurs at the anode when aluminium is purified by electrolysis.

Ans. (i) Sodium hydroxide

(ii) $2Al(OH)_3 \xrightarrow{\Delta} Al_2O_3 + 3H_2O$

(iii) Carbon

(iv) $Al^{3+} + 3e^- \longrightarrow Al$

(v) $O^{2-} - 2e^- \longrightarrow O$

14. (i) Give the name and formula of the ore of zinc containing its sulphide.

(ii) Write equations for the following steps in the extraction of zinc.

(a) Roasting of the ore.

(b) Reduction of the zinc compound which is the product of the above reaction.

(iii) In the process of extracting zinc, the above named ore is roasted. Write the equation for the reaction which takes place when the sulphide ore is roasted.

What in addition to a zinc compound, is put into the blast furnace?

(iv) State one large scale use of zinc.

(v) "Iron is removed from a blast furnace as a liquid". State how zinc leaves a furnace.

Ans. (i) Zinc blende (ZnS)

(ii) (a) $2ZnS + 3O_2 \longrightarrow 2ZnO + 2SO_2$

(b) $ZnO + C \longrightarrow Zn + CO$

(iii) Powdered coke.

(iv) For galvanising iron sheets to prevent rusting.

(v) Iron is removed from a blast furnace as a liquid because of the maximum liberated heat, coke reduces zinc oxide to zinc vapours to volatize a brilliant glow. These received in a condenser to liquifies, called spelter.

15. (i) (a) What is the function of adding the lime stone in the "Blast Furnace"?

(b) Give two uses of slag.

(ii) How is zinc extracted from zinc blende [ZnS]?

Ans. (i) (a) Lime stone decomposes at a higher temperature of blast furnace, to form calcium oxide and carbon dioxide.

$$CaCO_3 \longrightarrow CaO + \underset{\text{Carbon dioxide}}{CO_2} \uparrow$$

Calcium oxide combines with sand, present as impurities, to form calcium silicate known as the slag.

$$CaO + SiO_2 \longrightarrow \underset{\text{Calcium Silicate [Slag]}}{CaSiO_3}$$

The slag, is formed at the surface of blast furnace from where, it is removed from time to time. Thus, lime stone is used to remove the impurities of silicon dioxide present in ore.

(b) (1) It is used in the manufacture of cement.

(2) It is used as a fertilizer.

(ii) Zinc blende is strongly heated in the presence of air in a 'reverberatory furnace', where it is oxidized to zinc oxide and sulphur dioxide gas is liberated.

$$2ZnS + 3O_2 \longrightarrow \underset{\text{Zinc oxide}}{2ZnO} + 2SO_2 \uparrow$$

Zinc oxide so obtained is mixed with carbon and is strongly heated, where it is reduced to metallic zinc and carbon monoxide are formed.

$$\underset{\text{Zinc oxide}}{ZnO} + C \longrightarrow Zn + CO\uparrow$$

$$ZnO + CO \longrightarrow Zn + CO_2\uparrow$$

$$CO_2 + C \longrightarrow \underset{\text{Carbon monoxide}}{2CO}$$

Carbon monoxide is regenerated.

16. (i) The following is a list of metals : Na, Mg, Al, Zn, Fe, Pb, Cu and Au.

Name an ore of one of the above metals and state, how the metal is produced from it commercially by the process of :

(a) Electrolysis of the molten ore (b) Reduction by coke

(c) Roasting followed by reduction (d) Metal found in native state

(ii) How would you prove that the alloy solder contains lead?

Ans. (i) (a) Aluminium–Bauxite (b) Iron–Hematite

(c) Lead–Galena (d) Gold

(ii) Take the given sample of solder and dissolve it in concentrated nitric acid. Dilute the solution with water and add dilute hydrochloric acid. A white precipitate of lead chloride is obtained, which is soluble in hot water. To this solution add potassium chromate solution, a golden yellow precipitate, confirms the presence of lead.

$$Pb + 4HNO_3 \longrightarrow Pb(NO_3)_2 + 2H_2O + 2NO_2$$

Lead present in solder is confirmed as below :

$$Pb(NO_3)_2 \longrightarrow 2HCl \longrightarrow \underset{\text{White ppt.}}{PbCl_2 \downarrow} + 2HNO_3$$

$$PbCl_2 + K_2CrO_4 \longrightarrow \underset{\text{(Golden yellow ppt.)}}{\underset{\text{Lead chromate}}{PbCrO_4\downarrow}} + 2KCl$$

17. (i) The basic materials needed for production of iron in the blast furnace are lime stone, coke and air in addition to the iron ore.

(a) Name one iron ore and write its formula.

(b) Hot air is blown in, at the base of the furnace; where it reacts with coke. Give the equation for the reaction that takes place.

(c) Higher up in the furnace, the iron ore is reduced to iron by one of the gases produced in the furnace. Give the chemical equation for the reaction, by which this gas is produced and give a balanced equation to show, how the ore is reduced to iron.

(ii) (a) Which compound produced from the limestone takes part in forming the slag? Give the formula of the slag.

(b) What is the principal use of slag?

(iii) (a) Iron forms two series of slags ferrous and ferric. Using caustic soda solution, how would you distinguish between these two salts?

(b) How zinc is extracted from calamine ore?

Ans. (i) (a) Haematite (Ferric oxide) Fe_2O_3.

(b)
$$\underset{\text{Coke}}{C + O_2 \longrightarrow CO_2}$$

(c)
$$CO_2 + C \longrightarrow 2CO$$
$$\underset{\text{Haematite}}{Fe_2O_3} + 3CO \longrightarrow \underset{\text{Iron}}{2Fe} + 3CO_2$$

(ii) (a) The limestone decomposes at 1000°C, to form lime and carbon dioxide. Lime is the compound which takes part in the formation of slag.

$$CaCO_3 \xrightarrow{1000°C} \underset{\text{Lime}}{CaO} + CO_2$$

$$CaO + SiO_2 \longrightarrow \underset{\text{Slag}}{CaSiO_3}$$

Slag is calcium silicate with the formula $CaSiO_3$.

(b) The slag is used in the manufacture of cement.

(iii) (a) When caustic soda solution is added to a ferrous salt, a dirty green precipitate of ferrous hydroxide is obtained.

$$FeSO_4 + 2NaOH \longrightarrow \underset{\text{Dirty green ppt.}}{Fe(OH)_2\downarrow} + Na_2SO_4$$

On the other hand, ferric salts give a reddish brown precipitate of ferric hydroxide with caustic soda.

$$FeCl_3 + 3NaOH \longrightarrow \underset{\text{Reddish brown}}{Fe(OH)_3\downarrow} + 3NaCl.$$

(b) Calamine ore is chemically zinc carbonate [$ZnCO_3$]. Purified ore is roasted in a reverberatory furnace, where it is decomposed to form zinc oxide and carbon dioxide gas is liberated.

$$ZnCO_3 \longrightarrow ZnO + CO_2$$

Zinc oxide so obtained is mixed with carbon and heated in a specially designed clay retorts, placed in a furnace, where it is reduced to zinc metal and carbon monoxide is formed, which is further used to reduce the ore.

$$ZnO + C \longrightarrow Zn + CO$$
$$ZnO + CO \longrightarrow Zn + CO_2$$

18. (i) (a) Give the common name of an ore, from which aluminium is extracted.

(b) Name the process of its extraction and describe it.

(c) Name another metal, which is extracted by this process.

(ii) (a) Give a reason, why aluminium cannot be obtained from aluminium oxide; by the "Blast Furnace Process" or "Carbon Reduction Process."

(b) Name the properties for which aluminium is used in :
(i) Cooking utensils, (ii) Overhead electric transmission wire.

(c) Aluminium is more active than iron and yet there is less corrosion of the aluminium, when both are exposed to air. Explain this fact.

(iii) (a) What is pig-iron?

(b) What is the main difference in chemical composition of cast iron and steel?

Ans. (i) (a) Aluminium metal is extracted from Bauxite ore [$Al_2O_3.2H_2O$].

(b) Aluminium metal is extracted by electrolytic reduction process, in which electric current is passed through fused aluminium oxide, the purified Bauxite. Aluminium metal is collected at cathode, while oxygen gas is liberated at anode according to the following reactions.

$$Al_2O_3 \rightleftharpoons 2Al^{3+} + 3O^{2-}$$

At cathode : $2Al^{3+} + 6e^- \longrightarrow 2Al$

At anode : $3O^{2-} - 6e^- \longrightarrow 3O$

$$3O + 3O \longrightarrow 3O_2$$

(c) Calcium metal is also extracted by electrolytic reduction process.

(ii) (a) Aluminium metal is more electropositive than iron. It has the characteristic property of all the oxides of metals that are highly electropositive oxides of metals cannot be reduced by heating with carbon to form free metal. Hence aluminium is not extracted by the "Blast Furnace Process".

(b) (i) Aluminium is a good conductor of heat, light in weight, strong and can be made passive by nitric acid.

(ii) Aluminium is a good conductor of electricity, light in weight, tough and possess high tensile strength and is resistant to corrosion. It is a cheaper metal than copper.

(c) Aluminium metal is protected by a thin film of aluminium oxide, which sticks firmly with the metal and prevents further corrosion.

(iii) (a) The iron, which is obtained directly from the blast furnace is known as pig iron. Besides iron, it contains about 4% of carbon and small quantities of silicon, phosphorus, manganese and sulphur. When pig iron is heated to melting and is casted into desired moulds, then it is known as cast iron.

(b) Cast iron contains 2·5% to 4% of carbon, while steel contains 0·5% to 1·5% of carbon. Due to the higher percentage of carbon in cast iron, it resists corrosion.

19. (i) What is froth floatation process and for, what purpose it is used?

(ii) How is the metal sodium extracted? Write the equations for the reactions involved.

(iii) Name two other metals, which can be extracted by electrolytic reduction method.

Ans. (i) In this process, the heavy material containing metal, is floated upward with froath to separate it from waste material present in ore or mineral. Hence it is called froth floatation process.

(ii) Sodium metal is extracted by the electrolysis of fused sodium chloride. Sodium is collected at cathode, while chlorine gas is liberated at anode; as an important by product.

$$NaCl \rightleftharpoons Na^+ + Cl^-$$

At cathode : $Na^+ + e^- \longrightarrow Na$

At anode : $Cl^- - e^- \longrightarrow Cl$

$$Cl + Cl \longrightarrow Cl_2$$

(iii) Calcium, and magnesium are other two metals, which can be extracted by electrolytic reduction method.

20. (i) Give the reactions taking place in the blast furnace, when haematite, coke and lime stone are added to it.

(ii) What is the purpose of adding lime stone in the extraction of iron from haematite?

(iii) What is the use of galvanized iron?

(iv) Which two products are taken out from the blast furnace during the extraction of iron?

Ans. (i) (a) Carbon burns in the presence of oxygen of air to form carbon dioxide which is reduced to carbon monoxide in the presence of coke.

$$C + O_2 \longrightarrow CO_2$$
$$CO_2 + C \longrightarrow 2CO$$

(b) Carbon monoxide so produced, reduces haematite ore [Ferric oxide] to metallic iron.

$$Fe_2O_3 + 3CO \longrightarrow 2Fe + 3CO_2 \uparrow$$

(c) Lime stone, at a higher temperature of blast furnace decomposes into calcium oxide and carbon dioxide.

$$CaCO_3 \longrightarrow CaO + CO_2 \uparrow$$

(d) Calcium oxide so formed reacts with sand [silicon dioxide] to form calcium silicate, the slag and thus the impurity of sand is removed.

$$CaO + SiO_2 \longrightarrow CaSiO_3$$

(ii) Lime stone is added to "blast furnace" in the extraction of iron to remove the impurities of silicon dioxide, *i.e.*, the sand.

(iii) Galvanized iron is used in making different varieties of tools for industries, scientific apparatus, and household fittings.

(iv) Pig iron and slag are the two products, which are taken out from the blast furnace during the extraction of iron.

21. (i) The following represents a summary of the reaction which occur in the "Blast furnace", leading to the production of molten iron.

$$\text{Coke} + O_2 \longrightarrow \text{Gas Y}$$
$$\text{Gas Y} + \text{Coke} \longrightarrow \text{Gas Z}$$
$$\text{Gas Z} + \text{Iron oxide} \longrightarrow \text{Gas Y} + \text{Iron}$$

Identify the gases, Y and Z. Explain why gas Z is said to act as a reducing agent in the last step in the above equations.

(ii) Calcium, Copper, Lead, Aluminium, Zinc, Chromium, Magnesium, Iron, Choose the major metals from the list given above to make the following alloys : (a) Stainless steel (b) Brass.

Ans. (i) The gas Y, carbon dioxide and the gas Z, is carbon monoxide. The gas Z is a reducing agent because it removes oxygen from iron oxide and converts it into metallic iron and itself gets oxidized to carbon dioxide.

(ii) (a) Iron, Chromium (b) Zinc, Copper

22. (i) (a) Name two ores of iron.

(b) Name three raw materials used in the extraction of iron.

(c) Write equations that occur in the "Blast Furnace".

(ii) What are the main constituents of steel?

(iii) What is tempering of steel?

Ans. (i) (a) Two main ores of iron are : Haematite [Fe_2O_3] and Magnetite [Fe_3O_4].

(b) Iron ore, lime stone, and coke are used in the extraction of iron.

(c) Following are the reactions, which take place in the "Blast Furnace", during the extraction of iron.

$$C + O_2 \longrightarrow CO_2$$
$$CO_2 + C \longrightarrow 2CO$$
$$Fe_2O_3 + 3CO \longrightarrow 2Fe + 3CO_2$$
$$Fe_3O_4 + 4CO \longrightarrow 3Fe + 4CO_2$$
$$CaCO_3 \longrightarrow CaO + CO_2$$
$$CaO + SiO_2 \longrightarrow \underset{\text{Calcium Silicate [Slag]}}{CaSiO_3}$$

(ii) Steel is an alloy of iron and carbon containing very small amounts of impurities, that are present in the cast iron. The carbon content varies from 0·5 to 1·5%. The variety containing lower percentage of carbon is called "mild steel" and the variety which contains higher percentage of carbon is known as "hard steel".

(iii) The hardened steel is brittle in nature and when it is heated upto a definite temperature and for certain time and then allowed to be cooled down slowly, then it loses its brittleness. This process is known as tempering of steel and is employed for bringing the steel into a suitable state of hardness and elasticity. The temperature required is generally judged from the colour of a thin film of oxide which is formed on the surface and varies from yellow to brown to blue as the temperature rises from 200°C to 300°C.

23. (i) Most of pig iron obtained from blast furnace is converted into steel. Suggest a reason for this.

(ii) The element X, has the electronic configuration 2, 8, 8, 1. Describe the symbol for an ion of X. Give a reason for your answer and deduce whether the element X would be expected to have oxidizing or reducing properties.

(iii) How is the metal X, extracted from its given salt?

Ans. (i) Pigiron does not possess a high tensile strength, hence it cannot withstand with high stress and strain. Pigiron is brittle in nature and cannot be welded. Due to these disadvantages, pigiron cannot be put in different varieties of uses hence it is converted into steel.

(ii) The element X, contains one electron in the outermost orbit of its atom, which can be easily donated and one unit of positive charge is gained on its ion.

$$X - e^- \longrightarrow X^+$$

The symbol of X ion is X^+. The atom of element X is a donor of electron hence X is expected to have reducing properties.

(iii) It is clear from the properties of X, that the metal X is supposed to be highly positive in nature, hence it is extracted by electrolytic reduction method.

$$XCl \rightleftharpoons X^+ + Cl^-$$

At cathode : $X^+ + e^- \longrightarrow X$

At anode : $Cl^- - e^- \longrightarrow Cl$

$$Cl + Cl \longrightarrow Cl_2$$

24. A to F below relate to the source and extraction of either Zinc or Aluminium.

(A) Bauxite (B) Coke

(C) Cryolite (D) Froth floatation

(E) Sodium hydroxide solution. (F) Zinc blende.

(i) Write down the three letters each from the above list which are relevant to :

(1) Zinc (2) Aluminium.

(ii) (1) Metals are generally solid at room temperature. Name the metal which is liquid at room temperature [say 25°C].

(2) Which allotrope of the non-metal conducts electricity?

(iii) How many valence electrons are present in :

(a) Metals, (b) Non-metals.

Ans. (i) (1) B, D, F (2) A, C, E

(ii) (1) Mercury metal exists in liquid state at room temperature.

(2) Graphite, an allotrope of carbon is a good conductor of electric current.

(iii) Atom of metals contain 1, 2 and 3 valence electrons, while the atom of non-metals contain 4, 5, 6 and 7 valence electrons.

25. The following substances are put into the blast furnace while manufacturing iron :
Iron ore, coke, limestone and hot air. In this context—

(i) What is the name of the most common ore of iron and what is its chemical formula?

(ii) What is the purpose of using :

(a) coke (b) limestone

(iii) Write the equation for the reduction reaction which produces iron.

(iv) Name the substance which is collected along with cast iron at the bottom of the furnace.

(v) Write the chemical equation for the formation of the substance named in (iv) above.

Ans. (i) Haematite – Fe_2O_3.

(ii) (a) Coke acts as a reducing agent. (b) Limestone acts as a flux.

(iii) $Fe_2O_3 + 3CO \longrightarrow \underset{\text{Spongy iron}}{2Fe} + 3CO_2$

Short Answers

(iv) Fusible slag (calcium silicate)

(v) $CaO + SiO_2 \longrightarrow CaSiO_3$

26. Bayer's process is used to concentrate bauxite ore to alumina.*

Give balanced chemical equations for the reaction taking place for its conversion from bauxite to alumina.

Ans. Chemical reactions related to Bayer's process for the conversion from bauxite to alumina:

1. $Al_2O_3.2H_2O(s) + 2NaOH \xrightarrow{150° - 200° C} 2NaAlO_2 + H_2O$

2. $NaAlO_2 + 2H_2O \xrightarrow{50° - 60° C} Al(OH)_3 + NaOH$

3. $2Al(OH)_3 \xrightarrow[\text{Alumina}]{\text{Heat}} Al_2O_3 + 3H_2O$

27. Answer the following question based on the extraction of aluminium from alumina by **Hall-Heroult's Process :***

(i) What is the function of cryolite used along with alumina as the electrolyte ?

(ii) Why is powdered coke sprinkled on top of the electrolyte ?

(iii) Name the electrode, from which aluminium is collected.

Ans. (i) Cryolite (Na_3AlF_6) is mixed with alumina to bring down the melting point of electrolyte mixture and to increase its electrical conductivity.

(ii) The electrolytic mixture is sprinkled with coke to prevent the anode's oxidation by the oxygen evolved.

(iii) Aluminium is collected at cathode which is carbon lining covering the inside portion of vessel.

28. (i) Name an alloy used in aircraft construction and give a reason for its use.

(ii) What is rusting of iron?

(iii) (a) How are the following protected from rust?

(1) A car bumper and (2) A food can.

(b) How can iron or steel be prevented from rusting, when used for?

(c) What is galvanized iron and for what purposes it is used?

(d) To protect iron from rusting it is coated with a thin layer of zinc. Name this process.

Ans. (i) Duralumin, an alloy of aluminium, is used in the construction of aircraft; because it is light, resistant to corrosion and has great tensile strength.

(ii) The rusting of iron is a chemical process of atmospheric corrosion, *i.e.*, slow destruction of iron by moisture and atmospheric oxygen. Rust is a reddish-brown powdery deposit and consists of a mixture of ferric hydroxide and hydrated ferric oxide.

(iii) (a) (1) Nickel Plating and (2) Galvanizing.

(b) By painting and by coating with nickel.

(c) Iron coated with zinc is called galvanized iron. Galvanization is a process of depositing a thin layer of zinc, over the surface of iron to protect iron from rusting. Zinc is more electropositive and would be attacked first and thus iron is protected from any corrosion. Galvanized iron is used in making different varieties of tools for industries, scientific apparatus and household fittings.

(d) Galvanisation.

29. (i) How are the alloys classified?

(ii) What are ferrous alloys? Give one example.

(iii) What are non-ferrous alloys? Give one example.

(iv) An alloy usually has some property which makes it particularly useful. What is the special property of : (a) Type metal, (b) Duralumin?

Ans. (i) Alloys are classified on the basis of their constituents. They are classified as follows :

(a) Ferrous alloys (b) Amalgams (c) Non-ferrous alloys

*** Frequently asked previous years Board Exam Questions.**

(ii) **Ferrous alloys :** It is an alloy having iron as one of the constituent, *e.g.*, nickel, steel.

(iii) **Non-ferrous alloys :** An alloy that does not contain iron as one of its constituents, is called a non-ferrous alloy, *e.g.*, brass.

(iv) (a) Type metal is hard and expands on cooling and is therefore used for making types.

(b) Duralumin is light and strong therefore it is used in the construction of aircraft.

30. State the main components of the following alloys :

(i) Brass (ii) Duralumin

(iii) Bronze (iv) Stainless steel

Ans. (i) Copper and zinc (ii) Aluminium and magnesium

(iii) Copper and Tin (iv) Iron and Carbon

31. Give the composition and uses of the following alloys :

(i) Brass (ii) Bronze (iii) German silver

(iv) Type metal (v) Magnalium (vi) Duralumin

Ans. (i) It is an alloy of copper and zinc and is used for making utensils, condenser tubes, statues, and for making decoration pieces.

(ii) It is an alloy of copper, zinc and tin and is used for making statues, utensils and coins.

(iii) It is an alloy of copper, zinc and nickel and is used for making ornaments and utensils and also used for decoration.

(iv) It is an alloy of lead, antimony and tin and is used for making printing type.

(v) It is an alloy of aluminum and magnesium and is used for making light instruments, parts of machines and balance beams.

(vi) It is an alloy of aluminium, copper, magnesium and manganese and is used for making aeroplanes, space crafts, sea ship and pressure cookers.

32. (i) Write a note on thermite welding.

(ii) What is meant by electrochemical series of metal?

Ans. (i) Thermite is a mixture of 3 parts of ferric oxide and one part of aluminium powder. On the top of the thermite mixture, an ignition mixture of potassium chlorate and magnesium powder is placed. A burning Mg ribbon is inserted into the ignition mixture, which catches the fire and ignite the thermite. During this reaction ferric oxide is converted into iron and a large amount of heat is evolved and about 3000°C temperature is achieved.

$$Fe_2O_3 + 2Al \longrightarrow 2Al_2O_3 + 2Fe + Heat$$

The formed liquid iron is allowed to drop over the gap between the broken piece and thus they join together. This process is termed as Goldschmidt aluminothermic process.

(ii) The arrangement of metals in the order of their decreasing activity, in which the most reactive at the top and the least reactive at the bottom are placed in the series. This arrangement of metals in a series is called electrochemical or metal activity series.

Chapter 8. (a) Study of Compounds : Hydrogen Chloride

1. The following questions are pertaining to the laboratory preparation of hydrogen chloride gas :

(i) Write the equation for its preparation mentioning the condition required.

(ii) Name the drying agent used and justify your choice.

(iii) State a safety precaution you would take during the preparation of hydrochloric acid.

Ans. (i)

$$\underset{\text{Sodium chloride}}{NaCl} + \underset{\text{Sulphuric acid (Conc.)}}{H_2SO_4} \xrightarrow{< 200°C} \underset{\text{Sodium hydrogen Sulphate}}{NaHSO_4} + \underset{\text{Hydrogen chloride}}{HCl\uparrow}$$

(ii) It is dried by passing through conc. sulphuric acid because it does not react with hydrogen chloride gas.

(iii) During the process, back suction might occur in the inverted funnel, connected to hydrogen gas supply. In order to prevent an accident, empty flask (anti-suction device) is kept between the flask and water trough. This water in the empty flask collects the acid, reverted due to back suction.

2. (i) State one condition under which chlorine and hydrogen react to form hydrogen chloride gas.
(ii) Give balanced chemical equation for the above reaction.
(iii) Name the gas which is a covalent compound but becomes electrovalent when dissolved in water?
(iv) For which gas, ammonia fountain experiment can be used?

Ans. (i) Presence of diffused sunlight.
(ii) $H_2 + Cl_2 \longrightarrow 2HCl$
(iii) Hydrogen chloride (HCl) gas.
(iv) Hydrogen chloride gas.

3. State three uses of hydrochloric acid.

Ans. (i) It is used in the manufacture of silver chloride, which is used widely in photography.
(ii) It is used in the manufacture of dyes, drugs and paints.
(iii) It is used for cleaning metal surface before painting, electroplating, galvanising, soldering etc.

4. Hydrogen chloride gas is prepared in the laboratory using concentrated sulphuric acid and sodium chloride. Answer the questions that follow based on this reaction :*
(i) Give the balanced chemical equation for the reaction with suitable conditions(s) if any.
(ii) Why is concentrated sulphuric acid used instead of concentrated nitric acid ?
(iii) How is the gas collected ?
(iv) Name the drying agent not used for drying the gas.

Ans. (i) 1. $NaCl + \underset{conc.}{H_2SO_4} \xrightarrow[\text{below } 200°C]{} NaHSO_4 + HCl\uparrow$

2. $2NaCl + \underset{conc.}{H_2SO_4} \xrightarrow[\text{above } 200°C]{\text{Heated}} Na_2SO_4 + 2HCl\uparrow$

(ii) Concentrated nitric acid is not used for preparation of HCl gas as it is very strong oxidising agent and will end up oxidising HCl gas.
(iii) HCl is collected by the upward displacement of air.
(iv) Calcium oxide.

5. (i) (a) Name the oxidising agent in the reaction between manganese dioxide and conc. hydrochloric acid.
(b) State your observation when a rod dipped in ammonium hydroxide solution is brought near a gas jar containing hydrogen chloride gas.
(ii) Manganese(IV) oxide, lead(IV) oxide and red lead (Pb_3O_4) react with concentrated hydrochloric acid liberating chlorine.
(a) What is the common property being shown by these metal oxides?
(b) Write the equation for the reaction of concentrated hydrochloric acid with Pb_3O_4.
(c) What kind of compound can be added to bleaching powder to obtain chlorine?

Ans. (i) (a) Manganese dioxide acts as an oxidising agent.
(b) Dense white fumes appear in the jar on account of formation of fine particles of ammonium chloride which get suspended in the gas.
(ii) (a) Oxidizing agents
(b) $Pb_3O_4 + 8HCl \longrightarrow 3PbCl_2 + 4H_2O + Cl_2$
(c) Dilute acid (Hydrochloric acid)

6. Answer the following questions, stating your answer only to compounds in the following list : Silver nitrate, hydrochloric acid, chlorine, ammonia, bleaching powder.
(i) What is water sterilizer?
(ii) Which compound forms curdy white precipitate with hydrogen chloride?

* Frequently asked previous years Board Exam Questions.

(iii) Name the gas which produces dense white fumes with ammonia, write the balanced chemical equation.

Ans. (i) Chlorine is water sterilizer. It kills harmfull pathogens present in water.

(ii) Silver nitrate and hydrochloric acid forms white ppt.

$$AgNO_3 + HCl \longrightarrow \underset{\text{White ppt.}}{AgCl} + HNO_3$$

(iii) Hydrogen chloride (HCl)

$$NH_3 + HCl \longrightarrow NH_4Cl \text{ (Dense white fumes)}$$

7. (i) When moist chlorine reacts with hydrogen sulphide, two products are formed :

(a) A gas which fumes in moist air

(b) A yellow solid

Name these products.

(ii) What type of reaction is taking place when chlorine acts as a bleaching agent?

Ans. (i) (a) Hydrogen chloride gas (b) Sulphur

(ii) Oxidation reaction.

8. From the gases-ammonia, hydrogen chloride, hydrogen sulphide, sulphur dioxide. Select the following :

(i) The gas which gives a white precipitate when reacted with silver nitrate solution acidified with dilute nitric acid.

(ii) A solution of hydrogen chloride in water is prepared. The following substances are added to separate portions of the solution.

Substances added	Gas evolved	Odour
Calcium carbonate	—	—
Magnesium ribbon	—	—
Manganese(IV) oxide with heating	—	—
Sodium sulphide	—	—

Complete the table by writing the gas evolved in each case and its odour.

Ans. (i) Hydrogen chloride

(ii)

Substances added	Gas evolved	Odour
Calcium carbonate	Carbon dioxide	Odourless
Magnesium ribbon	Hydrogen	Odourless
Manganese(IV) oxide with heating	Chlorine	Pungent irritating
Sodium sulphide	Hydrogen sulphide	Rotten eggs

9. A colourless gas G fumes strongly in the air. The gas gives dense white fumes when a glass rod dipped in ammonia solution is held near the gas.

Answer the following questions :

(i) Name the gas G.

(ii) Name two chemicals used in the preparation of the gas G.

(iii) Write the chemical equations for the reaction of the chemicals named in (ii) when :

(a) The reaction mixture is not heated.

(b) The reaction mixture is heated above 100°C.

(iv) Why does the gas G fume strongly in air ?

(v) Why does the gas G form dense white fumes with ammonium hydroxide?

Ans. (i) The gas G is hydrogen chlorine gas.

(ii) The chemicals are (i) sodium chloride, (ii) concentrate sulphuric acid.

(iii) (a) $NaCl + H_2SO_4 \text{ (conc.)} \longrightarrow NaHSO_4 + HCl \text{ (g)}$

(b) $NaCl + NaHSO_4 \xrightarrow{\text{Heat}} Na_2SO_4 + HCl \text{ (g)}$

Short Answers

(iv) It is because the HCl gas is extremely soluble in water. Thus, the gas dissolves in water vapour present in the air to form tiny droplets of hydrochloric acid, which appear in the form of fumes.

(v) The HCl gas reacts with vapours of ammonium hydroxide to form very fine solid particles of ammonium hydroxide which are white in colour. These white particles of solid ammonium hydroxide appear in the form of white fumes.

10. (i) How will you dry HCl acid gas.

(ii) Give three tests of hydrogen chloride.

(iii) Which two colourless gases combine to form a white solid.

Ans. (i) HCl gas can be dried by passing it over conc. H_2SO_4, which acts as a powerful dehydrating agent.

(ii) Tests for hydrogen chloride.

(1) It gives dense white fumes with a rod dipped in NH_4OH solution.

(2) It produces white ppt. with $AgNO_3$ solution.

$$AgNO_3 + HCl \longrightarrow \underset{\text{White ppt.}}{AgCl \downarrow} + HNO_3$$

(3) It turns moist blue litmus red.

(iii) NH_3 and HCl gases combine to form a white solid NH_4Cl

$$\underset{\text{Gas}}{NH_3} + \underset{\text{Gas}}{HCl} \longrightarrow \underset{\text{Solid}}{NH_4Cl}$$

11. (i) (a) What must be added to sodium chloride to obtain hydrogen chloride?

(b) Write the equation for the reaction which takes place in (a) above.

(c) What would you see when hydrogen chloride is mixed with ammonia?

(ii) Hydrogen chloride dissolve in water forming an acidic solution :

(a) Name the experiment which demonstrates that hydrogen chloride is very soluble in water.

(b) Give three distinct tests (apart from using an indicator) you would carry out with this solution to illustrate the typical properties of an acid.

Ans. (i) (a) Concentrated sulphuric acid.

(b) $NaCl + H_2SO_4 \xrightarrow{< 200°C} NaHSO_4 + HCl\,(g)$

$NaHSO_4 + NaCl \xrightarrow{> 200°C} Na_2SO_4 + HCl\,(g)\uparrow$

(c) When aqueous solution of ammonia is taken in the jar of hydrogen chloride, it forms dense white fumes of ammonium chloride.

$$NH_3 + HCl \longrightarrow NH_4Cl$$

(ii) (a) Fountain experiment.

(b) An acid reacts with :

(1) Metal carbonates and bicarbonates with effervescence to liberate CO_2.

(2) Acids react with metal sulphides to liberate H_2S gas which has smell of rotten eggs.

(3) Acids react with metal sulphites to liberate SO_2 gas.

12. What is aqua regia? How does it help in dissolving gold or platinum?

Ans. A mixture of 1 part of conc. nitric acid and 3 parts of conc. hydrochloric acid by weight is called aqua regia.

The conc. HCl and conc. HNO_3 reacts to form nascent chlorine which reacts with gold or platinum to form their respective soluble chlorides.

$$HNO_3 + \underset{\text{(Conc.)}}{3HCl} \longrightarrow 2H_2O + NOCl + 2\,(Cl)$$

$$Au + 3[Cl] \longrightarrow \underset{\text{Gold(II) chloride}}{AuCl_3}$$

$$Pt + 4\,[Cl] \longrightarrow \underset{\text{Platinium(IV) chloride}}{PtCl_4}$$

Chapter 8. (b) Study of Compounds : Ammonia and Nitric Acid

1. How liquor ammonia is different from liquid ammonia?

Ans. A very strong solution of ammonia in water is called liquor ammonia. When ammonical liquor is boiled with milk of lime, the ammonia formed is passed into ice or cold water to form liquor ammonia.

When cooled under pressure, ammonia condenses to a colourless liquid which boils at – 33·4°C. This is liquid ammonia which when further cooled freezes to a white crystalline snow like solid, which melts at – 77·7°C. Liquid ammonia is used as a refrigerant because its heat of vaporisation is 5700 calories per gram molecule. Liquid ammonia is liquified ammonia and is available in cylinders. Laboratory bench reagent is a diluted solution of liquor ammonia.

2. The questions below are related to the manufacture of ammonia.

(i) Name the process.

(ii) In what ratio must the reactants be taken?

(iii) Name the catalyst used.

(iv) Give the equation for the manufacture of ammonia.

(v) Ammonia can act as a reducing agent — write a relevant equation for such a reaction.

Ans. (i) Haber's process.

(ii) Nitrogen one part, hydrogen three parts.

(iii) Iron powder.

(iv) $N_2 + 3H_2 \longrightarrow 2NH_3$

(v) $2NH_3 + \underset{\text{Copper oxide}}{3CuO} \longrightarrow \underset{\text{Copper}}{3Cu} + 3H_2O + N_2$

3. The following questions are based on the preparation of ammonia gas in the laboratory :

(i) Explain why ammonium nitrate is not used in the preparation of ammonia.

(ii) Name the compound normally used as a drying agent during the process.

(iii) How is ammonia gas collected?

(iv) Explain why it is not collected over water.

(v) Give the name of a hydride of nitrogen.

(vi) Which reactants are used in laboratory preparation of ammonia?

(vii) What is the vapour density and nature of ammonia?

(viii) Which feature of the ammonia molecules leads to the formation of the ammonium ion when ammonia dissolves in water. Name the other ion formed when ammonia dissolves in water.

Ans. (i) Ammonium nitrate is not used in the preparation of ammonia because ammonium nitrate is explosive in nature and dissociate into nitrous oxide and water on heating.

(ii) Quick lime.

(iii) By downward displacement of air.

(iv) It is highly soluble in water.

(v) Ammonia.

(vi) Ammonium chloride and slaked lime [$Ca(OH)_2$].

(vii) Vapour density of ammonia is 8·5, and it is alkaline in nature.

(viii) In ammonia molecule there is one lone pair of electrons available on nitrogen atom. This lone pair of electron leads to the formation of ammonium ion. Hydroxide ion (OH^-), is formed when ammonia dissolves in water.

4. What are the necessary conditions for getting maximum yield of ammonia?

Ans. The necessary conditions for getting maximum yield of ammonia are as follows :

(i) **Low temperature :** As the reaction is exothermic in nature, so the temperature should be low. When temperature is lowered, the rate of the reaction slows down and the yield is maximum. It is found for maximum yield, temperature should be between 450° to 500°C (optimum temperature).

(ii) **Catalyst :** A catalyst is used to accelerate the reaction at optimum temperature. Following are the catalysts used :

(a) Finely divided iron, mixed with molybdenum as promoter.

(b) A better catalyst is ferric oxide (Fe_2O_3) containing 1% of potassium oxide (K_2O) and 3% of aluminium oxide (Al_2O_3) which acts as promoters.

(iii) **High pressure :** When 4 volumes of a mixture of nitrogen and hydrogen are reduced to 2 volumes of ammonia, the pressure drops. In order to have maximum yield the pressure should be very high. The optimum pressure should be between 200 atmospheres to 900 atmospheres.

Pressure (in atm.)	Temperature (in C)	Yield (in %)
200 atms.	400°C	36·3%
1000 atms.	400°C	79·8%

(iv) **Purity of gases :** Any kind of impurity tends to slow down the reaction or poisons catalyst. So, the gases that are to be used should be pure and dry.

5. (i) What are the products formed when ammonia is oxidised with copper oxide?

(ii) What is the difference between chemical nature of an aqueous solution of hydrogen chloride and an aqueous solution of ammonia.

Ans. (i) Nitrogen, copper and water.

(ii) Aqueous solution of HCl is acidic while aqueous solution of ammonia is basic.

6. How would you prepare :

(i) Nitrogen from ammonia. (ii) Ammonia from nitrogen.

(iii) Nitric oxide gas and nitrous oxide gas.

Ans. (i) Ammonia gas burns in the presence of oxygen in a glass jar with a greenish yellow flame to form nitrogen and steam.

$$4NH_3 + 3O_2 \longrightarrow 2N_2 + 6H_2O$$

(ii) Nitrogen combines with heated metals; for example, aluminium to form aluminium nitride, which is readily decomposed by water to form ammonia gas which has a characteristic pungent odour and aluminium hydroxide is also formed.

$$2Al + N_2 \longrightarrow \underset{\text{Aluminium nitride}}{2AlN}$$

$$AlN + 3H_2O \longrightarrow \underset{\text{Aluminium hydroxide}}{Al(OH)_3} + NH_3$$

(iii) Introduce the given gas in the atmosphere of oxygen and if reddish brown fumes are produced, then the given sample of gas is nitric oxide and if no brown fumes are produced and the mixture remains colourless, then the gas is nitrous oxide.

7. Outline the conditions of the synthetic process for producing ammonia gas. Mention the catalyst used.

Ans. A mixture of nitrogen and hydrogen in the ratio of 1 : 3 by volume is passed over heated iron at 450°C in the presence of molybdenum, which acts as catalyst (promoter) under 200 atmospheric pressure. The reaction of nitrogen and hydrogen is a reversible reaction hence ammonia is removed from the reaction vessel from time to time by liquefaction under high pressure, so that the reaction proceeds in the forward direction.

$$N_2 + 3H_2 \rightleftharpoons \underset{\text{Ammonia}}{2NH_3} + 22 \text{ k cal.}$$

8. Name one element in each case, to which the following description would apply.

(i) The burning metal which combines directly with nitrogen.

(ii) Define fixation of nitrogen and state two ways in which fixation of nitrogen occurs, naturally.

(iii) Give two reasons to show that the solution of ammonia in water contains hydroxide ions.

Ans. (i) Magnesium, calcium and aluminium.

(ii) Conversion of atmospheric nitrogen into soluble nitrogenous compounds which can be easily absorbed by the roots of the plants.

(a) During thunder and lightning

(b) By symbiotic bacteria

(iii) (a) It turns red litmus solution to blue.

(b) It reacts with acids to form salt and water.

9. (i) Of the two gases, ammonia and hydrogen chloride, which is more dense? Name the method of collection of this gas.

(ii) Give one example of a reaction between the above two gases which produces a solid compound.

Ans. (i) HCl collected by upward displacement of air.

(ii) $NH_3 + HCl \longrightarrow NH_4Cl$

10. (i) Does ammonia burn in oxygen? If yes, give equation.

(ii) What is catalytic oxidation of ammonia?

(iii) What is the special feature of the apparatus that is used in the laboratory preparation of nitric acid ?

(iv) Why should the temperature of the reaction mixture of nitric acid not be allowed to rise above 200°C?

Ans. (i) Yes. Ammonia burns in oxygen with a greenish yellow flame producing water vapour and nitrogen.

$$4NH_3 + 3O_2 \longrightarrow 2N_2 + 6H_2O$$

(ii) Ammonia, when reacts with oxygen in presence of platinum (as catalyst) at 800°C, ammonia is oxidised to nitric oxide (NO). This is called catalytic oxidation of ammonia.

(iii) It is an all glass retort.

(iv) Because above 200°C nitric acid will decompose.

11. Oxidation of ammonia under certain conditions is represented by the following equation.

$$4NH_3 + 5O_2 \longrightarrow 4NO + 6H_2O$$

(i) Give the conditions required for this reaction.

(ii) Briefly describe the importance of this reaction in industry.

Ans. (i) Pure and dry ammonia gas mixed with air in the ratio of 1 : 8 by volume is first compressed and then passed over heated platinum gauze at 800°C.

(ii) Nitric oxide is obtained by the oxidation of ammonia, which is further oxidised to form nitrogen dioxide. Nitrogen dioxide is dissolved in water, in the presence of oxygen, to form nitric acid. Thus, nitric acid can be prepared by the catalytic oxidation of ammonia.

$$4NH_3 + 5O_2 \xrightarrow{\text{Platinum}} \underset{\text{Nitric Oxide}}{4NO} + 6H_2O$$

$$2NO + O_2 \xrightarrow{800°C} \underset{\text{Nitrogen dioxide}}{2NO_2}$$

$$4NO_2 + 2H_2O + O_2 \longrightarrow \underset{\text{Nitric acid}}{4HNO_3}$$

12. (i) How would you obtain the compound magnesium nitride?

(ii) What property of ammonia is illustrated by the reaction between ammonia and copper oxide?

(iii) What important industrial process starts with reaction between ammonia and oxygen. Name the catalyst uses.

(iv) During laboratory preparation how is ammonia dried and collected?

Ans. (i) By burning magnesium in the atmosphere of nitrogen by direct combination of elements.

$$3Mg + N_2 \longrightarrow Mg_3N_2$$

(ii) Reducing property of ammonia, as it reduces CuO to Cu in reaction.

(iii) Ostwald process for the manufacture of nitric acid and the catalyst used is platinum gauze.

(iv) Ammonia is dried by passing it through quick lime (CaO) and is collected by downward displacement of air.

13. Gas B turns moist red litmus paper blue.

(i) What is the name of gas B?

(ii) Write the equation for the reaction that takes place when gas B is passed over heated copper oxide.

Ans. (i) Gas B is ammonia.

(ii) $3CuO + 2NH_3 \longrightarrow 3Cu + 3H_2O + N_2\uparrow$

14. (i) What would you see during burning of ammonia in oxygen?

(ii) Name the catalyst used in the catalytic oxidation of ammonia.

(iii) In the reaction of the catalytic oxidation of ammonia the catalyst glows red hot, Why?

(iv) What is the name of the industrial process, which starts with the reaction of catalytic oxidation of ammonia.

(v) How is the temperature maintained in the Ostwald's process?

Ans. (i) The burning of ammonia in oxygen produces a bluish green flame. The reaction is irreversible and highly exothermic.

(ii) Platinum.

(iii) The catalyst glows red hot without external heating because much heat is liberated during the reaction.

(iv) Ostwald's process.

(v) In the Ostwald's process, the catalytic oxidation of ammonia to nitric oxide is an exothermic reaction. Once the reaction starts, the heat released maintains the temperature of the catalyst.

15. In what way does the action of dilute nitric acid on a metal like zinc differ from the action of dil. sulphuric acid.

OR

Dilute nitric acid is generally considered a typical acid except for its reaction with metals. In what way is dilute nitric acid different from other acids when it reacts with metals?

Ans. Nitric acid is a strong oxidising agent and oxidises the liberated hydrogen to water and is itself reduced to nitrogen dioxide. On the other hand dilute sulphuric acid can not oxidise the liberated hydrogen to water.

16. The reaction of nitric acid with metals depends upon the concentration of the acid. Give examples, alongwith equations to illustrate the reaction of the acid.

Ans. Copper reacts differently with nitric acid at different concentrations and temperature.

(i) Copper reacts with cold and dilute nitric acid to form copper nitrate, water and nitric oxide gas is liberated.

$$3Cu + 8HNO_3 \longrightarrow Cu(NO_3)_2 + 4H_2O + \underset{\text{Nitric oxide}}{2NO}$$

(ii) Copper reacts with hot and concentrated nitric acid to form copper nitrate, water and nitrogen dioxide gas, which has reddish brown fumes and a pungent smell is liberated.

$$Cu + 4HNO_3 \longrightarrow \underset{\text{Copper nitrate}}{Cu(NO_3)_2} + 2H_2O + \underset{\text{Nitrogen dioxide}}{2NO_2}$$

17. (i) Among the elements chlorine, nitrogen and sulphur select :

(a) The least reactive (b) Obtained from the atmosphere

(ii) Why commercial nitric acid is brown?

(iii) How can you get 'liquid air'?

Ans. (i) (a) Nitrogen (b) Nitrogen

(ii) Commercial nitric acid is brown in colour because it contains dissolved nitrogen dioxide.

(iii) CO_2 and moisture-free air is passed through a condenser. Finally, this air is compressed, cooled and suddenly allowed to expand. By repeating this process, the air can be liquified.

18. Explain the following :

(i) Dilute nitric acid is generally considered a typical acid but not so in its reaction with metals.

(ii) Concentrated nitric acid appears yellow when it is left standing in a glass bottle.

(iii) An all glass apparatus is used in the laboratory preparation of nitric acid.

Ans. (i) Because it does not liberate hydrogen. It is a powerful oxidising agent and the nascent oxygen formed oxidises the hydrogen to water.

(ii) Because when nitric acid is left standing in a glass bottle, it decompases to give reddish brown NO_2 gas which dissolves in undecomposed nitric acid to give a yellow colour.

(iii) Because nitric acid vapours are corrosive and may attack rubber, cork or metal.

19. (i) Sodium hydroxide solution is added to solution A, a white precipitate is formed which is soluble in excess of sodium hydroxide. What metal ion is present in A?

(ii) Ammonium hydroxide solution is added to solution B, when a pale blue precipitate is formed. This pale blue precipitate dissolves in excess of ammonium hydroxide to giveinky blue colouration. Name the cation present in B. Name the probable colour of solution B.

(iii) When an ammonium salt is warmed with sodium hydroxide solution, ammonia gas is evolved ? State three ways in which you can identify the gas.

Ans. (i) Ammonium ion is present in A.

(ii) The cation present in B is copper. Solution B is blue in colour.

(iii) The three ways in which the gas can be identified are as follows :

(a) It gives dense white fumes when a rod dipped in HCl is held in it.

(b) It turns moist red litmus paper blue.

(c) It turns phenolphthalein solution pink.

20. (i) What is the type of salt formed when the reactants are heated at a suitable temperature for the preparation of nitric acid ?*

(ii) State why for the preparation of nitric acid, the complete apparatus is made up of glass.

Ans. (i) Sodium sulphate is formed if the reactants (sulphuric acid and sodium nitrate) for the preparation of nitric acid are heated above 200°C. The sodium sulphate formed deposits as a hard crust and is difficult to remove.

$$NaNO_3 + NaHSO_4 \xrightarrow{200°C} Na_2SO_4 + HNO_3$$

(ii) All glass apparatus should be used while preparing nitric acid as the nitric acid vapours are highly corrosive and they corrode the cork or rubber fittings used in the apparatus.

21. (i) Give the balanced equations for laboratory preparation of nitric acid and describe the reaction.

(ii) In laboratory preparation of nitric acid, the mixture of potassium nitrate and concentrated sulphuric acid should not be heated above 200°C. Explain?

Ans. (i) In laboratory, nitric acid is prepared by heating potassium nitrate with concentrated sulphuric acid at low temperature, potassium hydrogen sulphate and nitric acid are formed, and at a higher temperature, potassium sulphate and nitric acid are formed.

$$KNO_3 + H_2SO_4 \xrightarrow{\text{Low temp.}} KHSO_4 + HNO_3$$

$$2KNO_3 + H_2SO_4 \xrightarrow{\text{High temp.}} K_2SO_4 + 2HNO_3$$

(ii) In laboratory preparation of nitric acid, the mixture of potassium nitrate and concentrated sulphuric acid should not be heated above 200°C because at a higher temperature, potassium sulphate is formed. Potassium sulphate is deposited as a hard, solid mass in the retort, which is difficult to remove. Moreover, most of the nitric acid decomposes at a higher temperature to form reddish brown fumes of nitrogen dioxide and the acid acquire a yellow colouration.

22. (i) (a) What compounds are required for the laboratory preparation of nitric acid?

(b) The first step in the manufacture of nitric acid is the catalytic oxidation of ammonia. What is the name of the catalyst?

(ii) (a) How soluble is ammonia in water?

(b) Give two reasons to show that the solution of ammonia in water contains hydroxide ions.

(c) Name a simple method you would employ to prepare ammonium salts in your laboratory.

Ans. (i) (a) Nitre (Potassium nitrate) and conc. sulphuric acid.

(b) Platinum.

(ii) (a) Extremely soluble.

(b) Ammonia solution in water turns red litmus blue and phenolphthalein pink. Ferricchloride, aluminium chloride solutions treated with ammonia solution giveprecipitates of $Fe(OH)_3$ and $Al(OH)_3$ respectively.

$$FeCl_3 + 3NH_4OH \longrightarrow Fe(OH)_3\downarrow + 3NH_4Cl$$

$$AlCl_3 + 3NH_4OH \longrightarrow Al(OH)_3\downarrow + 3NH_4Cl$$

(c) By titration or neutralisation.

23. Give two reactions to show that nitric acid is :

(i) An acid (ii) An oxidizing agent

* **Frequently asked previous years Board Exam Questions.**

Short Answers

Ans. (i) (a) Dilute nitric acid reacts with magnesium metal to liberate hydrogen gas and magnesium nitrate is formed.

$$Mg + 2HNO_3 \longrightarrow Mg(NO_3)_2 + H_2\uparrow$$

(b) Dilute nitric acid reacts with marble pieces to liberate carbon dioxide gas and calcium nitrate is formed.

$$CaCO_3 + 2HNO_3 \longrightarrow Ca(NO_3)_2 + H_2O + CO_2\uparrow$$

(ii) (a) When carbon is heated with concentrated nitric acid, it is oxidized to carbonic acid and the acid is reduced to nitrogen dioxide and water.

$$C + 4HNO_3 \longrightarrow H_2CO_3 + 4NO_2 + H_2O$$

(b) When sulphur is heated with concentrated nitric acid, it is oxidized to sulphuric acid and the nitric acid is reduced to nitrogen dioxide and water. Carbonic acid further decomposes to carbon dioxide.

$$\underset{\text{(Conc.)}}{S + 6HNO_3} \longrightarrow H_2SO_4 + 6NO_2 + 2H_2O$$

24. Answer the questions given below, relating your answer only to compounds given in the following list :

Tetrammine copper(II) sulphate, Iron(III) chloride, conc. nitric acid, Ammonium hydroxide.

(i) Write name of the colours and action when concentrated nitric acid is heated with copper turnings ?

(ii) A compound X is dissolved in the ammonia solution to give a deep blue coloured solution, write the name of the compound having deep blue colours?

(iii) The concentrated nitric acid produces yellow stains and blisters on skin. Give the reason.

Ans. (i) Brown fumes of nitrogen dioxide are evolved.

(ii) Tetrammine copper(II) sulphate.

(iii) This is due to the chemical reaction that occurs between nitric acid and proteins of skin and flesh. In this reaction, a complex compound xanthoprotic acid is formed.

25. The action of heat on the blue crystalline solid L gives a reddish brown gas M, a gas which relights a glowing splint and leaves a black residue. When as N, which has a rotten egg smell, is passed through a solution of L a black precipitate is formed.

(i) Identify L, M and N (Name or formula).

(ii) Write the equation for the action of heat on L.

(iii) Write the equation for the reaction between the solution of L and the gas N.

Ans. (i) L is copper nitrate.

M is nitrogen dioxide gas.

N is hydrogen sulphide gas.

(ii) $2\,Cu(NO_3)_2 \longrightarrow 2CuO + 4NO_2 + O_2$

(iii) $Cu(NO_3)_2 + H_2S \longrightarrow CuS + 2HNO_3$.

26. How will you show that nitric acid contains?

(i) Hydrogen (ii) Nitrogen (iii) Oxygen

Ans. (i) Dilute nitric acid reacts with magnesium metal to liberate hydrogen gas which burns with a pale blue flame and a pop sound and magnesium nitrate is formed.

$$\underset{\text{(Dil.)}}{Mg + 2HNO_3} \longrightarrow Mg(NO_3)_2 + H_2\uparrow$$

(ii) Nitric acid when heated with copper turnings, produces brown fumes of nitrogen dioxide, which are passed over red hot iron filings. The residual gas collected over water is nitrogen.

$$Cu + 4HNO_3 \longrightarrow Cu(NO_3)_2 + 2H_2O + 2NO_2$$

$$6NO_2 + 8Fe \longrightarrow 4Fe_2O_3 + 3N_2\uparrow$$

(iii) Nitric acid, when heated alone, is decomposed into a reddish brown coloured gas nitrogen dioxide, steam and a colourless, odourless gas oxygen, which rekindles a glowing splinter.

$$4HNO_3 \longrightarrow 4NO_2 + 2H_2O + O_2$$

27. What is the ring test for nitric acid? Briefly describe this test.

Ans. To the nitrate solution, add freshly prepared iron(II) sulphate solution. Filter off the precipitate if formed. Now carefully add conc. H_2SO_4 along the sides of the test tube. A dark brown ring is formed at junctions of two liquids. Brown ring is due to the formation of nitrosonium pentaqua iron(I) sulphate $[Fe^+ (NO^+) (H_2O)_5] SO_4$, which is an octahedral complex compound.

Chapter 8. (c) Study of Compounds : Sulphuric Acid

1. Sulphuric acid is said to be dibasic acid. What is meant by the term 'dibasic'?

Ans. Basicity of an acid is the number of H^+ ions that one formula unit of an acid liberates, *e.g.,*

Monobasic= HCl, HNO_3 etc.

Dibasic= H_2SO_4, H_2SO_3 etc.

Each acid can form as many kinds of salts, as it has hydrogen ions. Sulphuric acid can form two kinds of salts, *i.e.*, SO_4^{2-} and HSO_4^-. It ionizes in water to form two hydrogen ions. Hence, it is said to be dibasic.

$$H_2SO_4 \rightleftharpoons H^+ + HSO_4^-$$
$$HSO_4^- \rightleftharpoons H^+ + SO_4^{2-}$$

These acids can yield two kinds of salts, *i.e.*, the normal salt and the acid salt.

2. While diluting concentrated sulphuric acid, the acid should be added to water and not water to the acid. Explain?

Ans. When equal volumes of the acid and water are mixed at room temperature, the temperature may reach up to 120°C. Therefore, dilution of the acid should be done by adding small quantity of acid into water.

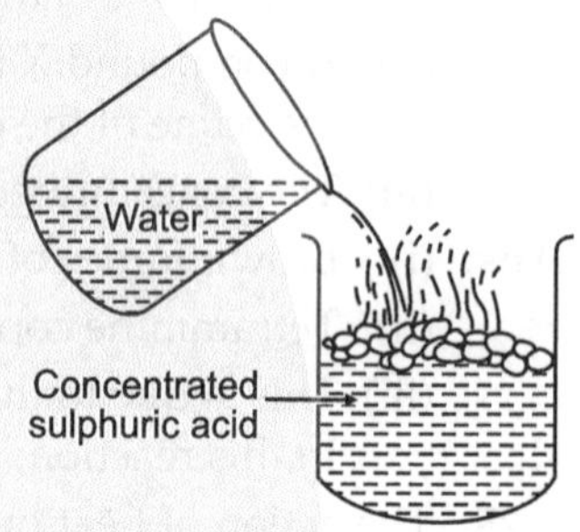

If water is added to concentrated sulphuric acid, the molecules of the acid try to grasp the molecules of water resulting in molecular tension, liberating heat and due to sudden rise in temperature, the acid starts splashing.

If a drop of concentrated acid is added to water, the molecules of acid go in different directions to pick up water which is available in plenty. Although the same amount of heat is formed but since the molecules are spread out, no splashing occurs.

3. (i) Which property of sulphuric acid is used to prepare the hydrochloric and nitric acid respectively.

(ii) What is the catalyst used in catalytic chamber and what is the temperature used?

Ans. (i) Non-volatile nature of sulphuric acid is responsible to prepare HCl and HNO_3 by using H_2SO_4.

(ii) The catalyst used is platinized asbestos or vanadium pentaoxide and the temperature used is about 450°C.

4. (i) (a) Name the acid formed when sulphur dioxide dissolves in water.

(b) What are the two necessary conditions for the direct combination of sulphur dioxide and chlorine forming sulphuryl chloride?

(c) State the property of sulphur dioxide which causes potassium permanganate to change its colour from purple to colourless.

(ii) Answer the following questions related to dilute and concentrated sulphuric acid.

(a) Which acid does not react with metals that are placed below hydrogen in activity series ?

(b) Which acid will give white precipitates with barium sulphate?

Ans. (i) (a) Sulphurous acid.

(b) Sunlight and absence of moisture.

(c) Reducing.

(ii) (a) Dilute sulphuric acid. (b) Dilute sulphuric acid.

5. Which property of sulphuric acid is used in the following :

(i) As a source of hydrogen when treated in dilute form with metals like Zn, Mg, Fe, etc.

(ii) Production of hydrogen chloride on treating concentrated acid with sodium chloride.

(iii) Production of sulphur dioxide on heating in concentrated form with copper turnings.

(iv) Liberation of sulphur from H_2S with concentrated form.

(v) Charring of sugar with hot concentrated acid.

(vi) Liberation of ethylene gas with hot concentrated acid.
(vii) Liberation of carbon monoxide with hot concentrated acid.

Ans. (i) Acidic property (ii) Non-volatile nature
(iii) Oxidising nature (iv) Oxidising property
(v) Dehydrating nature (vi) Dehydrating nature
(vii) Dehydrating nature

6. A, B, C and D summarize the properties of sulphuric acid depending on whether it is dilute or concentrated. Choose the property (A, B, C or D), depending on which is relevant to each of the preparations (i) to (iii) :
(A) Dilute acid (typical acid properties) (B) Non-volatile acid.
(C) Oxidizing agent. (D) Dehydrating agent
(i) Preparation of hydrogen chloride.
(ii) Preparation of ethene from ethanol.
(iii) Preparation of copper sulphate from copper oxide.

Ans. (i) B (non-volatile acid). (ii) D (dehydrating agent). (iii) A (dilute acid).

7. Name from the list of substances given below, the substances which you would use to prepare each of the following salts, named in parts (i) to (iv) :
The substances are :
Copper, Lead, Sodium, Zinc, Copper oxide, Lead carbonate, Sodium carbonate solution, Dilute hydrochloric acid, Dilute nitric acid and Dilute sulphuric acid :
(i) Zinc sulphate; (ii) Copper sulphate; (iii) Sodium sulphate; (iv) Lead sulphate.

Ans. (i) For zinc sulphate — Zinc and dilute sulphuric acid.
(ii) For copper sulphate — Copper oxide and dilute sulphuric acid.
(iii) For sodium sulphate — Sodium carbonate and dilute sulphuric acid.
(iv) For lead sulphate — Lead carbonate and add dil. nitric acid and then dil. sulphuric acid.

8. Give examples of the use of sulphuric acid as :
(i) An electrolyte in everyday use. (ii) A non-volatile acid. (iii) An oxidizing agent.

Ans. (i) In lead accumulators or in storage batteries.
(ii) In the manufacture of other acids like nitric acid, hydrochloric acid and phosphoric acid.
(iii) For cleaning metals before enameling, electroplating and galvanizing, as a pickling agent.

9. A, B, C and D summarize the properties of sulphuric acid depending on whether it is dilute or concentrated.
A = Typical acid property.
B = Non-volatile acid
C = Oxidizing agent
D = Dehydrating agent
Choose the property (A, B, C or D) depending on which is relevant to each of the following :
(i) Preparation of Hydrogen chloride gas.
(ii) Preparation of copper sulphate from copper oxide.
(iii) Action of conc. sulphuric acid on sulphur.

Ans. (i) B = Non-volatile acid.
(ii) A = Typical acid property.
(iii) C = Oxidizing agent.

10. Which property of sulphuric acid is shown by the reaction of concentrated sulphuric acid with :*
(i) Ethanol ? (ii) Carbon ?

Ans. (i) Reaction of concentrated sulphuric acid with ethanol leads to formation of ethene, which shows that it is dehydrating in nature.

$$\underset{\text{Ethanol}}{C_2H_5OH} \xrightarrow[\text{Dehydration}]{H_2SO_4} \underset{\text{Ethene}}{C_2H_4} + H_2O$$

(ii) Reaction of concentrated sulphuric acid with carbon shows its oxidizing nature, where it oxidizes carbon to carbon dioxide.

$$C + 2H_2SO_4 \rightarrow CO_2 + 2H_2O + 2SO_2$$

* **Frequently asked previous years Board Exam Questions.**

11. Some bacteria obtain their energy by oxidizing sulphur, producing sulphuric acid as a byproduct. In the laboratory, or industrially, the first step in the conversion of sulphur to sulphuric acid is to produce sulphur dioxide. Then sulphur dioxide is converted to sulphur trioxide which reacts with water, producing sulphuric acid.

(i) Name one catalyst used industrially which speeds up the conversion of sulphur dioxide to sulphur trioxide.

(ii) Write the equation for the conversion of sulphur dioxide to sulphur trioxide. Why does this reaction supply energy?

(iii) What is the name of the compound formed between sulphur trioxide and sulphuric acid?

Ans. (i) Platinum and Vanadium pentaoxide.

(ii) When conversion of SO_2 to SO_3 takes place according to the following reaction.

$$SO_2 + O_2 \longrightarrow 2SO_3 + 45 \text{ k cal}$$

The 45 k cal energy supplied by above reaction.

(iii) Oleum ($H_2S_2O_7$).

12. (i) Write balanced equations for the three chemical reactions that take place during the conversion of sulphur dioxide to sulphuric acid in the contact process.

(ii) Name the catalyst used in the contact process.

(iii) Name another ore which on roasting gives sulphur dioxide?

Ans. (i) The chemical reactions are summarised as below :

$$S + O_2 \longrightarrow SO_2$$
$$2SO_2 + O_2 \longrightarrow 2SO_3$$
$$SO_3 + H_2SO_4 \longrightarrow \underset{\text{(Oleum)}}{H_2S_2O_7}$$
$$H_2S_2O_7 + H_2O \longrightarrow 2H_2SO_4$$

(ii) Platinized asbestos or V_2O_5.

(iii) Zinc blende or ZnS :

$$2ZnS + 3O_2 \longrightarrow 2ZnO + 2SO_2 \uparrow$$

13. (i) With the help of equations, give an outline for the manufacture of sulphuric acid by the contact process.

(ii) What property of sulphuric acid is shown by the reaction of concentrated sulphuric acid when heated with

(a) potassium nitrate (b) carbon?

Ans. (i) Contact Tower :

$$\underset{\text{Sulphur dioxide}}{2SO_2} + O_2 \xrightarrow[450°C]{V_2O_5} \underset{\text{Sulphur trioxide}}{2SO_3}$$

Absorption Tower :

$$SO_3 + \underset{\text{Conc.}}{H_2SO_4} \longrightarrow \underset{\text{Oleum}}{H_2S_2O_7}$$

Dilution Tower :

$$H_2S_2O_7 + H_2O \longrightarrow \underset{\text{Sulphuric acid}}{2H_2SO_4}$$

(ii) (a) Non-volatile nature. (b) Oxidising property.

14. (i) Which two gases are combined during contact process?

(ii) Write the equation for the reaction between zinc and the final product of the contact process ?

(iii) What happens when sulphur trioxide gas is passed into concentrated sulphuric acid.

Ans. (i) SO_2 and O_2 (sulphur dioxide and oxygen)

(ii)
$$Zn + \underset{\text{(Conc.)}}{2H_2SO_4} \longrightarrow ZnSO_4 + 2H_2O + SO_2$$

(iii) Sulphur trioxide gas dissolves in concentrated sulphuric acid to form fuming sulphuric acid, commonly known as oleum.

$$SO_3 + H_2SO_4 \longrightarrow \underset{\text{Oleum}}{H_2S_2O_7}$$

Short Answers

15. Give one reaction in each case to illustrate the following properties of sulphuric acid :

(i) As an acid. (ii) As an oxidising agent.

(iii) As a dehydrating agent. (iv) As a less volatile acid.

Ans. (i) Dilute sulphuric acid reacts with metals above hydrogen in the activity series, for example, magnesium to liberate hydrogen gas and magnesium sulphate.

$$Mg + H_2SO_4 \longrightarrow MgSO_4 + H_2\uparrow$$

(ii) When sulphur is heated with concentrated sulphuric acid, it is oxidised to sulphur dioxide.

$$S + 2H_2SO_4 \longrightarrow \underset{\text{Sulphur dioxide}}{3SO_2} + 2H_2O$$

(iii) Add few drops of concentrated sulphuric acid to blue coloured crystals of copper(II) sulphate. After sometime, white anhydrous copper(II) sulphate is left, due to loss of water of crystallization.

$$\underset{\text{Hydrated copper sulphate (Blue)}}{CuSO_4.5H_2O} + [H_2SO_4] \longrightarrow \underset{\text{Anhydrous copper sulphate (White)}}{CuSO_4} + [H_2SO_4.5H_2O]$$

(iv) Concentrated sulphuric acid, when heated with sodium chloride, produces volatile hydrochloric acid.

$$NaCl + \underset{\text{(Conc.)}}{H_2SO_4} \xrightarrow{\text{Heat}} NaHSO_4 + HCl$$

16. What happens when conc. sulphuric acid reacts with :

(i) Potassium chloride

(ii) Zinc nitrate.

Ans. (i) Conc. sulphuric acid reacts with potassium chloride and forms hydrogen chloride and potassium hydrogen sulphate.

$$KCl + H_2SO_4 \longrightarrow KHSO_4 + HCl$$

(ii) Conc. sulphuric acid reacts with zinc nitrate to form zinc sulphate and nitric acid.

$$Zn(NO_3)_2 + H_2SO_4 \longrightarrow ZnSO_4 + 2HNO_3$$

17. State how you can obtain :

(i) Sulphur dioxide from sulphur

(ii) Hydrogen sulphide from iron(II) sulphide

(iii) Oxalic acid

(iv) Sodium hydroxide

(v) Hydrogen sulphide gas

Ans. (i) When sulphur is burnt in a deflagerating spoon, it melts to form a reddish brown liquid which catches fire. It burns with a blue flame forming an extremely pungent gas sulphur dioxide.

$$S + O_2 \longrightarrow SO_2$$

(ii) In the laboratory, hydrogen sulphide gas is prepared by the action of dilute sulphuric acid on ferrous sulphide.

$$FeS + H_2SO_4 \longrightarrow FeSO_4 + H_2S$$

(iii) Concentrated sulphuric acid, when heated with oxalic acid crystals, it absorbs water from oxalic acid and mixtures of carbon monoxide and carbon dioxide is formed.

$$\underset{\text{Oxalic acid}}{\begin{array}{c}COOH\\|\\COOH\end{array}} + H_2SO_4 \longrightarrow CO\uparrow + CO_2\uparrow + H_2O$$

In this reaction sulphuric acid acts as a dehydrating agent.

(iv) Sodium hydroxide solution reacts with dilute sulphuric acid to form sodium sulphate and water.

$$2NaOH + H_2SO_4 \longrightarrow Na_2SO_4 + 2H_2O$$

In this reaction sulphuric acid acts as an acid as it neutralizes sodium hydroxide to form salt and water only.

(v) When hydrogen sulphide gas is passed through concentrated sulphuric acid, it is oxidized to free sulphur. Sulphur dioxide and water are also formed.

$$2H_2S + H_2SO_4 \longrightarrow 2H_2O + SO_2\uparrow + \underset{\text{Sulphur}}{2S\downarrow}$$

In this reaction sulphuric acid acts as an oxidizing agent.

18. Some properties of sulphuric acid are listed below. Choose the property A, B, C or D which is responsible for the reactions (i) to (v)? Some properties may be repeated :

(A) Acid (B) Dehydrating agent

(C) Non-volatile acid (D) Oxidizing agent

(i) $C_{12}H_{22}O_{11} + nH_2SO_4 \longrightarrow 12C + 11H_2O + nH_2SO_4$

(ii) $S + 2H_2SO_4 \longrightarrow 3SO_2 + 2H_2O$

(iii) $NaCl + H_2SO_4 \longrightarrow NaHSO_4 + HCl$

(iv) $CuO + H_2SO_4 \longrightarrow CuSO_4 + H_2O$

(v) $Na_2CO_3 + H_2SO_4 \longrightarrow Na_2SO_4 + H_2O + CO_2$

(vi) $CuSO_4{\cdot}5H_2O \xrightarrow{\text{Conc. } H_2SO_4} CuSO_4 + 5H_2O$

(vii) $NaNO_3 + H_2SO_4\text{ (conc.)} \xrightarrow{< 200°C} NaHSO_4 + HCl$

(viii) $MgO + H_2SO_4 \longrightarrow MgSO_4 + H_2O$

(ix) $Zn + 2H_2SO_4\text{ (conc.)} \longrightarrow ZnSO_4 + SO_2 + 2H_2O$

Ans. (i) B (ii) D (iii) C (iv) A (v) A

(vi) B (vii) C (viii) A (ix) D

19. Some of the properties of six pure substances represented by A, B, C, D, E and F are given below :

A–when heated with concentrated sulphuric acid, it gives off a choking gas which dissolves in water giving an acid.

B–is a greenish-yellow gas which dissolves in water and when this aqueous solution is exposed to sunlight, bubbles of a gas are evolved, which rekindles a glowing splinter.

C– is a metal which when treated with concentrated nitric acid, gives off a brown gas and a blue solution is obtained.

D– is a white solid, which when heated, gives off a sweet smelling gas which rekindles a glowing splinter.

E– is a heavy oily liquid which when added to moist sugar, chars it into a black porous mass.

F– is a gas which turns moist red litmus to blue. When the gas is passed over heated copper oxide, an inactive gas is obtained.

(i) Name the substances A, B, C, D, E and F.

(ii) Write equations for the following reactions involving A, B, C, D, E and F.

(a) A is heated with concentrated sulphuric acid.

(b) An aqueous solution of B is exposed to bright sunlight.

(c) Concentrated nitric acid and the metal C are heated.

(d) The action of heat on D.

(e) Oily liquid E is added to sugar.

(f) The action of F on heated copper(II) oxide.

Ans. (i) A is sodium chloride, B is chlorine gas, C is copper, D is ammonium nitrate, E is concentrated sulphuric acid and F is ammonia gas respectively.

(ii) (a) When sodium chloride is heated with concentrated sulphuric acid, hydrogen chloride gas is liberated and sodium sulphate is also formed.

$$2NaCl + \underset{\text{(Conc.)}}{H_2SO_4} \xrightarrow{\text{Heat}} Na_2SO_4 + 2HCl.$$

(b) Chlorine gas reacts with water in the presence of sunlight to liberate oxygen gas and hydrochloric acid is formed.

$$2Cl_2 + 2H_2O \xrightarrow{\text{Sunlight}} 4HCl + \underset{\text{Oxygen}}{O_2\uparrow}$$

Short Answers

(c) Copper and concentrated nitric acid when heated, a brown gas, nitrogen dioxide is evolved and a blue coloured copper nitrate is formed.

$$\underset{\text{(Conc.)}}{Cu + 2HNO_3} \xrightarrow{\text{Heat}} \underset{\substack{\text{Copper nitrate}\\ \text{Blue}}}{Cu(NO_3)_2} + 2H_2O + \underset{\substack{\text{Nigrogen dioxide}\\ \text{Brown gas}}}{2NO_2\uparrow}$$

(d) Ammonium nitrate on heating gives off a sweet smelling gas nitrous oxide, commonly known as laughing gas and water is formed.

$$NH_4NO_3 \xrightarrow{\text{Heat}} \underset{\substack{\text{Nitrous oxide}\\ \text{(laughing gas)}}}{N_2O\uparrow} + 2H_2O$$

(e) Sulphuric acid acts as dehydrating agent and chars the sugar to black porous mass, *i.e.*, carbon.

$$\underset{\text{Sugar}}{C_{12}H_{22}O_{11}} + H_2SO_4 \longrightarrow 12C + \underset{\text{Hydrate}}{[H_2SO_4 + 11H_2O]}$$

(f) When ammonia is passed over heated copper (II) oxide, it is oxidized to form nitrogen and water. Copper(II) oxide itself reduces to metallic copper.

$$2NH_3 + 3CuO \longrightarrow 3Cu + 3H_2O + \underset{\text{Nitrogen gas}}{N_2\uparrow}$$

Chapter 9. Organic Chemistry

1. The melting point of three members X, Y and Z of a homologous series of hydrocarbons are – 80°C, –140°C and –30°C respectively.

(i) Which one of the three would have the lowest number of carbon atoms in its molecule? Justify your answer.

(ii) Which one of the three have the maximum number of carbon atoms in its molecule? Justify your answer.

Ans. (i) The homologue with lower number of C-atoms in its molecule has lower melting point. Therefore, compound X has the lowest number of carbon atoms in its molecule. It is clear from the given values of melting points in which – 180°C is lowest.

(ii) The homologue with maximum number of carbon atoms in its molecule has the highest melting point. Therefore, compound Z has the maximum number of carbon atoms in its molecules. It is clear from the given values of melting point in which 30°C is the highest of the three.

2. State two uses of ethane and methane.

Ans. Uses of ethane :

(i) Liquified ethane is used as a fuel in automobiles.

(ii) It is used in the manufacture of organic compounds such as acetic acid, ethyl alcohol etc.

Uses of Methane :

(i) It is used on industrial scale for the preparation of hydrogen by the process of pyrolysis.

(ii) It is used in the manufacture of industrial compounds such as carbon tetrachloride, chloroform, formic acid etc.

3. State two uses of ethylene and acetylene.

Ans. Uses of ethylene :

(i) It is used in the manufacture of polyethylene which is a valuable plastic.

(ii) It is used in the artificial ripening of fruits.

Uses of acetylene :

(i) It is used in oxyacetylene flame which is used for cutting and welding of metals.

(ii) It is used in the manufacture of acetaldehyde, acetic acid etc.

4. The list of some organic compound is given below :

Ethanol, ethane, methanol, methane, ethyne, and ethene.

From the list above, name a compound :

(i) Formed by the dehydration of ethanol by concentrated sulphuric acid.

(ii) Which will give red precipitate with ammonical cuprous chloride solution?

(iii) Which forms methanoic acid on oxidation in the presence of copper at 200°C?
(iv) Which has vapour density 14 and turns alkaline potassium permanganate green?
(v) Which forms chloroform on halogenation in the presence of sunlight?
(vi) Which decolourises bromine solution in carbon tetrachloride?

Ans. (i) Ethene (ii) Ethyne
(iii) Methane (iv) Ethene
(v) Methane (vi) Ethene

5. (i) Alkanes are called saturated hydrocarbons. Give a brief explanation, by taking the example of C_2H_6.
(ii) 'Alkenes are unsaturated hydrocarbons'. Illustrate it, by taking the example of ethene (C_2H_4).
(iii) A compound has number of H atoms just double that of C atoms. What types of hydrocarbon is it ?

Ans. (i) Alkane are called saturated hydrocarbons because tetra-valency of each carbon atom is satisfied by single covalent bond.

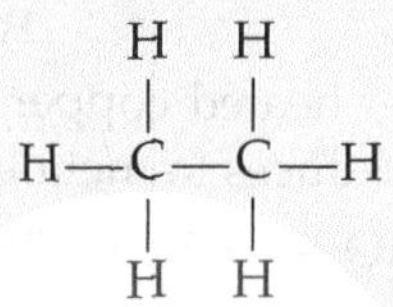

(ii) Ethene (C_2H_4) is an unsaturated hydrocarbon, commonly known as alkene. Ethene molecule contains two carbon atoms bonded by double bond.

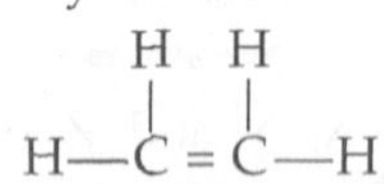

Double bond in ethane

(iii) The compound is an unsaturarted hydrocarbon having general formula C_nH_{2n}. So, this compound is an alkene.

6. (i) In the general formula C_nH_{2n+2}, write the meaning of n and $2n + 2$.
(ii) Write the formulae and names of the first four members of the alkane family.
(iii) Write the molecular formula of an alkane, which is composed of 16H atoms.
(iv) In a molecule of saturated hydrocarbon the number of C-atoms is 5, what is the number of H-atoms ?

Ans. (i) The number of alkane family represents the general formula C_nH_{2n+2}. In this formula :
n = number of carbon atoms in the same molecule of alkane.
$2n + 2$ = number of H-atoms in a molecule of alkane.
(ii) CH_4 (methane), C_2H_6 (ethane), C_3H_8 (propane), C_4H_{10} (butane).
(iii) C_7H_{16} (Heptane). [$\because 2n + 2 = 16, 2n = 16 - 2, 2n = 14, n = 14/2 = 7$].
(iv) According to general formula C_4H_{2n+2} when $n = 5$, $C_5H_{2\times5+2}$ or C_5H_{10+2} or C_5H_{12}.
Thus, the number of hydrogen atom is 12.

7. The molecules of alkene family are represented by a general formula C_nH_{2n}. Now answer the following :
(i) What do n and $2n$ signify?
(ii) What is the lowest value which can be assigned to n?
(iii) What is the molecular formula of alkene, when $n = 4$?
(iv) What is the structural formula of the first member of the alkene family?

Ans. (i) n = Number of C-atoms in a molecule of alkene.
$2n$ = Number of H-atoms in a molecule of alkene.
(ii) The lowest value of n is 2.
(iii) Butene (C_4H_8).
The first member of alkene family is C_2H_4.
(iv) It is called ethene, the structural formula is :

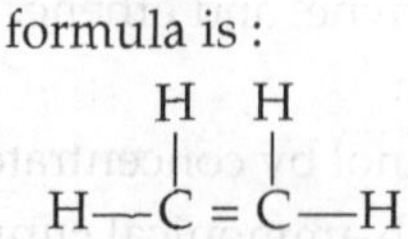

8. (i) Ethane and chlorine react together to form monochloroethane [ethyl chloride].
   (a) Write down the structural formula of ethane.
   (b) What type of reaction has taken place between ethane and chlorine?

   (ii) The type of reaction between ethene and chlorine is different from that between ethane and chlorine.
   (a) What is the type of reaction between ethene and chlorine.
   (b) What feature of the ethene structure makes such reaction possible?
   (c) Name the product of the reaction between ethene and chlorine.

   (iii) Ethane burns completely in air or oxygen to give carbon dioxide and water vapours. With a limited supply of air or oxygen, carbon monoxide is formed. The same gases are found in automobile exhaust gases. Both gases can be considered as atmospheric pollutants.
   (a) Write the equation for the complete combustion of ethane.
   (b) What danger is associated with carbon monoxide?
   (c) What effect is associated with too much carbon dioxide in the atmosphere?
   (d) Burning of acetylene [Ethyne] in oxygen, under appropriate conditions, produces a very hot flame. What is this hot flame used for?

Ans. (i) (a) The structural formula of ethane is given by :

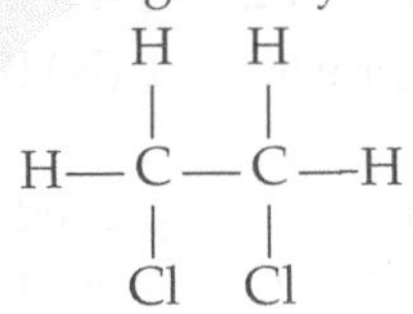

(b) Substitution reaction takes place in between ethane and chlorine.

$$\underset{\text{Ethane}}{C_2H_6} + Cl_2 \longrightarrow \underset{\text{Ethyl chloride}}{C_2H_5Cl} + HCl$$

(ii) (a) Addition reaction takes place in between ethene and chlorine.

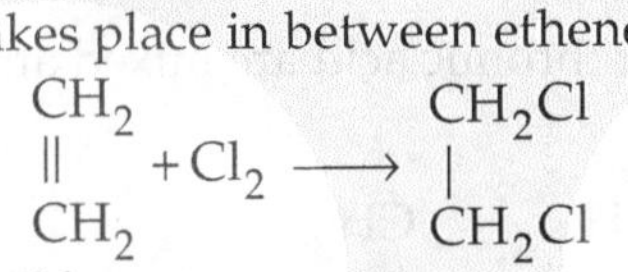

Ethene　　　　Dichloroethene

(b) Ethene is an unsaturated hydrocarbon [Alkene], containing double covalent bonds, which respond to the addition reaction.

(c) Ethene combines with chlorine to form 1 : 2 dichloroethane [Ethylene dichloride].

(iii) (a) Ethane burns in atmospheric oxygen to form carbon dioxide and steam [water].

$$2C_2H_6 + 7O_2 \longrightarrow 4CO_2 \uparrow + 6H_2O$$

(b) Carbon monoxide is a highly poisonous gas. It readily combines with haemoglobin of blood to form carboxy haemoglobin. Carboxy haemoglobin is a stable compound and is incapable of taking up the oxygen from the inhaled air and as a result, people die due to suffocation.

(c) Greenhouse effect or global warming is associated with too much carbon dioxide in the atmosphere.

(d) This hot flame is used for welding and cutting of steel.

9. Indicate the type of reaction that occurs when :
   (i) Ethane reacts with chlorine.
   (ii) Ethene reacts with chlorine.
   (iii) What type of reaction is common in C_2H_4 and C_2H_2?
   (iv) What is formed when ethene reacts with steam at 300°C in the presence of phosphoric acid as catalyst?
   (v) Name a solid which on reaction with water forms :
   (a) methane　　(b) ethyne (acetylene)
   (vi) Give the names of each of the following compounds :
   (a) $CH_3CH_2CH_2CH_3$　　(c) $CH \equiv CH$
   (b) $CH_2 = CH_2$

Ans. (i) Substitution reaction (ii) Addition reaction
(iii) Addition reaction (iv) Ethanol
(v) (a) Aluminium carbide, (b) Calcium carbide.
(vi) (a) Butane, (b) Ethylene (Ethene), (c) Acetylene (Ethyne).

10. (i) Experimentally, how can polychlorination of methane be minimized?
(ii) What are the conditions required for the addition of hydrogen to ethene?
(iii) Which catalyst is used for the addition of hydrogen to ethene at room temperature?
(iv) Write the names of all the possible organic products in the reaction of methane with chlorine.

Ans. (i) If excess of methane over chlorine is used, the chance of chlorine reacting with methane is greatest than with any other of the formed chloromethane.
(ii) Addition of hydrogen to ethene occurs at 300°C in the presence of nickel (Ni) catalyst.
(iii) Palladium (Pd) or Platinum (Pt) are used as catalyst at room temperature for the addition of hydrogen.
(iv) The main products is methyl chloride (CH_3Cl), (CH_2Cl_2) dichloro methane; ($CHCl_3$) trichloro methane, (CCl_4) and tetra chloromethane.

11. How does ethene gas react with the following?
(i) Hydrogen, (ii) Halogen acid, (iii) Sulphuric acid, (iv) Bromine, (v) Alkaline potassium permanganate, (vi) HCl gas.

Ans. (i) When a mixture of ethene and hydrogen are passed over heated catalyst (Ni, Pd or Pt), an addition reaction takes place with the formation of ethane, a saturated hydrocarbon.

$$\begin{matrix} CH_2 \\ \| \\ CH_2 \end{matrix} + H_2 \xrightarrow{\text{Ni–300°C}} \underset{\text{Ethane}}{\begin{matrix} CH_3 \\ | \\ CH_3 \end{matrix}}$$

(ii) When vapours of ethene and hydrobromic acid are mixed at room temperature, they react to form addition product, bromoethane.

$$\begin{matrix} CH_2 \\ \| \\ CH_2 \end{matrix} + \begin{matrix} H \\ | \\ Br \end{matrix} \longrightarrow \begin{matrix} CH_3 \\ | \\ CH_2Br \end{matrix}$$

(iii) When ethene is passed through conc. sulphuric acid, an addition reaction takes place at room temperature with the formation of ethyl hydrogen sulphate.

$$\begin{matrix} CH_2 \\ \| \\ CH_2 \end{matrix} + \begin{matrix} H \\ | \\ HSO_4 \end{matrix} \longrightarrow \underset{\text{Ethyl hydrogen sulphate}}{\begin{matrix} CH_3 \\ | \\ CH_2HSO_4 \end{matrix}}$$

(iv) When bromine is passed through the inert solution of ethene, an addition reaction takes place with the formation of 1, 2, dibromoethane.

$$\begin{matrix} CH_2 \\ \| \\ CH_2 \end{matrix} + \begin{matrix} Br \\ | \\ Br \end{matrix} \xrightarrow[\text{inert solvent}]{CCl_4} \underset{\text{1, 2-dibromoethane}}{\begin{matrix} CH_2Br \\ | \\ CH_2Br \end{matrix}}$$

(v) Ethene reacts with alkaline potassium permanganate solution to form glycol.

$$\begin{matrix} CH_2 \\ \| \\ CH_2 \end{matrix} + 2KMnO_4 + 4H_2O \longrightarrow \begin{matrix} CH_2OH \\ | \\ CH_2OH \end{matrix} + 2MnO_2 + 2KOH$$

(vi) Ethene reacts with HCl to form ethane glycol monochloride.

$$\begin{matrix} CH_2 \\ \| \\ CH_2 \end{matrix} + \begin{matrix} H \\ | \\ Cl \end{matrix} \longrightarrow \underset{\text{Ethane monochloride}}{\begin{matrix} CH_3 \\ | \\ CH_2Cl \end{matrix}}$$

12. (i) The alkenes having how many carbon atoms are in liquid state at normal temperature?
(ii) The alkenes having how many carbon atoms are in solid state at normal temperature?

Ans. (i) The alkenes having six to seventeen carbon atoms are in liquid state at normal temperature.
(ii) The alkenes having eighteen or more carbon atoms are in solid state at normal temperature.

13. (i) What is the type of reaction taking place between ethane and chlorine to form monochlorethane?
(ii) The reaction between ethene and chlorine forms only one product. Name the type of this reaction.
(iii) (1) Draw the structural formula of ethene.
(2) What is the feature of the ethene structure, which allows ethene to react with chlorine in the way it does?

Ans. (i) Substitution reaction.
(ii) Addition reaction.
(iii) (1)

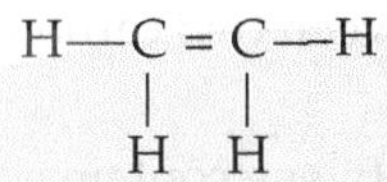

(2) Unsaturated hydrocarbon with double bond.

14. Compound A is bubbled through bromine dissolved in carbon tetrachloride and the product is

$$A \xrightarrow{Br_2/CCl_4} CH_2Br-CH_2Br$$

(i) Draw the structural formula of A.
(ii) What type of reaction has A undergone?
(iii) What is your observation?
(iv) Name (not formula) the compound formed when steam reacts with A in the presence of phosphoric acid.
(v) What is the procedure for converting the product of (b) (iv) back to A?

Ans. (i)

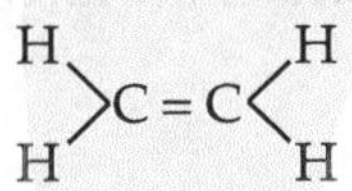

(ii) Addition reaction.
(iii) Bromine solution gets decolourised.
(iv) Ethanol
(v) By heating it (ethanol) with concentrated sulphuric acid at 170°C.

15. (i) Give the names and structural formulae of :
(a) An alkane with a carbon to carbon single bond.
(b) An unsaturated hydrocarbon with a double bond and triple bond.
(ii) (a) Write the equation, for the laboratory preparation of ethyne (acetylene) from calcium carbide.
(b) What is the special feature of the structure of ethyne?
(iiii) Name the addition product formed between ethene and water.

Ans. (i) (a) Saturated hydrocarbon – methane and ethane.

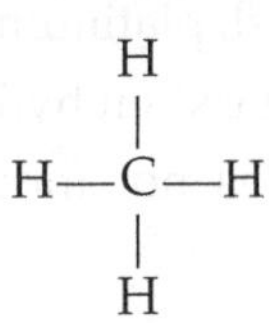

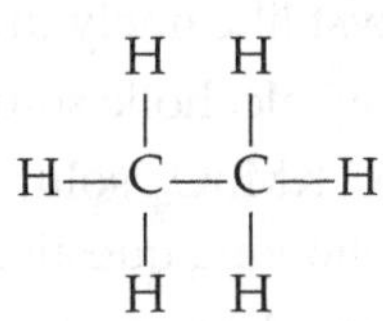

(b) Unsaturated hydrocarbon – ethene, ethyne.

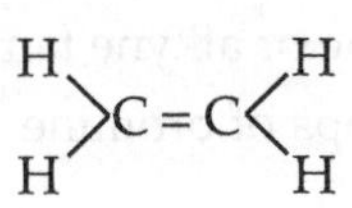

$$H-C\equiv C-H$$

(ii) (a) $$CaC_2 + 2H_2O \longrightarrow Ca(OH)_2 + \underset{\text{Acetylene}}{C_2H_2} \uparrow$$

(b) Ethyne is highly reactive because of presence of one triple bond between two carbon atoms.

(iii) Ethyl alcohol.

16. (i) What word is used to describe these three compounds taken together? CH_4, C_2H_4, C_2H_5OH

(ii) What is the special feature of the structure of :

(a) C_2H_4 (b) C_2H_2

(iii) What type of reaction is common in both of these compounds?

(iv) How is acetylene filled in commercial gas cylinders?

Ans. (i) Organic compounds .

(ii) (a) C_2H_4 contains a double bond between two carbon atoms.

(b) C_2H_2 contains a triple bond between two carbon atoms.

(iii) Addition reaction.

(iv) The commercial gas cylinders of acetylene contain a solution of acetylene in acetone. The cylinder contains a porous material into which the acetone and acetylene are absorbed. The pressure in a freshly filled cylinder of acetylene is about 15 atmosphere.

17. Acetylene can be converted to benzene by suitable temperature and catalyst.

(i) State the temperature and catalyst.

(ii) What type of reaction has taken place?

(iii) Write down the equation involved.

Ans. (i) 600° C in presence of copper tube.

(ii) Polymerisation reaction.

(iii) $$3C_2H_2 \xrightarrow[\text{Cu}]{\text{At } 600°\text{C}} C_6H_6 \text{ (Benzene)}$$

18. (i) (a) A compound has triple bond in its molecule and has only two carbon atoms with two hydrogen atoms. Name the compound.

(b) What is hydrogenation?

(c) What is halogenation?

(d) What 'substitution reaction' and 'substitution product'.

(e) What is 'pyrolysis'? What is the other term signifying the same?

(ii) State the conditions required for the following reactions to take place :

(a) Catalytic hydrogenation of ethyne.

(b) Preparation of ethyne from ethylene dibromide.

Ans. (i) (a) Acetylene $H-C\equiv C-H$.

(b) Addition of hydrogen to some unsaturated hydrocarbons is called hydrogenation.

(c) Addition of halogens (Cl, Br, I) to some unsaturated hydrocarbons is called halogenation.

(d) A substitution reaction is one in which one atom in a molecule is replaced by another atom (or a group of atoms). The product of a substitution reaction is known as a substitution product.

(e) Decomposition of alkanes by heat is called pyrolysis. Another term signifying the same is cracking.

(ii) (a) In presence of catalyst like finely divided nickel, platinum, heating upto 473 K.

(b) Hot and concentrated alcoholic solution of potassium hydroxide.

19. A hydrocarbon decolourises $KMnO_4$ solution but does not form any precipitate with ammonical $AgNO_3$. Now answer the following questions :

(i) Is the hydrocarbon saturated or unsaturated?

(ii) What is the type of bonds between two carbon atoms?

(iii) Does the hydrocarbon belong to alkane, alkene or alkyne family?

(iv) What will be the change on adding a few drops of bromine solution in a test tube filled with this hydrocarbon?

Ans. (i) Unsaturated.

(ii) Double bond between two carbon atoms.

(iii) Alkene family.

(iv) Ethene decolourises the solution of bromine in carbon tetrachloride and dibromo-ethane is formed

$$\underset{\text{Brownish red}}{C_2H_4} + Br_2 \longrightarrow \underset{\substack{\text{Dibromo ethane}\\ \text{(Colourless)}}}{CH_2Br.\ CH_2Br}$$

20. A compound **X** (having vinegar like smell) when treated with ethanol in the presence of the acid **Z,** gives a compound **Y** which has a fruity smell. The reaction is :*

$$C_2H_5OH + X \xrightarrow{Z} Y + H_2O$$

(i) Identify **Y** and **Z.**

(ii) Write the structural formula of **X.**

(iii) Name the above reaction.

Ans. (i) Compound X is acetic acid (CH_3COOH) as it has vinegar like smell. Compound Y is a ester *i.e.,* $CH_3COOC_2H_5$ ethyl ethanoate.

Z is a protic acid for example HCl (aq) or conc. H_2SO_4.

(ii) The structural formula of X is CH_3COOH acetic acid.

(iii) The above reaction is known as 'Esterification' reaction.

$$\underset{\text{Ethanol}}{C_2H_5OH} + \underset{\text{Acetic acid}}{CH_3COOH} \xrightarrow[\text{Esterification}]{H_3O^+} \underset{\text{Ethyl acetate}}{CH_3COOC_2H_5}$$

21. (i) Write the equation, for the preparation of ethylene from ethyl alcohol.

(ii) Write the general formula of a saturated hydrocarbon and give one example of a saturated hydrocarbon with its structural formula.

(iii) Name a compound, which will give acetylene gas, when treated with water.

Ans. (i) $C_2H_5OH(l) \xrightarrow[160\text{-}170°C]{\text{conc. } H_2SO_4} C_2H_4(g) + H_2O(l)$

(ii) General formula C_nH_{2n+2}.

Example : Methane – CH_4

Structural formula
$$H-\overset{\displaystyle H}{\underset{\displaystyle H}{\overset{|}{\underset{|}{C}}}}-H$$

(iii) Calcium carbide.

22. (i) A compound 'X' reacts with compound 'Y' in presence of lime to form a 'Z'. It is insoluble in water.

(a) Name the compound X and Y.

(b) Name the gas Z.

(c) Write only balanced chemical equation.

(ii) (a) Name the product of the reaction between ethene and chlorine.

(b) Burning acetylene in oxygen, under appropriate conditions, produces a very hot flame. For what purpose, this hot flame is used?

Ans. (i) (a) X – Sodium ethanoate.

Y – Sodium hydroxide.

(b) Z – Methane (CH_4).

(c) $CH_3COONa + NaOH \xrightarrow{CaO} CH_4 + Na_2CO_3$.

(ii) (a) Ethene dichloride.

(b) For welding.

* Frequently asked previous years Board Exam Questions.

23. Name the functional group of each of CH_3OH, CH_3COOH, CH_3CHO.

Ans. Alcoholic – OH group present in CH_3OH.

Carboxylic – COOH group present in CH_3COOH.

Aldehydic – CHO group present in CH_3CHO.

24. Illustrate the following :

(i) Thermal cracking

(ii) Catalytic cracking

(iii) Substitution reactions.

Ans. (i) **Thermal cracking :** Breaking of higher hydrocarbons into smaller hydrocarbons on heating in absence of air, is called thermal cracking.

$$\underset{}{C_{12}H_{26}} \xrightarrow{273K} \underset{\text{Heptane}}{C_7H_{16}} + \underset{\text{Pentene}}{C_5H_{10}}$$

(ii) **Catalytic cracking :** The breaking of higher hydrocarbons into smaller hydrocarbons, by heating under pressure, in the presence of suitable catalysts, is called catalytic cracking :

$$CH_3CH_2CH_2CH_3 \xrightarrow[450\,K]{Cr_2O_3 + SiO_2} \underset{\text{Propylene}}{CH_3CH = CH_2} + \underset{\text{Methane}}{CH_4}$$

$$CH_3CH_2CH_2CH_3 \xrightarrow[450\,K]{Cr_2O_3 + SiO_2} \underset{\text{Ethylene}}{CH_2 = CH_2} + \underset{\text{Ethane}}{C_2H_6}$$

(iii) In the presence of diffused light or ultraviolet rays, an atom of hydrogen from the hydrocarbon is substituted by an atom of halogen (Cl, Br, I). These are called substitution reactions.

e.g. : When ethane reacts with chlorine, all the hydrogen atoms are replaced by chlorine atom and substitution products are obtained.

$$e.g.: \underset{\text{Ethane}}{C_2H_6} \xrightarrow[-HCl]{Cl_2} \underset{\text{Ethyl chloride}}{C_2H_5Cl} \xrightarrow[-HCl]{+HCl} \underset{\text{Dichloro ethane}}{C_2H_4Cl} \xrightarrow[-HCl]{+Cl_2} \underset{\text{Trichloro ethane}}{C_2H_3Cl_3}$$

$$\xrightarrow[-HCl]{+Cl_2} \underset{\text{Tetra chloroethane}}{C_2H_2Cl_2} \xrightarrow[-HCl]{+Cl_2} \underset{\text{Penta chloroethane}}{C_2HCl_5} \xrightarrow[-HCl]{+Cl_2} \underset{\text{Hexachloroethane}}{C_2Cl_6}$$

25. (i) Which compound should be heated with sodalime to obtain ethane gas in the laboratory?

(ii) Write the equation for the reaction in (i) above.

(iii) Write a balanced equation for the complete combustion of ethane.

(iv) Name a solid which can be used instead of concentrated sulphuric acid to prepare ethylene by the dehydration of ethanol.

(v) Ethylene forms an addition product with chlorine. Name this addition product and write its structural formula.

Ans. (i) Sodium propionate.

(ii) $$\underset{\text{Sodium propionate}}{C_2H_5COONa} + \underset{\text{Sodium hydroxide}}{NaOH} \xrightarrow[CaO]{\Delta} \underset{\text{Ethane}}{C_2H_6} + Na_2CO_3$$

(iii) $$\underset{\text{Ethane}}{2C_2H_6} + 7O_2 \longrightarrow 4CO_2 + 6H_2O$$

(iv) Alumina (Al_2O_3).

$$C_2H_5OH \xrightarrow[\Delta]{Al_2O_3} C_2H_5 + H_2O$$

(v) $$\underset{\text{Ethylene}}{\begin{matrix} CH_2 \\ \| \\ CH_2 \end{matrix}} + Cl_2 \longrightarrow \underset{\text{Ethene dichloride}}{\begin{matrix} CH_2Cl \\ | \\ CH_2Cl \end{matrix}}$$

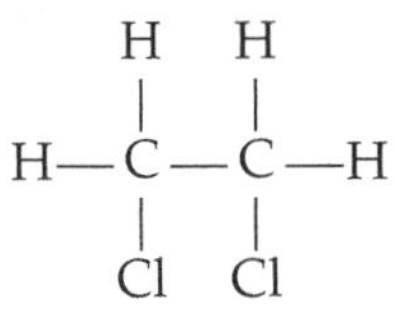

1, 2-dichloroethane

Chapter 10. Practical Chemistry

1. Name the three ions which can not be identified by $K_4[Fe(CN)_6]$ solution.

Ans. Fe^{3+}, Z^{2+}, Cu^{2+}

2. State the colour of the following :
   (i) Lead monoxide (ii) Hot zinc oxide
   (iii) Copper carbonate (iv) Copper nitrate crystals

Ans. (i) Lime yellow (ii) Yellow
(iii) Emerald green (iv) Deep blue

3. Mention the colour changes observed when the following indicators are added to :
   (i) Alkaline phenolphthalein solution (ii) Methyl orange solution
   (iii) Neutral litmus solution

Ans. (i) From pink to colourless (ii) From orange to pink (red)
(iii) From colourless to red

4. Identify the anion present in the following compounds :
   (i) Compound X on heating with copper turnings and concentrated sulphuric acid liberates a reddish brown gas.
   (ii) When a solution of compound Y is treated with silver nitrate solution a white precipitate is obtained which is soluble in excess of ammonium hydroxide solution.
   (iii) Compound Z which on reacting with dilute sulphuric acid liberates a gas which turns lime water milky, but the gas has no effect on acidified potassium dichromate solution.
   (iv) Compound L on reacting with barium chloride solution gives a white precipitate insoluble in dilute hydrochloric acid or dilute nitric acid.

Ans. (i) Nitrate ion, NO_3^- (ii) Chloride ion, Cl^-
(iii) Carbonate ion, CO_3^{2-} (iv) Sulphate ion, SO_4^{2-}

5. Identify the substances P, Q, R, S and T in each case based on the information given below :
   (i) The deliquescent salt P, turns yellow on dissolving in water, and gives a reddish brown precipitate with sodium hydroxide solution.
   (ii) The white crystalline solid Q is soluble in water. It liberates a pungent smelling gas when heated with sodium hydroxide solution.
   (iii) The pale green solid R turns reddish brown on heating. Its aqueous solution gives a white precipitate with barium chloride solution. The precipitate is insoluble in mineral acids.
   (iv) The reddish brown liquid S is dissolved in water. When ethyne gas is passed through it, turns colourless.

Ans. (i) Hydrogen chloride gas (ii) Nitric oxide
(iii) Oxygen (iv) Ammonia

6. Select from the list given (A to E) one substance in each case which matches the description given in parts (i) to (v). (Note : Each substance is used only once in the answer.)
   (A) Nitroso Iron(II) sulphate, (B) Iron(III) chloride, (C) Chromium sulphate, (D) Lead(II) chloride, (E) Sodium chloride.
   (i) A compound which is deliquescent.
   (ii) A compound which is insoluble in cold water, but soluble in hot water.
   (iii) The compound responsible for the brown ring during the brown ring test of nitrate ion.
   (iv) A compound whose aqueous solution is neutral in nature.

(v) The compound which is responsible for the green colouration when sulphur dioxide ispassed through acidified potassium dichromate solution.

Ans. (i) B, Iron(III) chloride (ii) D, Lead(II) chloride
(iii) A, Nitroso Iron(II) sulphate (iv) E, Sodium chloride
(v) C, Chromium sulphate

7. The questions (i) to (v) refer to the following salt solutions listed A to F :
(A) Copper nitrate (B) Iron(II) sulphate.
(C) Iron(III) chloride (D) Lead nitrate
(E) Magnesium sulphate (F) Zinc chloride.
(i) Which two solutions will give a white precipitate when treated with dilute hydrochloric acid followed by barium chloride solution?
(ii) Which two solutions will give a white precipitate when treated with dilute nitric acid followed by silver nitrate solution?
(iii) Which solution will give a white precipitate when either dilute hydrochloric acid or dilute sulphuric acid is added to it?
(iv) Which solution becomes a deep or inky blue colour when excess of ammonium hydroxide is added to it?
(v) Which solution gives a white precipitate with excess ammonium hydroxide solution?

Ans. (i) B and E (Iron(II) sulphate and magnesium sulphate)
(ii) C and F (Iron(III) chloride and zinc chloride)
(iii) D (lead nitrate)
(iv) A (copper nitrate)
(v) F (zinc chloride)

8. A white crystalline solid A on dissolving in water forms a neutral solution. When solid A is heated with calcium hydroxide, it gives off colourless gas B having a sharp biting smell. The gas fumes very strongly in air and turns red litmus blue. The aqueous solution of A on treating with lead acetate solution forms a white precipitate C. The precipitate is insoluble in all acids.
(i) Name the cation present in solid A.
(ii) Name the anion present in solid A.
(iii) Name the solid A and write its chemical formula.
(iv) Name the colourless gas B.
(v) Write one more chemical test for the identification of gas B.
(vi) Name the white precipitate C.
(vii) Write fully balanced chemical equations for :
(a) Solid A and calcium hydroxide.
(b) Solution of solid A and lead acetate solution.

Ans. (i) Cation in A is ammonium (NH_4^+)
(ii) Anion in A is sulphate (SO_4^{2-})
(iii) Solid A is ammonium sulphate $[(NH_4)_2SO_4]_2$
(iv) The colourless gas B is ammonia gas.
(v) Bring a rod dipped in HCl sol. near the gas. The dense white fumes of ammonium chloride are formed.
(vi) White precipitate C is lead sulphate.
(vii) (a) $(NH_4)_2SO_4 + Ca(OH)_2 \longrightarrow CaSO_4 + 2NH_3 + 2H_2O$
(b) $(NH_4)_2SO_4 + (CH_3COO)_2Pb \longrightarrow PbSO_4 + 2CH_3COONH_4$

9. Identify the following solids.
(i) A silvery white solid which floats on the surface of water and reacts violently to give tiny bubbles of a colourless gas. The gas burns in air with a pop sound. The chloride of solid gives a non-persistent lilac colour to non-luminous bunsen flame.
(ii) A pale yellow compound gives a persistent golden yellow colour in non-luminous bunsen burner flame. When treated with water, it liberates a colourless gas, which burns with a pop sound.

Short Answers

(iii) A white solid on treating with water gives off a colourless gas which has a sharp biting smell and turns red litmus blue. The white solid formed after the reaction is soluble in conc. sodium hydroxide solution.

Ans. (i) The solid is potassium metal (ii) The solid is sodium hydride

(iii) The solid is aluminium nitride

10. Identify the cations and anions in each case and write relevant equations wherever necessary.

(i) A white crystalline solid on heating swells and gives off colourless vapours, which condense to form a colourless liquid X. The X turns anhydrous copper sulphate solution blue. The flame test shows the flame as persistent golden yellow. When the solution of solid is treated with HCl, it gives off a colourless gas Y which turns lime water milky.

(ii) A blue solid gives a non-persistent green flame during the flame test. When the solid is heated strongly, it leaves behind a white residue. The residue on treating with few drops of water changes to blue colour.

(iii) A white solid gives a non-persistent brick red flame. On treating with dilute sulphuric acid, it gives off a colourless gas which turns acidified potassium dichromate solution green.

(iv) A white solid gives a non-persistent lilack colour. On treating with dilute sulphuric acid, it gives off a highly offensive smell. The gas turns lead acetate paper black.

Ans. (i) Cation ⇒ Sodium (Na^+);

Anion ⇒ Carbonate (CO_3^{2-})

X is water of crystallisation.

$$Na_2CO_3.10H_2O \xrightleftharpoons{\text{Heat}} Na_2CO_3 + 10H_2O \text{ (Steam)}$$

Y is carbon dioxide gas.

(ii) Cation ⇒ Copper (Cu^{2+});

Anion ⇒ Sulphate (SO_4^{2-}) Salt is $CuSO_4{\cdot}5H_2O$

$$CuSO_4 \cdot 5H_2O \xrightarrow{\text{Heat}} CuSO_4 + 5H_2O$$

$$CuSO_4 + 5H_2O \longrightarrow CuSO_4{\cdot}5H_2O$$

(iii) Cation ⇒ Calcium (Ca^{2+});

Anion ⇒ Sulphite (SO^{2-}) Salt is $CaSO_3$.

$$CaSO_3 + H_2SO_4 \text{ (dil.)} \longrightarrow CaSO_4 + H_2O + SO_2$$

(iv) Cation ⇒ Potassium (K^+);

Anion ⇒ Sulphide (S^{2-})

Salt is potassium sulphide.

$$K_2S + H_2SO_4 \text{ (dil.)} \longrightarrow K_2SO_4 + H_2S$$

11. Identify the salts **P**, **Q**, **R** from the following observations : *

(i) Salt **P** has light bluish green colour. On heating, it produces a black coloured residue. Salt **P** produces brisk effervescence with dil. HCl and the gas evolved turns lime water milky, but no action with acidified potassium dichromate solution.

(ii) Salt **Q** is white in colour. On strong heating, it produces buff yellow residue and liberates reddish brown gas. Solution of salt **Q** produces chalky white insoluble precipitate with excess of ammonium hydroxide.

(iii) Salt **R** is black in colour. On reacting with concentrated HCl, it liberates a pungent greenish yellow gas which turns moist starch iodide paper blue black.

Ans. (i) P is Copper carbonate.

$CuCO_3(s) \rightarrow CuO(s) + CO_2(g)$

Copper II carbonate → Copper (II) oxide

Bluish green Black + Carbon dioxide(g)

$CuCO_3 + HCl \rightarrow CuCl_2 + CO_2 + H_2O$

(P) Effervescence

$CuCO_3 + K_2Cr_2O_7 \rightarrow$ No reaction

* **Frequently asked previous years Board Exam Questions.**

(ii) The salt Q is Lead nitrate

$$2Pb(NO_3)_2(s) \xrightarrow{heat} 2PbO(s) + 4NO_{2(g)} + O_2(g)$$

Lead nitrate Colourless (Q) — Lead monoxide Yellow residue — Nitrogen dioxide Reddish brown — Oxygen

When reacted with ammonium hydroxide solution it gives a chalky white precipitate of lead hydroxide :

$Pb(NO_3)_2 + 2NH_4OH \rightarrow 2NH_4NO_3 + Pb(OH)_2\downarrow$

(Q) — Chalky white ppt.

(iii) The salt R is MnO_2, which is black in colour and reacts with HCl to give Cl_2 gas

$$MnO_2(s) + 4HCl(aq) \rightarrow MnCl_2(aq) + 2H_2O(l) + Cl_2(g)$$

Chlorine gas is pungent and greenish yellow in colour. The chlorine gas oxidises some of the iodide ions in the starch iodide paper to create iodine diatomic molecules. These molecules react with the iodide ions and the starch to form a charge-transfer complex of blue colour.

12. Write balanced chemical equations, for the preparation of the given salts :*
(i) to (iii) by using the methods A to C respectively :
A : Neutralization B : Precipitation C : Titration
(i) Copper sulphate
(ii) Zinc carbonate
(iii) Ammonium sulphate

Ans. (i) Copper sulphate by neutralization

$Cu(OH)_2 + H_2SO_4 \rightarrow CuSO_4\downarrow + 2H_2O$

Copper hydroxide — Sulphuric acid — Copper sulphate

(ii) Zinc carbonate by precipitation

$Zn(NO_3)_2\ (aq) + 2NaCO_3 \rightarrow Zn(CO_3)_2 + 2NaNO_3\ (aq)$

Zinc nitrate — (aq) Sodium carbonate — Zinc carbonate

(iii) Ammonium sulphate can be prepared by titration of NH_3 solution with dil. H_2SO_4 solution.

$2NH_4OH\ (aq) + H_2SO_4\ (aq) \rightarrow (NH_4)_2SO_4 + 2H_2O\ (l)$

White powder

13. Choose the method of preparation of the following salts, from the methods given in the list.*
[List : A. Neutralization B. Precipitation
C. Direct combination D. Substitution]
(i) Lead chloride
(ii) Iron (II) sulphate
(iii) Sodium nitrate
(iv) Iron (III) chloride

Ans. (i) (B) Lead chloride can be prepared by precipitation.

$$Pb(NO_3)_2(aq) + 2NaCl(aq) \rightarrow PbCl_2(s) + 2NaNO_3(aq)$$

(ii) (D) Iron (II) Sulphate can be prepared by substitution.

(iii) (A) Sodium nitrate can be prepared by Neutralization.

$$2HNO_3 + Na_2CO_3 \rightarrow 2\ NaNO_3 + H_2O + CO_2$$

(iv) (C) Iron (III) chloride can be prepared by direct combination.

$$2Fe + 3Cl_2 \rightarrow 2FeCl_3$$

* **Frequently asked previous years Board Exam Questions.**

Numericals

Set 14

Chapter 5. Mole Concept and Stoichiometry

1. Calculate the volume occupied by 3.4 g of ammonia at S.T.P.

Sol. Gram molecules of NH_3 = (1 × 14) + (3 × 1)

= 14 + 3 = 17 g

Mass of one mole of ammonia = 17 g

Molar volume = 22.4 litre

Thus, 17 grams of NH_3 occupied 22.4 litre

$$3.4 \text{ gram} = \frac{3.4 \times 22.4}{17} = 4.48 \text{ litre}$$

Volume occupied by 3.4 g of NH_3 at S.T.P. is 4.48 litre.

2. A cylinder contains 68 g of ammonia gas at S.T.P.

(i) What is the volume occupied by this gas ?

(ii) How many moles of ammonia are present in the cylinder ?

(iii) How many molecules of ammonia are present in the cylinder ?

[N–14, H–1]

Sol. Molecular weight of NH_3 = (14 + 3 × 1) = 17

Number of moles of NH_3 in 68 g = $\frac{\text{Wt.}}{\text{Mol. wt.}} = \frac{68}{17}$ = 4 moles

At S.T.P. one mole of NH_3 occupies 22.4 lit.

(i) ∴ Volume occupied by 4 moles of the gas

= 4 × 22.4 lit. = 89.6 lit. at S.T.P.

(ii) 4 moles of ammonia are present in the cylinder.

(iii) Number of molecules = No. of moles × N_A

= 4 × 6.023 × 10^{23}

= 24.092 × 10^{23}

3. What weight of zinc is needed to produce 100 ml of dry hydrogen at S.T.P. from dilute sulphuric acid solution ?

Sol. The chemical equation representing the reaction is

$$\underset{(65\ g)}{Zn} + H_2SO_4 \longrightarrow ZnSO_4 + \underset{(22400\ ml)}{H_2}$$

∵ 22400 ml of hydrogen is liberated at S.T.P. from 65 g of zinc

∴ 100 ml of hydrogen is liberated at S.T.P. from

$$= \frac{65}{22400} \times 100 = 0.29 \text{ g of zinc.}$$

4. What weight and volume of oxygen at S.T.P. will be given, when 18 g of water is electrolysed ?

Sol. The decomposition of water can be represented as :

$$\underset{(36\ g)}{2H_2O} \longrightarrow \underset{(4\ g)}{2H_2} + \underset{(32\ g)}{O_2}$$

∵ 36 g of water yields = 32 g of oxygen.

∴ 18 g of water will yield = $\frac{32}{36}$ × 18 = 16 g oxygen.

Again, 32 g oxygen occupies 22.4 litre volume at S.T.P.

∴ 16 g oxygen will occupy = $\frac{22.4}{32}$ × 16 = 11.2 litre.

5. Calculate the volume of propane burnt for every 200 cm^3 of oxygen used in the reaction.

$$C_3H_8 + 5O_2 \longrightarrow 3CO_2 + 4H_2O$$

Sol. From the above reaction it is clear that for every 5 volumes of oxygen, 1 volume of propane is burnt.

Hence, volume of propane burnt for every 200 cm^3 of oxygen $= \frac{1}{5} \times 200$

$$= 40\ cm^3.$$

6. Calculate the volume of methane gas that must be burnt completely to produce 100 lit of CO_2 at S.T.P.

Sol. Let x be the volume of methane to be burnt to produce 100 lit CO_2

$$\underset{\substack{1\text{ mol}\\22.4\text{ lit}}}{\overset{x}{CH_4(g)}} + 2O_2(g) \longrightarrow \underset{\substack{1\text{ mol}\\22.4\text{ lit}}}{\overset{100\text{ lit}}{CO_2(g)}} + 2H_2O(l)$$

$$\text{Ratio proportion} = \frac{x\,CH_4}{22.4\text{ lit }CH_4} = \frac{100\text{ lit }CO_2}{22.4\text{ lit }CO_2}$$

$$\therefore \quad x = \frac{100\text{ lit}}{22.4\text{ lit}} \times 22.41 = 100\text{ lit}$$

Volume of methane to be burnt is 100 lit.

7. Calculate the volume of carbon dioxide formed when 8 g of methane gas burns completely as represented by the equation :

$$CH_4(g) + 2O_2(g) \longrightarrow CO_2(g) + 2H_2O(l)$$

Sol. Let x be the volume of CO_2 produced when 8 g CH_4 is burnt.

$$\underset{\substack{1\text{ mole}\\16\text{ g} = 22.4\text{ litre}}}{CH_4(g)} + 2O_2(g) \longrightarrow \underset{\substack{1\text{ mole}\\44\text{ g} = 22.4\text{ litre}}}{CO_2(g)} + 2H_2O(l)$$

$$\text{Ratio proportion} = \frac{8\text{ g }CH_4}{16\text{ g }CH_4} = \frac{x\text{ lit }CO_2}{22.4\text{ lit }CO_2}$$

$x = \frac{8\text{ g}}{16\text{ g}} \times 22.4$ lit = 11.2 lit is the volume of CO_2 produced at S.T.P.

8. What volume of oxygen would be required for the complete combustion of 100 litre of ethane according to the following equation ?

$$2C_2H_6 + 7O_2 \longrightarrow 4CO_2 + 6H_2O$$

Sol.

$$2C_2H_6 + 7O_2 \longrightarrow 4CO_2 + 6H_2O$$

2 volume of ethane = 7 volume of oxygen

2 volume of ethane = 100 litre

$\therefore$ 1 volume of ethane = 50 litre

7 volume of O_2 = 50 × 7 = 350 litre

$\therefore$ 100 litre of ethane require = 350 litre of oxygen.

9. Ammonia may be oxidised to nitrogen monoxide in the presence of a catalyst according to the following equation.

$$4NH_3 + 5O_2 \longrightarrow 4NO + 6H_2O$$

If 27 litre of reactants are consumed, what volume of nitrogen monoxide is produced at the same temperature and pressure ?

Sol. According to equation,

$$4NH_3 + \underset{5\text{ vol.}}{5O_2} \longrightarrow \underset{4\text{ vol.}}{4NO} + 6H_2O$$

$\because$ 5 volumes of reactants are consumed to give 4 volumes of nitrogen monoxide.

$\therefore$ 27 litre of reactants are consumed to give $= \frac{27 \times 4}{5} = \frac{108}{5}$

$$= 21.6\text{ litre.}$$

10. Hydrogen sulphide gas burns in oxygen to yield 12.8 g of sulphur dioxide according to the equation :

$$2H_2S + 3O_2 \longrightarrow 2H_2O + 2SO_2$$

Calculate the volume of hydrogen sulphide at S.T.P. Also calculate the volume of oxygen required at S.T.P., which will complete the combustion of hydrogen sulphide. [S = 32; O = 16; H = 1]

Sol. According to equation,

$$\underset{\substack{2 \times 22.4 \\ = 44.8\ cm^3 \text{ at S.T.P.}}}{2H_2S} + \underset{\substack{3 \times 22.4 \\ = 67.2\ dm^3 \text{ at S.T.P.}}}{3O_2} \longrightarrow 2H_2O + \underset{\substack{2[32+32] \\ = 128\ g}}{2SO_2}$$

128 g of SO_2 is produced from H_2S = 44.8 dm^3

∴ 12.8 g of SO_2 is produced from H_2S = $\frac{44.8 \times 12.8}{128}$

= 4.48 dm^3

128 g of SO_2 is produced from O_2 = 67.2 dm^3

∴ 2.8 g of SO_2 is produced from O_2 = $\frac{67.2 \times 12.8}{128}$

= 6.72 dm^3

11. Ammonia burns in oxygen and the combustion, in the presence of a catalyst; may be represented as :

$$2NH_3 + 2\frac{1}{2}\ O_2 \longrightarrow 2NO + 3H_2O$$

(i) What mass of steam is produced when 1.5 g of nitrogen monoxide is formed ?

(ii) What volume of oxygen at S.T.P. is required to form 10 moles of products ?

Sol. (i)

$$2NH_3 + 2\frac{1}{2}O_2 \longrightarrow 2NO + 3H_2O$$

Wt. of 2NO = 2[14 + 16]

= 60 g

and Wt. of $3H_2O$ = 3[2 (1) + 16]

= 54 g

When 60 g of NO is formed, mass of steam produced = 54 g

∴ For 1.5 g of NO mass of steam produced = $\frac{54 \times 15}{60}$ = 13.5 g

(ii) For 5 moles of products, oxygen required = 2.5 × 22.4 litre

∴ For 10 moles of products, oxygen required = $\frac{2.5 \times 22.4 \times 10}{5}$

= 112 litre.

12. $3Cu + 8HNO_3 \longrightarrow 3Cu(NO_3)_2 + 4H_2O + 2NO$

(H = 1, N = 14, O = 16, Cu = 64)

Calculate from the equation :

(i) The mass of copper needed to react with 63 g of HNO_3.

(ii) The volume of nitric oxide that can be collected at S.T.P. (the gram molecular volume of a gas at STP = 22.4 litre).

Sol. (i)

$$\underset{\substack{3 \times 64 \\ = 192\ g}}{3Cu} + \underset{\substack{8(1 \times 14 + 48) \\ = 504\ g}}{8HNO_3} \longrightarrow 3Cu(NO_3)_2 + 4H_2O + 2NO$$

∵ 504 g of HNO_3 reacts with 192 g of Cu.

∴ 1 g of HNO_3 reacts with = $\frac{192}{504}$ g of Cu.

63 g of HNO_3 reacts with = $\frac{192}{504} \times 63$ = 24 g copper.

(ii) 192 g of Cu with HNO_3 gives out = 2 × 22.4 litre of NO at S.T.P.

∴ 24 g of Cu with HNO_3 = $\frac{2 \times 22.4}{192} \times 24$

= 5.6 litre of NO.

13. From the equation :

(i) $(NH_4)_2Cr_2O_7 \longrightarrow Cr_2O_3 + 4H_2O + N_2$

Calculate :

(a) The volume of nitrogen at S.T.P. evolved when 63 g of ammonium dichromate is heated.

(b) The mass of chromium(II) oxide (Cr_2O_3) formed at the same time.
(N = 14; H = 1; Cr = 52; O = 16)

(ii) The reaction $4N_2O + CH_4 \longrightarrow CO_2 + 2H_2O + 4N_2$ takes place in the gaseous state. If all volumes are measured at the same temperature and pressure, calculate the volume of dinitrogen oxide (N_2O) required to give 150 cm^3 of steam. (N = 14; O = 16; C = 12; H = 1)

Sol. (i) The reaction is represented as :

(a)
$$\underset{252\text{ g}}{(NH_4)_2Cr_2O_7} \longrightarrow Cr_2O_3 + 4H_2O + \underset{22.4\text{ litre at S.T.P.}}{N_2}$$

$$\therefore \quad \text{Volume of } N_2 \text{ at STP} = \frac{22.4 \times 63}{252}$$
$$= 5.6 \text{ litre at S.T.P.}$$

(b) 252 g $(NH_4)_2Cr_2O_7$ gives 152 g of Cr_2O_3

$$\therefore \quad 63 \text{ g } (NH_4)_2Cr_2O_7 \text{ gives } = \frac{152 \times 63}{254} = 37.70 \text{ g}$$

(ii)
$$\underset{\substack{4\text{ vol.}\\300\text{ cm}^3}}{4N_2O} + CH_4 \longrightarrow CO_2 + \underset{\substack{2\text{ vol.}\\150\text{ cm}^3}}{2H_2O} + 4N_2$$

300 cm_3 of N_2O will be required to yield 150 cm^3 of steam.

14. (i) Rewrite the following equation in terms of moles of reactants and products :

$$2NH_4Cl + Ca(OH)_2 \longrightarrow CaCl_2 + 2H_2O + 2NH_3$$

(ii) Use the given equation to calculate the weight of calcium hydroxide needed to decompose 2.14 g NH_4Cl.

(iii) Find out the volume of ammonia at S.T.P. obtained from 4.28 g of NH_4Cl.

Sol. (i) In term of moles, the equation is as follows :

$2NH_4Cl$	+	$Ca(OH)_2$	$\longrightarrow$	$CaCl_2$	+	$2H_2O$	+	$2NH_3$
2 moles		1 moles		1 mole		2 moles		2 moles
2 (14 + 4 + 35.5)		40 + 32 + 2		40 + 71		2 (2 + 16)		2 (14 + 3)
= 107 g		= 74 g		= 111 g		= 36 g		= 34 g

So, 2 moles of ammonium chloride combines with 1 mole of calcium hydroxide to form 1 mole of calcium chloride, 2 moles of water and 2 moles of ammonia gas.

(ii) 107 g of NH_4Cl needed to react with 74 g of $Ca(OH)_2$

$$\therefore \quad 2.14 \text{ g of } NH_4Cl \text{ needed} = \frac{74 \times 2.14}{107} \text{ g of } Ca(OH)_2$$
$$= 1.48 \text{ gm.}$$

(iii) 107 g of NH_4Cl will liberate with slaked lime

$$= 2 \times 22.4 \text{ litre of } NH_3.$$

$\therefore$ 4.28 g of NH_4Cl liberates with slaked lime

$$= \frac{2 \times 22.4 \times 4.28}{107}$$
$$= 1.792 \text{ litre.}$$

15. Solve the following :

(i) What volume of oxygen is required to burn completely 90 dm^3 of butane under similar conditions of temperature and pressure ?

$$2C_4H_{10} + 13O_2 \longrightarrow 8CO_2 + 10H_2O$$

(ii) The vapour density of a gas is 8. What would be the volume occupied by 24.0 g of the gas at STP ?

(iii) A vessel contains X number of molecules of hydrogen gas at a certain temperature and pressure. How many molecules of nitrogen gas would be present in the same vessel under the same conditions of temperature and pressure ?

Sol. (i)

$$2C_4H_{10} + 13O_2 \longrightarrow 8CO_2 + 10H_2O$$
$$\text{2 vol.} \quad \text{13 vol.}$$

∵ 2 vol. of butane require 13 vol. of oxygen (according to Gay-Lussac's law)

∴ $90\ dm^3$ of butane require $= \frac{13 \times 90}{2} = 585\ dm^3$

Ans. $585\ dm^3$ of oxygen is required.

(ii) Given : Vapour density (VD) = 8

∴ Mol. wt = 2 × VD = 2 × 8 = 16

No. of moles in 24.0 g of gas $= \frac{\text{Wt.}}{\text{Mol. wt.}} = \frac{24}{16}$

= 1.5 moles

At S.T.P. 1 mole of a gas occupies 22.4 l.

∴ 1.5 moles (or 24.0 g) of the gas will occupy $= \frac{22.4 \times 1.5}{1} = 33.6$ litres

(iii) 'X' number of molecules.

16. O_2 is evolved by heating $KClO_3$ using MnO_2 as a catalyst

$$2KClO_3 \xrightarrow{MnO_2} 2KCl + 3O_2$$

(i) Calculate the mass of $KClO_3$ required to produce 6.72 litre of O_2 at S.T.P. [atomic masses of K = 39, Cl = 35.5, O = 16]

(ii) Calculate the number of moles of oxygen present in the above volume and also the number of molecules.

(iii) Calculate the volume occupied by 0.01 mole of O_2 at S.T.P.

Sol. (i) Molecular wt. of $KClO_3$ = 39 + 35.5 + 16 × 3

= 122.5

$$2\ KClO_3 \longrightarrow 2\ KCl + 3O_2$$
$$\text{2 moles} \qquad \text{3 moles}$$
$$2 \times 122.5\ g \qquad 3 \times 22.4\ \text{lit at S.T.P.}$$

∵ 3 × 22.4 lit of oxygen is produced from 2 × 122.5 g of $KClO_3$

∴ 6.72 lit of oxygen is produced from $\frac{2 \times 122.5 \times 6.72}{3 \times 22.4} = 24.5$ g

Ans. 24.5 g of $KClO_3$ is required to produce 6.72 lit of O_2 at S.T.P.

(ii) At S.T.P. 22.4 lit of a gas = 1 mole

∴ 6.72 lit O_2 gas $= \frac{1 \times 6.72}{22.4} = 0.3$ moles

One mole contains = 6×10^{23} molecules

∴ 0.3 mole contains = $6 \times 10^{23} \times 0.3$

= 1.8×10^{23} molecules.

Ans. 6.72 l of oxygen contains 0.3 moles and hence 1.8×10^{23} molecules.

(iii) At STP one mole of O_2 occupies 22.4 lit

∴ 0.01 mole of O_2 occupies = 22.4 × 0.01

= 0.224 lit

Ans. 0.01 mole of O_2 will occupy 0.224 l at S.T.P.

17. 4.5 moles of calcium carbonate are reacted with dilute hydrochloric acid.

(i) Write the equation for the reaction.

(ii) What is the mass of 4.5 moles of calcium carbonate ? (Relative molecular mass of calcium carbonate is 100).

(iii) What is the volume of carbon dioxide liberated at S.T.P. ?

(iv) What mass of calcium chloride is formed ? (Relative molecular mass of calcium chloride is 111).

(v) How many moles of HCl are used in this reaction ?

Sol. (i)

$$CaCO_3 + 2HCl \longrightarrow \underset{\text{Calcium chloride}}{CaCl_2} + H_2O + CO_2$$

(ii) 4.5 moles of $CaCO_3 = 4.5 \times 100 = 450$ g

(iii) $\because$ 1 mole $CaCO_3$ liberates 1 mole of CO_2

$\therefore$ 4.5 moles of $CaCO_3$ liberates 4.5 moles of CO_2

$\therefore$ Volume of CO_2 liberated at STP = 22.4 × 4.5 = 100.8 lit

(iv) $\because$ 1 mole of $CaCO_3$ gives 1 mole of $CaCl_2$.

$\therefore$ 4.5 moles of $CaCO_3$ gives 4.5 moles of $CaCl_2$ = 4.5 × 111 = 499.5 g

(v) $\because$ 1 mole of $CaCO_3$ requires 2 moles of HCl

$\therefore$ 4.5 moles of $CaCO_3$ requires = 2 × 4.5 = 9 moles of HCl

18. (i) Oxygen oxidises ethyne to carbon dioxide and water as shown by the equation :

$$2C_2H_2 + 5O_2 \longrightarrow 4CO_2 + 2H_2O$$

What volume of ethyne gas at S.T.P. is required to produce 8.4 dm^3 of carbon dioxide at S.T.P. ? [H = 1, C = 12, O = 16]

(ii) A compound made up of two element X and Y has an empirical formula X_2 Y. If the atomic weight of X is 10 and that of Y is 5 and the compound has a vapour density 25, find its molecular formula.

Sol. (i) Given :

$$\underset{\substack{\text{2 vol.} \\ \text{1 vol.}}}{2C_2H_2} + 5O_2 \longrightarrow \underset{\substack{\text{4 vol} \\ \text{2 vol.}}}{4CO_2} + 2H_2O$$

According to Gay-Lussac's law :

2 volume of CO_2 is produced from 1 vol. of C_2H_2

$\therefore$ 8.4 dm^3 of CO_2 at S.T.P. produced from

$$= \frac{1 \times 8.4}{2}$$

$$= 4.2\ dm^3 \text{ of } C_2H_2$$

Ans. At S.T.P. 4.2 dm^3 of ethyne is required.

(ii)

$$\text{Molecular formula} = (\text{Empirical formula})_n$$

$$n = \frac{\text{Molecular formula weight}}{\text{Empirical formula weight}}$$

$$= \frac{2 \times \text{V.D.}}{(2 \times 10 + 5)} \quad \text{(V.D. = Vapour density)}$$

$$= \frac{2 \times 25}{25} = 2$$

$\therefore$

$$\text{Molecular formula} = (X_2Y)_2$$

$$= X_4Y_2$$

19. (i) Calculate the volume of 320 g of SO_2 at S.T.P. (Atomic mass : S = 32 and O = 16).

(ii) Calculate the volume of oxygen required for the complete combustion of 8.8 g of propane (C_3H_8). (Atomic mass : C = 14, O = 16, H = 1, Molar volume = 22.4 dm^3 at S.T.P.)

Sol. (i) Gram molar mass of SO_2 = 32 + (2 × 16) = 64 g.

$\therefore$ No. of moles in 320 g of $SO_2 = \frac{320}{64}$ = 5 moles.

At S.T.P. 1 mole of SO_2 occupies 22.4 dm^3.

$\therefore$ 5 moles of SO_2 will occupy 5 × 22.4 = 112 dm^3

(ii) Chemical equation for the complete combustion of propane is :

$$\underset{\substack{\text{1 mole} \\ \text{1 mole} \\ (12 \times 3) + (1 \times 8) = 44\text{ g}}}{C_3H_8} + \underset{\substack{\text{5 mole} \\ 5 \times 22.4 \text{ lit at S.T.P.}}}{5O_2} \longrightarrow 3CO_2 + 4H_2O$$

$$44\text{g} = 5 \times 22.4 \text{ lit S.T.P.}$$

$\therefore$ 8.8 g of propane would require $\dfrac{5 \times 22 \times 8.8}{44} = 22$ lit of oxygen at S.T.P.

20. (i) How many grams of sulphuric acid will be required to completely neutralise 16.0 g of caustic soda ?

(ii) What volume of H_2 be evolved at S.T.P. ?

Sol. (i) The complete reaction between H_2SO_4 and NaOH is represented by :

$$2NaOH + H_2SO_4 \longrightarrow Na_2SO_4 + 2H_2$$

$\therefore$ 2 [23 + 16 + 1] g of NaOH react with [2 × 1 + 32 + 4 × 16] g of H_2SO_4

or 80 g of NaOH react with 98 g of H_2SO_4

$\therefore$ 16 g of NaOH react with $\dfrac{98 \times 16}{80} = 19.6$ g of H_2SO_4.

(ii) 80 g of NaOH formed at S.T.P. = 2 × 22.4 litre of H_2

$\therefore$ 16 g of NaOH formed at S.T.P. $= \dfrac{2 \times 22.4 \times 16}{80}$

$= 8.96$ litre of H_2.

21. (i) How many molecules are present in (the numerical value of Avogadro's number can be used as 6×10^{23}) ?

(a) 2.2 grams of carbon dioxide

(b) 16 grams of sulphur dioxide

(c) 2 grams of oxygen

(d) 8.5 grams of sodium chloride

(C = 12, O = 16, Na = 23, S = 32, Cl = 35.5)

(ii) How many moles are present in ?

(a) 100 g of calcium carbonate, (b) 2.3 g of sodium

(c) 80 g of sodium hydroxide.

Sol. (i) (a) 44 g of CO_2 contain 6×10^{23} molecules at S.T.P.

$\therefore$ 2.2 g of $CO_2 = \dfrac{6 \times 10^{23}}{44} \times 2.2$

$= 3 \times 10^{22}$ molecules

(b) 64 g of SO_2 contain 6×10^{23} molecules at S.T.P.

$\therefore$ 16 g of S.T.P. $= \dfrac{6 \times 10^{23}}{64} \times 16$

$= 1.5 \times 10^{22}$ molecules

(c) 32 g of O_2 contain 6×10^{23} molecules at S.T.P.

$\therefore$ 2 g of $O_2 = \dfrac{6 \times 10^{23}}{32} \times 2$

$= 3.75 \times 10^{22}$ molecules

(d) 58.5 g of NaCl contain 6×10^{23} molecules at S.T.P.

$\therefore$ 58.5 g of NaCl $= \dfrac{6 \times 10^{23}}{58.5} \times 8.5$

$= 0.871 \times 10^{23}$ molecules

(ii) (a) Number of moles $= \dfrac{\text{Weight of the substance (W)}}{\text{Molecular weight (M)}}$

Molecular weight of $CaCO_3 = 40 + 12 + 48 = 100$

No. of moles $= \dfrac{100}{100} = 1$ mole

(b) No. of moles of sodium $= \dfrac{2.3}{23}$ = molecular wt. of sodium = 23

$= 0.1$ mole

(c) No. of moles of NaOH $= \frac{80}{40} = 2$

Molecular wt. of NaOH $= 23 + 16 + 1 = 40 = 2$ moles

22. The gases chlorine, nitrogen, ammonia and sulphur dioxide are collected under the same conditions of temperature and pressure. Copy the following table which gives the volumes of gases collected and the number of molecules (X) in 20 litre of nitrogen. You have to complete the table giving the number of molecules in the other gases in terms of X.

Gas	Volume (litres)	Number of molecules
Chlorine	10	
Nitrogen	20	X
Ammonia	20	
Sulphur dioxide	5	

Sol. In accordance with Avogadro's law which states that, under similar conditions of temperature and pressure, equal volumes of all the gases contain the same number of molecules.

Hence, the table can be represented as :

Gas	Volume (litres)	Number of molecules
Chlorine	10	$\frac{10}{20}X = \frac{X}{2}$
Nitrogen	20	$\frac{20}{20}X = X$
Ammonia	20	$\frac{20}{20}X = X$
Sulphur dioxide	5	$\frac{5}{20}X = \frac{X}{4}$

23. A sample of ammonium nitrate when heated yields 8.96 litres of steam (measured at S.T.P.).

$$NH_4NO_3 \longrightarrow N_2O + 2H_2O$$

(i) What volume of dinitrogen oxide is produced at the same time as 8.96 litres of steam ?

(ii) What mass of ammonium nitrate should be heated to produe 8.96 litres of steam ? (Relative molecular mass of ammonium nitrate is 80)

(iii) Determine the percentage of oxygen in ammonium nitrate (O = 16).

$$\underset{\substack{1\text{ mole}\\22.41}}{NH_4NO_3} \longrightarrow \underset{\substack{1\text{ mole}\\22.41}}{N_2O} + \underset{\substack{2\text{ mole}\\2\times 22.41 = 44.81}}{2H_2O}$$

Sol. (i) $44.8\text{ L of }H_2O \equiv 22.4\text{ L of }N_2O\text{ at S.T.P.}$

$$8.96\text{ L of }H_2O \equiv \frac{22.4\text{ L} \times 8.96\text{ L}}{44.8\text{L}}\text{ at S.T.P.}$$

$$= 4.48\text{ L of }N_2O\text{ at S.T.P.}$$

(ii) 44.8 L steam is liberated by 80 g NH_4NO_3

∴ 8.96 L steam will be liberated by $\frac{80 \times 8.96}{44.8} = 16\text{ L}$

(iii) 80 g NH_4NO_3 contains 48 g O_2

Percentage of oxygen $= \frac{48}{80} \times 100 = 60\%$

24. (i) What volume of hydrogen sulphide at S.T.P. will burn in oxygen to yield 12.8 g of sulphur dioxide according to the equation ?

$$2H_2S + 3O_2 \longrightarrow 2H_2O + 2SO_2\ (H = 1, O = 16, S = 32)$$

Numericals

(ii) For the volume of hydrogen sulphide determined in (i) above, what volume of oxygen would be required for complete combustion ?

$$\underset{\substack{2 \times 22.4 \\ \text{litres at S.T.P.}}}{2H_2S} + \underset{\substack{3 \times 22.4 \\ \text{litres at S.T.P.}}}{3O_2} \longrightarrow 2H_2O + \underset{\substack{2(32 + 2 \times 16) \\ = 128 \text{ g}}}{2SO_2}$$

Sol. (i) Volume of H_2S at S.T.P. required to form 128 g of $SO_2 = 2 \times 22.4$ litres

$\therefore$ Volume of H_2S at S.T.P. required to form 1.28 g of $SO_2 = \frac{2 \times 22.4}{128} \times 12.8$ litres

$= 44.8$ litres

(ii) Volume of O_2 required for complete combustion of 2×22.4 litres of H_2S

$= 3 \times 22.4$ litres at S.T.P.

$\therefore$ Volume of O_2 required for combustion of 4.48 litres of H_2S

$= \frac{3 \times 22.4}{2 \times 22.4} \times 4.48 = 6.72$ litres

25. (i) Iron pyrites has formula FeS_2. What mass of sulphur is contained in 30 g iron pyrites ?

(ii) When roasted, iron pyrites gives sulphur dioxide according to the equation ?

$$4FeS_2 + 11O_2 \longrightarrow 2Fe_2O_3 + 8SO_2$$

What volume of SO_2 at S.T.P. would be liberated by roasting 30 g of iron pyrites [S = 32, Fe = 56, Molar volume of gas is 22.4 litre at S.T.P.]

Sol. (i) Mass of sulphur in iron pyrites.

Gram molecular mass of $FeS_2 = 56 + 2 \times 32 = 120$ g

Mass of sulphur $= 2 \times 32 = 64$ g

Sulphur contained in 120 g of $FeS_2 = 64$ g

$\therefore$ Sulphur contained in 30 g of $FeS_2 = \frac{64}{120} \times 30 = 16$ g.

(ii) Volume of SO_2.

$$4FeS_2 + H_2O \longrightarrow 2Fe_2O_3 + 8SO_2$$

Now 4 moles of FeS_2 give 8 moles of SO_2 or 1 mole of FeS_2 gives 2 moles of SO_2.

[Molecular mass of $FeS_2 = 56 + 32 \times 2 = 120$]

Thus, 120 g of FeS_2 gives 2×22.4 litres of SO_2 at S.T.P.

$\therefore$ Volume of SO_2 obtained from 120 g of $FeS_2 = 2 \times 22.4$ litres at S.T.P.

$\therefore$ Volume of SO_2 obtained from 30 g of $FeS_2 = \frac{2 \times 22.4}{120} \times 30 = 11.2$ litres at S.T.P.

26. 40 g of sulphur is taken in an enclosed vessel containing 22.4 litre of oxygen, and ignited. Calculate the volume of sulphur dioxide formed and the mass of uncombined sulphur.
(1 mol. of S = 32 g; molar volume = 22.4 litre)

Sol.

$$\underset{32\text{ g}}{S} + \underset{1\text{ vol.}}{O_2} \longrightarrow \underset{1\text{ vol.}}{SO_2}$$

Oxygen : Sulphur dioxide

1 : 1

22.4 : x

Volume of SO_2 produced $= 22.4$ litres (Equal to the volume of O_2)

Sulphur : Oxygen 32 gm : 1 vol.

$\therefore$ $x = 22.4$ lit

$x = 32$ g

$\therefore$ Mass of uncombined sulphur $= 40 - 32 = 8$ g.

27. Solid ammonium dichromate (Relative molecular mass 252) decomposes according to the following equation :

$$(NH_4)2Cr_2O_7 \longrightarrow N_2 + Cr_2O_3 + 4H_2O$$

(i) What volume of nitrogen at S.T.P., will be evolved when 63 g of ammonium dichromate is decomposed ? (H = 1, N = 14, O = 16, Cr = 52)

Numericals

(ii) If 63 g of ammonium dichromate is heated above 1000°C, what will be the loss of mass ?

$$(NH_4)_2 Cr_2O_7 \longrightarrow N_2 + Cr_2O_3 + 4H_2O$$

$(NH_4)_2 Cr_2O_7$	N_2	Cr_2O_3	$4H_2O$
1 mole	1 mole	1 mole	4 mole
22.41	22.41	22.41	4 × 22.41
252 g	28 g		4 × 18 = 72 g

Sol. (i) 252 g of ammonium dichromate evolves 22.4 litre N_2 at S.T.P.

∴ 1 g of ammonium dichromate evolves $= \frac{22.4}{252}$ litre N_2

∴ 63 g of ammonium dichromate evolves $= \frac{22.4 \times 63}{252}$ litre N

$= 5.6$ litre of N_2.

(ii) Loss of mass due to evolution of nitrogen according to the reaction.

∵ 252 g of ammonium dichromate loses 28 g of N_2 gas on heating

∴ 63 g of ammonium dichromate loses $= \frac{28 \times 63}{252}$

$= 7$ g of N_2 gas

Loss due to water vapours :

∵ 252 g of ammonium dichromate loses 72 g of H_2O

∴ 63 g of ammonium dichromate loses $= \frac{72 \times 63}{252}$

$= 18$ g of H_2O

∴ Total loss of mass = Loss of N_2 + Loss of water vapours

$= 7\text{ g} + 18\text{ g}$

$= 25\text{ g}$

28. Solid ammonium dichromate (relative molecular mass = 252) on heating decomposes as follows :

$$(NH_4)_2 Cr_2 O_7 \xrightarrow{\text{Heat}} N_2 + Cr_2O_7 + 4H_2O$$

(i) Calculate the volume of nitrogen at S.T.P., that will be evolved when 31.5 g of ammonium dichromate is heated and

(ii) The mass of chromium(III) oxide formed at the same time

(H = 1; N = 14; O = 16; Cr = 52)

$(NH_4)_2 Cr_2 O_7$	$\longrightarrow$	N_2	+	$Cr_2 O_7$	+	$4H_2$
2(14+4) + (52 × 2) + 7 × 16 = 252 g		22.4 litres at S.T.P.		(52 × 2) + (7 × 16) = 216 g		

Sol. (i) Volume of N_2 evolved at S.T.P. when 31.5 g of $(NH_4)_2Cr_2O_7$ is decomposed

$= \frac{22.4 \text{ litre}}{252 \text{ g}} \times 31.5 \text{ gm}$

$= 2.80$ litre

(ii) Amount of Cr_2O_3 formed $= \frac{216 \text{ g}}{252 \text{ g}} \times 31.5 \text{ g}$

$= 27.0$ g

29. (i) From the balanced chemical equation

$$CH_4 + 2O_2 \longrightarrow CO_2 + 2H_2O.$$

Compute the mass of oxygen gas that will combine with 8 g of methane.

(Relative atomic mass : H = 1; O = 16; C = 12)

(ii) (a) Iron pyrites has the formula FeS_2. What mass of sulphur is contained in 30 g of pyrites ?

(b) When roasted, iron pyrite gives sulphur dioxide according to the following equation :

$$4FeS_2 + 11O_2 \longrightarrow 2Fe_2 O_3 + 8SO_2$$

What volume of sulphur dioxide (at S.T.P.) would be liberated by roasting 30 g of pyrites ?

(S = 32; Fe 56; molar volume of a gas is 22.4 litres at S.T.P.)

Numericals

Sol. (i) Let x be the mass of oxygen that will combine with 8 g of methane.

$$\underset{\substack{1\text{ mol}\\ \\1\text{ molar mass}\\12+4\times1\\=16\text{ g}}}{CH_4} + \underset{\substack{2\text{ mol}\\ \\2\text{ molar mass}\\2\times(2\times16)\\=64\text{ g}}}{2O_2} \longrightarrow CO_2 + 2H_2O.$$

$$\text{Ratio proportion} = \frac{8\text{ g }CH_4}{16\text{ g }CH_4} = \frac{x\text{ g }O_2}{64\text{g }O_2}$$

$$\therefore \quad x = \frac{8\text{ g}}{16\text{ g}} \times 64\text{ g} = 32\text{ g}$$

Mass of oxygen gas that will combine with 8 g of methane is 32 g.

(ii) (a)

$$FeS_2 : 2S$$

$$(56 + 64) = 120 : 64 \;(= 2 \times 32)$$

$$120\text{ g pyrites contain} = 64\text{ g S}$$

$$30\text{ g pyrites contain} = \frac{64}{120} \times 30 = 16\text{ g}$$

(b)

$$\underset{4\times120=480\text{ g}}{4FeS_2} + 11O_2 \longrightarrow 2Fe_2O_3 + \underset{8\times22.4\text{ litres}}{8SO_2}$$

$$480\text{ g pyrites liberate} = 8 \times 22.4\text{ litres of }SO_2$$

$$30\text{ g pyrites liberate} = \frac{8\times22.4}{480} \times 30 = 11.2\text{ litres}$$

30. Water can be split into hydrogen and oxygen under suitable conditions. The equation representing the change is :

$$2H_2O(l) \longrightarrow 2H_2(g) + O_2(g)$$

(i) In a given experiment if 2500 cm^3 of hydrogen is being produced, what volume of oxygen is liberated at the same time, under the same conditions of temperature and pressure ?

(ii) The 2500 cm^3 of hydrogen is subjected to $2\frac{1}{12}$ times increase in pressure (temperature remaining constant). What volume will the hydrogen now occupy ?

(iii) Taking the volume of hydrogen calculated in (ii), what change must be made in the Kelvin temperature to return the volume to 2500 cm^3 (P remaining constant) ?

Sol. (i) The reaction is represented as :

$$2H_2O(l) \longrightarrow 2H_2(g) + O_2(g)$$

From the above reaction, it is clear that the volume of O_2 evolved is half that of hydrogen. Hence,

$$\text{Volume of }O_2\text{ evolved} = \frac{2500\text{ cm}^3}{2} = 1250\text{ cm}^3$$

(ii)

$$\text{Let initial pressure of }H_2 = P_i$$

$$\text{Final pressure} = 2.5\text{ P}$$

$$P_iV_i = P_fV_f$$

$$P \times 2500 = 2.5\text{ P} \times V_f$$

$$V_f = \frac{P\times2500}{2.5\text{ P}} = 1000\text{ cm}^3.$$

(iii)

Let initial temperature $= T_i$

and Final temperature $= T_f$

According to Charles law,

$$\frac{V_i}{T_i} = \frac{V_f}{T_f}$$

or

$$T_f = \frac{V_f}{V_i} \times T_i$$

$$= \frac{2500 \times T_i}{1000} = 2.5\ T_i$$

Thus, the Kelvin temperature must increase by 2.5 times.

31. Ammonia and oxygen combine to produce water vapour and nitric oxide as per chemical equation :

$$4NH_3\,(g) + 5O_2(g) \longrightarrow 6H_2O + 4NO.$$

(i) Write the chemical equation in terms of Gay-Lussac's law of combining volumes and the Avogadro's law.

(ii) How many moles of oxygen are required to burn 85 g of ammonia ?

(iii) How many moles of nitric oxide will be produced in reaction (i) ?

(iv) What is the volume of NH_3 at S.T.P. that will combine with oxygen in reaction (ii) ?

Sol. (i) The chemical equation can be written as below :

$4NH_3\,(g)$	+	$5O_2\,(g)$	$\longrightarrow$	$6H_2O$	+	$4NO(g)$	
4 volumes		5 volumes		6 volumes		4 volumes	(In terms of Gay-Lussac's law)
4 mol		5 mol		6 mol		4 mol	(In terms of Avogadro's law)

(ii)
$$\text{Molar mass of ammonia} = \frac{\text{Mol. wt}}{\text{No. of moles}}$$

$$\text{Moles wt.} = 4\,(14 + 3 \times 1) = 68\text{ g}$$

$$\text{No. of moles} = 4$$

$$\therefore \quad \text{Molar mass of ammonia} = \frac{68}{4} = 17\text{ g/mol.}$$

$$\text{Moles of ammonia} = \frac{\text{Mass of ammonia}}{\text{Molar mass of ammonia}}$$

$$= \frac{85\text{ g}}{17\text{ g/mol}} = 5\text{ mol}$$

Let x be the moles of O_2 required to combine with 5 moles of NH_3

$$\underset{4\text{ mol}}{\overset{5\text{ mol}}{4NH_3\,(g)}} + \underset{5\text{ mol}}{\overset{x\text{ mol}}{5O_2(g)}} \longrightarrow 6H_2O + 4NO(g)$$

$$\text{Ratio proportion} = \frac{5\text{ mol }NH_3}{4\text{ mol }NH_3} = \frac{x\text{ mol }O_2}{5\text{ mol }O_2}$$

$$x = \frac{5\text{ mol} \times 5\text{ mol}}{4\text{ mol}} = 6.25\text{ mol }O_2$$

(iii) Let y be the moles of nitric oxide formed when 5 moles of ammonia are completely burnt in oxygen.

$$\underset{4\text{ mol}}{\overset{5\text{ mol}}{4NH_3\,(g)}} + 5O_2(g) \longrightarrow 6H_2O + \underset{4\text{ mol}}{\overset{y\text{ mol}}{4NO(g)}}$$

$$\text{Ratio proportion} = \frac{5\text{ mole }NH_3}{4\text{ mole }NH_3} = \frac{y\text{ mol NO}}{4\text{ mol NO}}$$

$$y = \frac{5\text{ mole }NH_3}{4\text{ mole }NH_3} \times 4\text{ mole NO}$$

On burning ammonia completely = 5 mole NO will be produced.

(iv) $\because$ 1 mole NH_3 occupies 22.4 litre at S.T.P.

$\therefore$ 5 mole NH_3 will occupy 5 × 22.4 = 112 litre.

Volume of ammonia at S.T.P. is 112 litre.

32. The equations given below relate to the manufacture of sodium carbonate (Molecular weight of Na_2CO_3 = 106).

$$NaCl + NH_3 + CO_2 + H_2O \longrightarrow NaHCO_3 + NH_4Cl$$

$$2NaHCO_3 \longrightarrow Na_2CO_3 + H_2O + CO_2$$

Questions (i) and (ii) are based on the production of 21.2 g of sodium carbonate.

(i) What mass of sodium hydrogen carbonate must be heated to give 21.2 g of sodium carbonate (Molecular weight of $NaHCO_3 = 84$) ?

(ii) To produce the mass of sodium hydrogen carbonate calculated in (a), what volume of carbon dioxide, measured at S. T. P., would be required ?

Sol. (i)
$$\underset{\substack{2(23+1+12+48)\\=2\times 84\\=168\text{ g}}}{2NaHCO_3} \longrightarrow \underset{\substack{(23\times 2+12+48)\\=106\\=106\text{ g}}}{Na_2CO_3} + H_2O + CO_2$$

∵ 106 g of Na_2CO_3 is obtained from 168 g of $NaHCO_3$

∴ 21.2 g of Na_2CO_3 is obtained from $= \dfrac{168\times 21.2}{106}$

$= 33.6$ g $NaHCO_3$.

(ii)
$$NaCl + NH_3 + \underset{\substack{1\text{ mole}\\22.4\text{ lit at S.T.P.}}}{CO_2} + H_2O \longrightarrow \underset{\substack{1\text{ mole}\\84\text{ g}}}{NaHCO_3} + NH_4Cl$$

∵ 84 g of $NaHCO_3$ requires 22.4 lit of CO_2

∴ 33.6 g of $NaHCO_3$ requires $= \dfrac{22.4\times 33.6}{84}$

$= 8.96$ lit of CO_2.

33. (i) Concentrated nitric acid oxidizes phosphorus to phosphoric acid according to the following equation :

$$P + 5HNO_3 \longrightarrow H_3PO_4 + H_2O + 5NO_2$$

(a) What mass of phosphoric acid can be prepared from 6.2 g of phosphorus ?

(b) What mass of nitric acid will be consumed at the same time ?

(c) What will be the volume of steam at the same time if measured as 760 mm Hg pressure and 273°C ?

[H = 1; N = 14; O = 16; P = 31]

(ii) Ammonia may be oxidised to nitrogen monoxide in the presence of a catalyst according to the following equation :

$$4NH_3 + 5O_2 \longrightarrow 4NO + 6H_2O$$

If 27 litres of reactants are consumed, what volume of nitrogen monoxide is produced at the same temperature and pressure ?

Sol. (i)
$$\underset{(31\text{ g})}{P} + \underset{(315\text{ g})}{5HNO_3} \longrightarrow \underset{(98\text{ g})}{H_3PO_4} + H_2O + \underset{(22.4\text{ lit at S.T.P.})}{5NO_2}$$

(a) 31 g of phosphorus produces 98 phosphoric acid

($H_3PO_4 = 3\times 1 + 31 + 4\times 16 = 98$ g)

6.2 g phosphorus produces $= \dfrac{98}{31}\times 62 = 196$ g

(b) 31 g of P consumes 5×63 gm of HNO_3

Mass of HNO_3 consumed by 6.2 g of phosphorus $= \dfrac{5\times 63}{31}\times 6.2 = 63$ g

(c) 31 g of phosphorus at S.T.P. produces = 22.4 litres of steam.

Vol. of steam S.T.P. formed by 6.2 g of phosphorus $= \dfrac{22.4}{31}\times 6.2$

$= 4.48$ litres

As we know, $\dfrac{P_1V_1}{T_1} = \dfrac{P_2V_2}{T_2}$

At S.T.P. $P_1 = 760$ mm; $V_1 = 4.48$ lit; $T_1 = 273$ K

$P_2 = 760$ mm; $T_2 = 273°C + 273 = 546$ K; $V_2 = ?$

$$V_2 = \frac{760 \times 4.48 \times 546}{760 \times 273} = 8.96 \text{ litres}$$

(ii) The balanced equation for the oxidation of ammonia is

$$\underset{4 \text{ vols.}}{4NH_3(g)} + \underset{5 \text{ vols.}}{5O_2(g)} \longrightarrow \underset{4 \text{ vols.}}{4NO(g)} + 6H_2O(l)$$

Total volume of reactant consumed = 27 litres

Ratio by volume of $NH_3 : O_2$ is 4 : 5

$\therefore$ $$\text{Volume of } NH_3 = \frac{27 \times 4}{9} = 12 \text{ litres}$$

$\therefore$ $$\text{Volume of } O_2 = \frac{27 \times 5}{9} = 15 \text{ litres}$$

$$4NH_3 + 5O_2 \longrightarrow 4NO + 6H_2O$$

9 litres of reactants give 4 litres of NO

$$27 \text{ litres would give } \frac{27 \times 4}{9} = 12 \text{ litres}$$

12 litres, of NH_3 produces 4 litres, of NO

$\therefore$ 12 litres of NH_3 produces of 12 litres, of NO

$\therefore$ Volume of NO produced = 12 litres

34. What is the mass of nitrogen in 1000 kg of urea $[CO(NH_2)_2]$? (Answer correct to the nearest kg) [H = 1; C = 12; N = 14; O = 16]

Sol. Molecular formula of urea is NH_2CONH_2.

Molecular mass of urea $[CO(NH_2)]_2$ = 12 + 16 + 2 (14 + 1 × 2)
= 28 + 32
= 60 g

60 g of urea contains nitrogen = 2 × 14
= 28 g

$\therefore$ 1000 kg of urea contains nitrogen = 60 × 1000
= 467 kg.

35. Calculate the atomicity of oxygen molecule from the following information :

Vapour density of oxygen = 16
Relative atomic mass of oxygen = 16

Show all the calculations.

Sol. Vapour density of oxygen = 16

$\therefore$ Molecular weight of oxygen = 2 × V.D.
= 2 × 16 = 32

Relative atomic mass of oxygen = 16

$\therefore$ No. of atoms in one molecule of oxygen $= \frac{\text{Molecular wt.}}{\text{Atomic wt.}}$

$$= \frac{32}{16} = 2$$

$\therefore$ Atomicity of oxygen = 2

36. (i) 67.2 litres of hydrogen combines with 44.8 litres of nitrogen of form ammonia under specific conditions as :

$$N_2(g) + 3H_2(g) \longrightarrow 2NH_3(g)$$

Calculate the volume of ammonia produced. What is the other substance, if any, that remains in the resultant mixture ?

(ii) The mass of 5.6 dm^3 of a certain gas at S.T.P. is 12.0 g. Calculate the relative molecular mass of the gas.

(iii) Find the total percentage of Magnesium in magnesium nitrate crystals, $Mg(NO_3)_2.6H_2O$. [Mg = 24, N = 14; O = 16 and H = 1]

Numericals

Sol. (i)

$$\underset{\substack{1\text{ vol.}\\22.4\text{ lit}}}{N_2} + \underset{\substack{3\text{ vol.}\\67.2\text{ lit}}}{3H_2} \longrightarrow \underset{\substack{2\text{ vol.}\\44.8\text{ lit}}}{2NH_3}$$

According to Gay-Lussac's law :
3 volumes of H_2 will give 2 vol. of NH_3.

$\therefore$ 67.2 lit of H_2 will give $= \dfrac{2\times 67.2}{3} = 44.8$ lit of NH_3

At the same time
3 vol. of H_2 will react with 1 vol. of N_2

$\therefore$ 67.2 lit of H_2 will react with $= \dfrac{1}{3}\times 67.2 = 22.4$ lit of N_2

$\therefore$ $(44.8 - 22.4) = 22.4$ lit of N_2 will be left unreacted in the mixture.

(ii) According to molar volume concept, weight of 22.4 lit of a gas at S.T.P. is equal to molecular mass of the gas. Now,

$\because$ 5.6 lit of gas at S.T.P. weighs 12.0 g

$\therefore$ 22.4 lit of gas at S.T.P. weighs $= \dfrac{12\times 22.4}{5.6} = 48$ g

Hence, molecular mass of the gas is 48 g.

(iii) Relative molecular mass of $Mg(NO_3)_2.6H_2O$

$= (24) + (14 \times 2) + (16 \times 6) + (12 \times 1) + (6 \times 16)$

$= 24 + 28 + 96 + 12 + 96 = 256$ g

Amount of Mg in 256 g of magnesium nitrate = 24 g

$\therefore$ % of Mg in $Mg(NO_3)_2.6H O = \dfrac{24}{256}\times 100 = 9.37\%$

37. (i) What is the volume (measured in dm^3 or litres) occupied by one mole of a gas at S.T.P. ?

(ii) 112 cm^3 (at S.T.P.) of a gaseous fluoride of phosphorus has a mass of 0.63 g. Calculate the relative molecular mass of the fluoride. If the molecule of the fluoride contains only one atom of phosphorus, then determine the formula of the phosphorus fluoride. (F = 19; P = 31)

(iii) 2.24 litres of a gas weighs 4.4 g at S.T.P. Calculate the vapour density of the gas.

Sol. (i) The volume occupied by one mole of a gas at S.T.P. is 22.4 litre.

(ii) Mass of 112 cm^3 of phosphorus fluoride = 0.63 g

Mass of 22,400 cm^3 of phosphorus fluoride $= \dfrac{0.63}{112}\times 22400$

$= 126$ g

Relative molecular mass of fluoride = 126 g

Mass of fluoride in phosphorus fluoride = $126 - 31 = 95$ g.

Hence, Number of fluorine in phosphorus fluoride $= \dfrac{95}{19} = 5$

Thus, The formula of phosphorus fluoride = PF_5

(iii) $\because$ 2.24 litres of gas weighs 4.4 g at S.T.P.

$\therefore$ 22.4 litres of gas weighs $= \dfrac{4.4\times 22.4}{2.24} = 44$ g

$\therefore$ Vapour density $= \dfrac{\text{Molecular wt.}}{2} = \dfrac{44}{2} = 22$

38. Vapour density of a gas Z is 23. Calculate : (i) number of moles, (ii) weight in grams and (iii) number of molecules in 6.72 dm^3 of gas at S.T.P.

Sol. Molecular weight of gas Z = 2 × V.D.

= 2 × 23 = 46 a.m.u.

$\therefore$ 1 mole of gas = 46 g

(i) 22.4 dm^3 of gas Z at S.T.P. = 1 mole

$\therefore$ 6.72 dm^3 of gas Z at S.T.P. $= \dfrac{6.72}{22.4} = 0.3$ moles.

(ii) 1 mole of gas Z at S.T.P. weighs = 46 g

∴ 0.3 mole of gas Z at S.T.P. weighs = 46 × 0.3 = 13.8 g

(iii) 1 mole of gas Z at S.T.P. contains = 6×10^{23} molecules

∴ 0.3 mole of gas Z at S.T.P. contains = $6 \times 10^{23} \times 0.3$

= 1.8×10^{23} molecules.

39. The atomic weight of oxygen is 16.0 and the formula of oxygen molecule is O_2. Calculate the weight of : (i) 1 atom of oxygen, (ii) One molecule of oxygen.

Sol. (i) Molecular weight of oxygen = 16 × 2 = 32 g

∴ 32 g of oxygen = $2 \times 6.023 \times 10^{23}$ atoms of oxygen

or $2 \times 6.023 \times 10^{23}$ atoms weighs = 32 g

∴ Weight of one oxygen atom = $\frac{32}{2 \times 6.023 \times 10^{23}} = 2.656 \times 10^{-23}$ g

(ii) Similarly 32 g of oxygen contains = 6.023×10^{23} molecules of oxygen

∴ 1 molecule of oxygen will weighs = $\frac{32}{6.023 \times 10^{23}}$

= 5.312×10^{-23} g.

40. Calculate the relative molecular masses (or molecular weights) of the following compounds :

(i) Copper sulphate crystals, $CuSO_4.5H_2O$.

(ii) Ammonium sulphate, $(NH_4)_2SO_4$.

(iii) Cane sugar, $C_{12}H_{22}O_{11}$.

(Given atomic masses Cu = 63.5, S = 32, O = 16, N = 14, C = 12)

Sol. (i) $CuSO_4.5H_2O$ = 63.5 + 32 + (16 × 4) + 5 × (2 + 16)

= 159.5 + 90 = 249.5 a.m.u.

(ii) $(NH_4)_2SO_4 = N_2H_8SO_4$

= 14 × 2 + 1 × 8 + 32 + 16 × 4

= 28 + 8 + 32 + 64 = 132 a.m.u.

(iii) $C_{12}H_{22}O_{11}$ = 12 × 12 + 22 × 1 + 11 × 16

= 144 + 22 + 176 = 342 a.m.u.

41. Half-litre of carbon dioxide is passed over red hot carbon. The volume becomes 700 ml. Find the composition of the product, assuming that all the volumes of gases are measured at S.T.P.

Sol. The chemical equation representing the reaction is

$$\underset{\text{1 vol.}}{CO_2} + C \longrightarrow \underset{\text{2 vol.}}{2CO}$$

Let, x ml of CO_2 react with red hot carbon to form $2x$ ml of CO. Initial volume of carbon dioxide = 500 ml.

Final volume of the reaction mixture = 700 ml.

Thus, we have $(500 - x) + 2x = 700$

or $2x - x = 700 - 500$

$x = 200$

Volume of carbon monoxide formed = 2 × 200 = 400 ml

Volume of carbon dioxide remains after the reaction

= (500 – 200) = 300 ml.

42. Consider the following reaction and based on the reaction answer the questions that follow :

$$(NH_4)\,Cr_2O_7 \xrightarrow{\text{Heat}} N_2(g) + 4H_2O(g) + Cr_2O_3$$

Calculate :

(i) The quantity in moles of $(NH_4)_2Cr_2O_7$ if 63 g of $(NH_4)_2Cr_2O_7$ is heated.

(ii) The quantity in moles of nitrogen formed.

(iii) The volume in litres or dm^3 of N_2 evolved at S.T.P.

(iv) The mass in grams of Cr_2O_3 formed at the same time.

[Atomic masses : H = 1, Cr = 52, N = 14]

Sol.

$$\underset{\substack{2\,(14+4)+(52\times 2)+(16\times 7)\\ =36+104+112=252\text{ g}}}{(NH_4)Cr_2O_7} \xrightarrow{\text{Heat}} N_2(g) + 4H_2O(g) + \underset{\substack{(2\times 52)+(16\times 3)\\ =104+48=152}}{Cr_2O_3}$$

(i) 252 g $(NH_4)_2Cr_2O_7$ = 1 mole

$\therefore$ 63 g $(NH_4)_2Cr_2O_7 = \dfrac{63}{252} = 0.25$ mole

Hence, 0.25 mole of $(NH_4)_2Cr_2O_7$ is heated.

(ii) From the chemical equation

1 mole of $(NH_4)_2Cr_2O_7$ liberates 1 mole of N_2

$\therefore$ 0.25 mole of $(NH_4)_2Cr_2O_7$ liberates 0.25 moles of N_2.

(iii) Volume of 1 mole of N_2 at S.T.P. is 22.4 lit.

$\therefore$ 0.25 mole of N_2 at S.T.P. has volume = 22.4 × 0.25 = 5.6 lit

(iv) 252 g $(NH_4)_2Cr_2O_7$ gives 152 g Cr_2O_3

$\therefore$ 63 g $(NH_4)_2Cr_2O_7$ gives = $\dfrac{152\times 63}{252}$ g Cr_2O_3 = 38 g Cr_2O_3

Hence, the mass of Cr_2O_3 formed is 38 g.

43. (a) (i) Propane burns in air according to the following equation :

$$C_3H_8 + 5O_2 \longrightarrow 3CO_2 + 4H_2O.$$

What volume of propane is consumed on using 1000 cm^3 of air, considering only 20% of air contains oxygen ?

(ii) The mass of 11.2 litres of a certain gas at S.T.P. is 24 g. Find the gram molecular mass of the gas.

(b) A gas cylinder can hold 1 kg of hydrogen at room temperature and pressure :

(i) Find the number of moles of hydrogen present.

(ii) What weight of CO_2 can the cylinder hold under similar conditions of temperature and pressure ? (H = 1, C = 12, O = 16)

(iii) If the number of molecules of hydrogen in the cylinder is X, calculate the number of CO_2 molecules in the cylinder under the same conditions of temperature and pressure.

(iv) State the law that helped you to arrive at the above result.

Sol. (a) (i) For every 5 moles of O_2, 1 mole of propane is burnt.

20% of 1000 = 20 × 1000/100 = 200 cm^3 of O_2.

Thus, volume of propane = 40 cm^3

(ii) Mass of gas = 24 g

Volume of gas = 11.6 litres

22.4 L of gas at S.T.P. = 1 mole

11.2 L of gas at S.T.P. = 11.2 / 22.4 = 0.5 moles

Mass of 0.5 moles of gas = 24 g

Mass of 1 mole of gas or molar mass = 24/0.5 = 48 g

(b) (i) 1 kg = 1000 grams

2 g of hydrogen molecules = 1 mole

1 g of hydrogen molecules = 1/2 mole

1000 g of hydrogen molecules = 1/2 × 1000 = 500 moles

(ii) Molecular weight of carbon dioxide = 44 g

$$\text{Vapour density} = \frac{44}{2} = 22$$

Now,

$$\text{Vapour density} = \frac{\text{Weight of carbon dioxide at certain temperature}}{\text{Weight of same volume of hydrogen at same temperature and pressure}}$$

$$22 = \text{Weight of carbon dioxide/1 kg}$$

Weight of carbon dioxide = 22 kg.

(iii) If the number of molecules of hydrogen is X, then number of molecules of carbon dioxide will also be X.

(iv) This is according to the **Avogadro's law** which states that, "Equal volumes of all gases under similar conditions of temperature and pressure contain equal number of molecules".

44. (i) The reaction of potassium permanganate(VII) with acidified iron(II) sulphate is given below :

$$2KMnO_4 + 10FeSO_4 + 8H_2SO_4 \longrightarrow K_2SO_4 + 2MnSO_4 + 5Fe_2(SO_4)_3 + 8H_2O$$

If 15.8 g of potassium permanganate was used in the reaction, calculate the mass of iron(II) sulphate used in the above reaction. (K = 39, Mn = 55, Fe = 56, S = 32, O = 16)

(ii) 20% nitric acid reacts with 4.11 g of lead carbonate of 65% purity. Calculate the weight of nitric acid to complete the reaction. [Pb = 207; C = 12; O = 16; N = 14]

Sol. (i)

$$\underset{2\times158\text{ g}}{2KMnO_4} + \underset{10\times152\text{ g}}{10FeSO_4} + 8H_2SO_4 \longrightarrow K_2SO_4 + 2MnSO_4 + 5Fe_2(SO_4)_3 + 8H_2O$$

(2 × 158) g $KMnO_4$ uses (10 × 152) g $FeSO_4$

$$15.8\text{ g } KMnO_4 \text{ uses } \frac{(10\times152)\text{g}}{(2\times158)\text{g}} \times 15.8\text{ g} = 76\text{ g } FeSO_4$$

Amount of $FeSO_4$ used by 15.8 g $KMnO_4$ = 76 g

(ii) Weight of impure lead carbonate = 4.11 g

$$\therefore \quad \text{Weight of pure lead carbonate} = \frac{4.11\times65}{100} = 2.6715\text{ g}$$

Writing gram-molecular weights

$$\underset{\substack{[207+12+3(16)]\\=267\text{ g}}}{PbCO_3} + \underset{\substack{2[1+14+3(16)]\\=126\text{ g}}}{2HNO_3} \longrightarrow Pb(NO_3)_2 + H_2O + CO_2$$

When pure $PbCO_3$ is 267 g pure nitric acid required = 126 g

When pure $PbCO_3$ is 2.6715 g pure nitric acid required

$$= \frac{126\times2.6715}{267}\text{g} = 1.260\text{ g}$$

∴ Wt. of pure nitric acid required = 1.260 g

$$\therefore \quad \text{Wt. of 20\% nitric acid required} = \frac{1.26\times100}{20} = 6.30\text{ g}$$

45. When heated, potassium permanganate decomposes according to the following equation :

$$2KMnO_4 \longrightarrow K_2MnO_4 + \underset{\text{Solid residue}}{MnO_2} + O_2$$

(i) Some potassium permanganate was heated in a test tube. After collecting one litre of oxygen at room temperature, it was found that the test tube had undergone a loss in mass of 1.32 g. If one litre of hydrogen under the same conditions of temperature and pressure has a mass of 0.0825 g, calculate the relative molecular mass of oxygen.

(ii) Given that the molecular mass of potassium permanganate is 158, what volume of oxygen (measured at room temperature) would be obtained by the complete decomposition of 15.8 of potassium permanganate (molar volume at room temperature is 24 litres) ?

Sol. (i) Mass of one litre of oxygen = 1.32 g

Mass of one litre of hydrogen under the same condition = 0.0825 g

$$\text{V.D. of oxygen} = \frac{\text{Mass of one litre of oxygen}}{\text{Mass of one litre of hydrogen}}$$

$$= \frac{1.32\text{ g}}{0.0825} = 16$$

Relative molecular mass of oxygen = 2 × V.D.

$$= 2 \times 16 = 32$$

(ii) $$\underset{(2 \times 158\text{ g})}{2KMnO_4} \longrightarrow K_2MnO_4 + MnO_2 + \underset{1\text{ mol (24 lit)}}{O_2}$$

(2 × 158) g $KMnO_4$ on heating liberates 24 litre O_2 at room temperature

15.8 g $KMnO_4$ on heating liberates $\frac{24\text{ litre}}{(2 \times 158)\text{g}} \times 15.8 = 1.2$ litre O_2

Volume of O_2 liberated at room temperature by heating 15.8 g potassium permanganate = 1.2 litres.

46. Iron forms three different forms of oxides :
(i) Ferrous oxide [FeO], (ii) Ferric oxide [Fe_2O_3], (iii) Tri-ferric tetraoxide [Fe_3O_4].
Calculate the percentage of iron in each of the above oxides.

Sol. (i) Molecular formula of ferrous oxide is FeO.

∵ Molecular weight of ferrous oxide = 56 + 16 = 72 g

∴ $$\% \text{ of Iron in ferrous oxide (FeO)} = \frac{56}{72} \times 100 = \frac{700}{9} = 77.78\%$$

(ii) Molecular formula of ferric oxide is Fe_2O_3.

∵ Molecular weight of ferric oxide = (2 × 56) + (3 × 16)
= 112 + 48 = 160 g

∴ $$\% \text{ of Iron in ferric oxide } (Fe_2O_3) = \frac{112}{160} \times 100 = 70\%.$$

(iii) Molecular formula of tri-ferric tetraoxide is Fe_3O_4.

Molecular weight of tri-ferric tetraoxide = (3 × 56) + (4 × 16)
= 168 + 64 = 232 g

∴ $$\% \text{ of Iron in tri-ferric tetraoxide } (Fe_3O_4) = \frac{168}{232} \times 100 = \frac{2100}{29} = 72.41\%$$

47. (i) If 16.4 gram of calcium nitrate is heated :
(a) Calculate the volume of nitrogen dioxide obtained at S.T.P.
(b) The weight of calcium oxide obtained.

(ii) Paddy crop removes 20 kg of nitrogen from soil per hectare. Calculate the amount of calcium nitrate [$Ca(NO_3)_2$] which should be added to the soil to provide nitrogen to a farm of 10 hectares ? State your answer in kg (approx.). [Ca = 40, N = 14, O = 16]

Sol. (i) The reaction is :

$$\underset{2[40 + 2(14 + 48)] = 328\text{ g}}{2Ca(NO_3)_2} \longrightarrow \underset{112\text{ g}}{2CaO} + \underset{4\text{ moles}}{4NO_2} + O_2$$

(a) 328 g of $Ca(NO_3)_2$ liberate 4 × 22.4 NO_2

∴ $$16.4 \text{ g will liberate} = \frac{4 \times 22.4}{328} \times 16.4$$
$$= 4.48 \text{ litre } NO_2 \text{ at S.T.P.}$$

(b) 328 g of $Ca(NO_3)_2$ yields 112 g CaO

∴ $$16.4 \text{ g will yields} = \frac{112}{328} \times 16.4 = 5.6 \text{ g CaO}$$

(ii) Total amount of nitrogen required for 10 hectare farm = 10 × 20 = 200 kg

Gram molecular wt. of $Ca(NO_3)_2$ = 1(Ca) + 2(N) + 6(O)
= 1 × (40) + 2 × (14) + 6 × (16)
= 40 + 28 + 96
= 164 kg.

28 kg of nitrogen can be obtained from

$Ca(NO_3)_2$ = 164 kg.

Numericals

∴ 200 kg of nitrogen can be obtained from $Ca(NO_3)_2$

$$= \frac{164 \times 200}{28}$$

$$= 1171.42 \text{ kg.}$$

48. (i) Find the total percentage of oxygen in magnesium nitrate crystals.

$(MgNO_3)_2.6H_2O$ [H = 1, N = 14, O = 16, Mg = 24].

(ii) Calculate the percentage of nitrogen in aluminium nitride.

(Al = 27, N = 14)

(iii) Find out the percentage, by weight of phosphorus present in calcium phosphate.

Sol. (i) $Mg(NO_3)_2.6H_2O$

$$\text{The molecular mass} = 24 + (2 \times 62) + 6 \times 18$$

$$= 24 + 124 + 108 = 256 \text{ g}$$

$$\text{Mass due to oxygen} = (2 \times 48) + (6 \times 16) = 96 + 96 = 192 \text{ g}$$

$$\therefore \quad \% \text{ of oxygen} = \frac{192}{256} \times 100 = 75\%.$$

(ii)

$$\text{Molecular weight of AlN} = 27 + 14 = 41 \text{ g}$$

$$\% \text{ of nitrogen} = \frac{14 \times 100}{41} = 34.14\%$$

(iii) Molecular formula of calcium phosphate is $Ca_3(PO_4)_2$.

Molecular weight of calcium phosphate

$$= 3 \times 40 + 2 \times [31 + (4 \times 16)]$$

$$= 120 + 2 \times (31 + 64)$$

$$= 310 \text{ g}$$

and molecular weight of phosphorus = 62 g

$$\therefore \quad \% \text{ of phosphorus in calcium phosphate} = \frac{62}{310} \times 100 = 20\%.$$

49. (i) If 112 cm^3 of hydrogen sulphide is mixed with 120 cm^3 of chlorine at S.T.P., what mass of sulphur is formed ?

$$H_2S + Cl_2 \longrightarrow 2HCl + S$$

(ii) Washing soda has the formula $Na_2CO_3.10H_2O$. What mass of anhydrous sodium carbonate is left when all the water of crystallization is expelled by heating 57.2 g of washing soda ?

(iii) When excess lead nitrate solution was added to a solution of sodium sulphate, 15.15 g of lead sulphate was precipitated. What mass of sodium sulphate was present in the original solution ?

$$Na_2SO_4 + Pb(NO_3)_2 \longrightarrow PbSO_4 + 2NaNO_3$$

(H = 1; C = 12; O = 16; Na = 23; S = 32; Pb = 207)

Sol. (i)

$$\underset{\text{1 vol.}}{H_2S} + \underset{\text{1 vol.}}{Cl_2} \longrightarrow 2HCl + S$$

From the above equation, it is clear that 1 vol. of H_2S reacts with 1 vol. of Cl_2. Thus, when 112 cm^3 of H_2S is mixed with 120 cm^3 of Cl_2 at S.T.P. only 112 cm^3 of Cl_2 will react.

Amount of gas in 22,400 cm^3 at S.T.P. = 1 mole

$$\text{Amount of gas in 112 cm}^3 \text{ at S.T.P.} = \frac{112}{22400}$$

$$= 0.005 \text{ mole}$$

$$112 \text{ cm}^3 \text{ of } H_2S \text{ at S.T.P.} = 0.005 \text{ mole } H_2S$$

$$112 \text{ cm}^3 \text{ of } Cl_2 \text{ at S.T.P.} = 0.005 \text{ mole } Cl_2$$

$$\underset{0.005}{H_2S} + \underset{0.005}{Cl_2} \longrightarrow 2HCl + \underset{\text{0.005 mol.}}{S}$$

Amount of sulphur formed = (0.005 × 32)

= 0.16 g

(ii) $$\underset{286\ g}{Na_2CO_3.10H_2O} \longrightarrow \underset{106\ g}{Na_2CO_3} + 10H_2O$$

∵ 286 g washing soda on heating, produce 106 g anhydrous sodium carbonate.

∴ 57.2 g washing soda produce $= \frac{57.2 \times 106}{286}$

= 21.20 g.

(iii) $$\underset{\substack{2 \times 23 + 32 + 16 \times 4 \\ = 142\ g}}{Na_2SO_4} + Pb(NO_3)_2 \longrightarrow \underset{\substack{207 + 32 + 64 \\ = 303\ g}}{PbSO_4} + 2NaNO_3$$

∴ 303 g $PbSO_4$ is precipitated by 142 g Na_2SO_4

∴ 15.15 g $PbSO_4$ will be precipitated by $= \frac{15.15 \times 142}{303}$

= 7.1 g Na_2SO_4

50. Concentrated nitric acid oxidizes phosphorus to phosphoric acid according to the following equation :

$$P + 5HNO_3 \longrightarrow H_3PO_4 + H_2O + 5NO_2$$

(i) What mass of phosphoric acid can be prepared from 6.2 g of phosphorus ?

(ii) What mass of nitric acid will be consumed at the same time ? [H = 1; N = 14; O = 16; P = 31]

Sol. $$\underset{31\ g}{P} + \underset{\substack{5[1 + 14 + 48] \\ 5 \times 63 - 315\ g}}{5HNO_3} \longrightarrow \underset{\substack{(3) + (31) + (64) \\ = 98\ g}}{H_3PO_4} + 5NO_2 + \underset{\substack{22.4\ dm^3 \\ at\ S.T.P.}}{H_2O}$$

(i) ∵ 31 g of phosphorus can produce phosphoric acid = 98 g

∴ 6.2 g of phosphorus can produce phosphoric acid $= \frac{98 \times 6.2}{31} = 19.6$ g

(ii) ∵ 31 g of phosphorus requires nitric acid = 315 g

∴ 6.2 g of phosphorus requires nitric acid $= \frac{6.2 \times 315}{31}$

= 63.0 g.

51. Following chemical equation is given :

$$2NH_4Cl + Ca(OH)_2 \longrightarrow CaCl_2 + 2H_2O + 2NH_3.$$

(i) Calculate the mass of ammonia obtained from 321 g of ammonium chloride.

(ii) Find the mass of ammonium chloride required to obtain 6 moles of H_2O.

(iii) Find the mass of ammonium chloride required to obtain 4 moles of NH_3.

Sol. The chemical equation is :

$$\underset{\substack{2\ (14 + 4\ (1) + 35.5) \\ = 107\ g}}{2NH_4Cl} + Ca(OH)_2 \longrightarrow CaCl_2 + 2H_2O + \underset{\substack{2\ moles\ or\ 2\ [14 + (1 \times 3)] \\ = 2\ (17) = 34\ g}}{2NH_3}$$

(i) ∵ 107 g of NH_4Cl gives 34 g of ammonia

∴ 1 g of NH_4Cl gives $= \frac{34}{107}$ g of ammonia

∴ 321 g of NH_4Cl gives $= \frac{34 \times 321}{107}$ g = 102 g of ammonia.

(ii) ∵ 2 moles of H_2O is obtained from 107 g NH_4Cl

∴ 6 moles of H_2O will be obtained from

$= \frac{107 \times 6}{2} = 321$ g of NH_4Cl

(iii) ∵ 2 moles of NH_3 is obtained from 107 g of NH_4Cl

∴ 1 mole of NH_3 is obtained from $= \frac{107}{2} = 53.5$ g of NH_4Cl

$\therefore$ 4 moles of NH_3 is obtained from = $\frac{107 \times 4}{2}$ g = 214 g of NH_4Cl

52. 10 g of a mixture of sodium chloride and anhydrous sodium sulphate is dissolved in water. An excess of barium chloride solution is added and 6.99 g of barium sulphate is precipitated according to the equation given below :

$$Na_2SO_4 + BaCl_2 \longrightarrow BaSO_4 + 2NaCl$$

Calculate the percentage of sodium sulphate in the original mixture.
(O = 16, Na = 23, S = 32, Ba = 137)

Sol.

$$\underset{\substack{2 \times 23 + 32 + 4 \times 16 \\ = 142 \text{ g}}}{Na_2SO_4} + BaCl_2 \longrightarrow \underset{\substack{137 + 32 + 4 \times 16 \\ = 233 \text{ g}}}{BaSO_4} + 2NaCl$$

From the above equation, it is clear that 233 g barium sulphate is precipitated from 142 g sodium sulphate.

$\therefore$ 6.99 g barium sulphate is precipitated from = $\frac{142 \text{ g}}{233 \text{ g}} \times 6.99$ = 4.26 g sodium sulphate

Amount of sodium sulphate in 10 g mixture = 4.26 g
Percentage of sodium sulphate in original mixture

$$= \frac{4.26}{10 \text{ g}} \times 100 = 42.6\%$$

53. (i) Determine the empirical formula of the compound whose composition by mass is : 42% nitrogen, 48% oxygen and 9% hydrogen. [H = 1; N = 14; O = 16]

(ii) Determine the empirical formula of a compound containing 47.9% potassium, 5.5% beryllium and 46.6% fluorine by mass.
(Atomic weight of Be = 9; F = 19; K = 39) Work to one decimal place.

Sol. (i)

Element	Rel. At. Mass	% age comp.	Relative No. of atoms	Simple ratio
N	14	42	$\frac{42}{14} = 3$	$\frac{3}{3} = 1$
O	16	48	$\frac{48}{16} = 3$	$\frac{3}{3} = 1$
H	1	9	$\frac{9}{1} = 9$	$\frac{9}{3} = 3$

Empirical formula = NOH_3

(ii)

Element	%	Atomic mass	Relative no. of atoms	Simplest ratio
K	47.9	39	$\frac{47.9}{39} = 1.2$	$\frac{1.2}{0.6} = 2$
Be	5.5	9	$\frac{5.5}{9} = 0.6$	$\frac{0.6}{0.6} = 1$
F	46.6	19	$\frac{46.6}{19} = 2.4$	$\frac{2.4}{0.6} = 4$

Empirical formula = K_2BeF_4

54. (i) Calculate the empirical formula of the compound having 37.6% of sodium, 23.1% of silicon and 39.3% of oxygen. [O = 16, N = 23, Si = 28]

(ii) The Empirical formula of a compound is C_2H_5. It has a vapour density of 29. Determine the relative molecular formula mass of the compound and hence its molecular formula.

Numericals

Sol. (i)

Element	Atomic Mass	% Comp.	Rel. No. of moles	Simple Ratio
Na	23	37.6	$\frac{37.6}{23} = 1.63$	$\frac{1.63}{0.82} = 1.98 = 2$
Si	28	23.4	$\frac{23.1}{28} = 0.82$	$\frac{0.82}{0.82} = 1$
O	16	39.3	$\frac{39.3}{16} = 2.45$	$\frac{2.45}{0.82} = 2.98 = 3$

The empirical formula is Na_2SiO_3.

(ii) $\because$ The empirical formula $= C_2H_5$

$\therefore$ Empirical formula wt. $= 2\,(C) + 5\,(H)$

$= 2\,(12) + 5\,(1)$

$= 24 + 5 = 29$

The vapour density of C_2H_5 is 29,

$\therefore$ Molecular weight $= 2 \times$ V.D.

$= 2 \times 29$

$= 58$

No. of molecules $n = \frac{\text{M.F.W.}}{\text{E.F.W.}}$

$n = \frac{58}{29}$

$n = 2$

Molecular formula $= n \times$ [Empirical formula]

$= 2\,[C_2H_5]$

$= C_4H_{10}$

The molecular formula $= C_4H_{10}$

The relative molecular formula mass of $C_4H_{10} = 4\,(C) + 10\,(H)$

$= 4\,(12) + 10\,(1)$

$= 48 + 10$

$= 58$ g

55. (i) Calculate the percentage of platinum in ammonium chloroplatinate $(NH_4)_2PtCl_6$ (Give your answer correct to the nearest whole number).

(ii) The percentage composition of sodium phosphate as determined by analysis, is 42.1% sodium, 18.9% phosphorus and 39% oxygen. Find the empirical formula of the compound (work to two decimal places). (H = 1, N = 14, O = 16, Na = 23, P = 31, Cl = 35.5, Pt = 195)

Sol. (i) Mol. wt. of $(NH_4)_2PtCl_6 = [14 + 4 \times 1] \times 2 + 195 + 6 \times 35.5$

$= 18 \times 2 + 195 + 213$

$= 444$

% of platinum $= \frac{195}{444} \times 100 = 43.9\%$

(ii)

Element	Percentage	At. wt.	%/At. wt	Simple ratio
Na	42.1	23	$\frac{42.1}{23} = 1.8$	$\frac{1.8}{0.6} = 3$
P	18.9	31	$\frac{18.9}{31} = 0.6$	$\frac{0.6}{0.6} = 1$
O	39	16	$\frac{39}{16} = 2.4$	$\frac{2.4}{0.6} = 4$

Empirical formula = Na_3PO_4

56. A compound contains 82.35% by mass of nitrogen and 17.64% by mass of hydrogen. Determine the empirical formula of this compound.

Sol.

Element	Percentage	Relative No. of Atoms	Simple Ratio
N	82.35	$\frac{82.35}{14} = 5.8$	1
H	17.64	$\frac{17.64}{1} = 17.64$	3

Hence, empirical formula = NH_3.

57. A compound X consists of 4.8% carbon and 95.2% bromine by mass.

(i) Determine the empirical formula of this compound working correct to one decimal place (C = 12; Br = 80).

(ii) If the vapour density of the compound is 252, what is the molecular formula of the compound ?

(iii) Name the type of chemical reaction by which X can be prepared from ethane.

Sol. (i) Empirical formula :

Element	% composition	At. mass	Relative number of atoms	Simple Ratio
Carbon	4.8	12	$\frac{4.8}{12} = 0.4$	$\frac{0.4}{0.4} = 1$
Bromine	95.2	80	$\frac{95.2}{80} = 1.19$	$\frac{1.19}{0.4} = 2.975$

Empirical fromula of substance is CBr_3

(ii)
$$\text{Empirical mass} = 12 + 80 \times 252$$
$$\text{Molecular mass} = 2 \times \text{V.D} = 2 \times 252 = 504$$
$$\therefore \quad n = \frac{\text{Molecular mass}}{\text{Empirical formula mass}} = \frac{504}{252} = 2$$
or
$$\text{Molecular formula} = n \times \text{Empirical formula}$$
$$2 \times CBr_3 = C_2Br_6$$

(iii) This substance can be prepared by substitution method.

58. A gaseous organic compound contains 3.6 g of carbon and 0.8 g of hydrogen. The vapour density of this compound is 22.

(i) Calculate the empirical formula.

(ii) Calculate the molecular formula of the compound.

(iii) If 4.4 g of the above compound are completely burnt in oxygen, calculate the volume of carbon dioxide formed at S.T.P. [C = 12; H = 1; O = 16]

Sol. (i)

Element	Wt. of atoms	At. wt	Relative No. of atoms	Simple Ratio
C	3.6	12	$\frac{3.6}{12} = 0.3$	3
H	0.8	1	$\frac{0.8}{1} = 0.8$	8

Empirical formula of the compound = C_3H_8

(ii)
$$\text{V.D. of the compound} = 22$$
$$\text{Empirical formula wt.} = 36 + 8 = 44$$
$$\text{Molecular wt. of compound} = 2 \times \text{V.D.} = 2 \times 22 = 44$$
$$\text{No. of Molecules} = \frac{\text{Molecular wt.}}{\text{Empirical wt.}} = \frac{44}{44} = 1$$

$$\text{Molecular formula} = n \times \text{Empirical formula} = 1 \times C_3H_8 = C_3H_8$$

(iii)
$$\underset{44\text{ g}}{C_3H_8} + 5O_2 \longrightarrow \underset{\substack{3 \times 22.4 \text{ dm}^3 \text{ at S.T.P.} \\ = 67.2 \text{ dm}^3 \text{ at S.T.P.}}}{3CO_2} + 4H_2O$$

44 g of C_3H_8 yields carbon dioxide at S.T.P. = 67.2 dm^3

∴ 4.4 g of C_3H_8 will produce carbon dioxide at S.T.P.

$$= \frac{67.2 \times 4.4}{44} = 6.72 \text{ dm}^3 \text{ at S.T.P.}$$

59. (i) A solid organic compound contained 2.15% of hydrogen, 26.8% of carbon, and the rest of oxygen. Take the molecular weight of the compound as 90. Find the empirical and molecular formula of the compound.

(ii) A metal M forms a volatile chloride containing 65.5% chlorine. If the density of the chloride relative to hydrogen is 162.5, find the molecular formula of the chloride. (M = 56; Cl = 35.5)

Sol. (i) Percentage of oxygen (by difference) = 100 – (26.8 + 2.15)
= 100 – 28.95 = 71.05%

Relative number of atoms	Simplest ratio
$C = \frac{26.8}{12} = 2.23$	$C = \frac{2.23}{2.15} = 1$
$H = \frac{2.15}{1} = 2.15$	$H = \frac{2.15}{2.15} = 1$
$O = \frac{71.05}{16} = 4.44$	$O = \frac{4.44}{2.15} = 2$

Empirical formula of the compound is CHO_2

Empirical formula weight = 12 + 1 + 2 × 16
= 12 + 1 + 32 = 45

Given, that the molecular weight = 90

$$\therefore \quad n = \frac{\text{Relative molecular weight}}{\text{Empirical formula weight}} = \frac{90}{45} = 2$$

Molecular formula = $(CHO_2) \times 2$

Therefore, molecular formula of the compound is $C_2H_2O_4$.

(ii) % of Metal = 100 – 65.5 = 34.5

Chlorine % = 65.5

Relative number of atoms	Simplest ratio
$\text{Metal} = \frac{34.5}{56} = 0.616$	$\frac{0.616}{0.616} = 1$
$\text{Chlorine} = \frac{65.5}{35.5} = 1.85$	$\frac{1.85}{0.616} = 3$

∴ Empirical formula = MCl_3

Molecular mass = 2 × V.D.
= 2 × 162.5 = 325.0

$$n = \frac{\text{Molecular mass}}{\text{Empirical formula weight}} = \frac{325}{162.5} = 2$$

∴ Molecular formula = $(\text{Empirical formula})_n$
= $(MCl_3)_2 = M_2Cl_6$

60. (i) The compound A has the following percentage composition by mass. Carbon = 26.7%, oxygen = 71.1%, hydrogen = 2.2%. Determine the empirical formula of compound A [work to one decimal place] [H = 1, C = 12, O = 16].

(ii) If the relative molecular mass of A is 90, what is the molecular formula of A ?

(iii) The compound A is weak acid. What is meant by this statement ?

Sol. (i) Relative number of atoms,

$$C = \frac{26.7}{12} = 2.2 \qquad O = \frac{71.1}{16} = 4.4 \qquad H = \frac{2.2}{1} = 2.2$$

Simplest ratio,

$$C = \frac{2.2}{2.2} = 1 \qquad O = \frac{4.4}{2.2} = 2 \qquad H = \frac{2.2}{2.2} = 1$$

Empirical formula of the compound A is CO_2H.

(ii) The empirical formula weight = [12 + 32 + 1] = 45

$$n = \frac{\text{Molecular weight}}{\text{Empirical weight}} = \frac{90}{45}$$

$$n = 2$$

∴ Molecular formula of the compound A is $[CO_2H]_2$

$= C_2O_4H_2$

$= \begin{matrix} COOH \\ | \\ COOH \end{matrix}$ (Oxalic acid)

(iii) It does not completely ionise into $C_2O_4^{2-}$ and H^+ ions.

61. (a) A gas cylinder contains 12×10^{24} molecules of oxygen gas. If Avogadro's number is 6×10^{23}. Calculate :

(i) The mass of oxygen present in the cylinder.

(ii) The volume of oxygen at S.T.P. present in the cylinder. [O = 16]

(b) A gaseous hydrocarbon contains 82.76% of carbon. Given that its vapour density is 29, find its molecular formula. [C = 12, H = 1]

(c) The equation $4NH_3 + 5O_2 \longrightarrow 4NO + 6H_2O$, represents the catalytic oxidation of ammonia. If 100 cm^3 of ammonia is used calculate the volume of oxygen required to oxidise the ammonia completely.

Sol. (a) (i) 12×10^{24} molecules of O_2

$$\text{Number of mole} = \frac{12 \times 10^{24}}{6 \times 10^{23}} = 20 \text{ mole}$$

∵ 1 mole of oxygen has the atomic weight ⟶ 32 g

∴ 20 moles of oxygen have the atomic weight = 32 × 20 = 640 g

(ii) The volume of one mole gas at S.T.P. = 22.4 lit

∴ 20 mole of gas at S.T.P. will have the volume of oxygen

= 20 × 22.4 lit

= 448 lit

(b)

Element	Percentage	Molecules	Simple ratio	Simple whole ratio
C	82.76	$\frac{82.76}{12} = 6.89$	1	2
H	17.24	$\frac{17.24}{1} = 17.24$	2.5	5

∴ Empirical formula = C_2H_5

Empirical formula mass = (12 × 2) + (1 × 5)

Numericals

$$= 24 + 5 = 29$$

$$\text{Vapour density} = 29 \text{ (Given)}$$

$$\text{Molecular mass} = \text{V.D.} \times 2 = 29 \times 2$$

$$= 58 \text{ gm}$$

$$\text{Molecular formula mass} = n \times \text{Empirical formula mass}$$

$$\Rightarrow \quad n = \frac{\text{Molecular formula mass}}{\text{Empirical formula mass}}$$

$$= \frac{58}{29} = 2$$

$$\text{Molecular formula} = n \times \text{Empirical formula}$$

$$= 2 \times C_2H_5$$

$$= C_4H_{10}$$

(c) $4NH_3 + 5O_2 \longrightarrow 4NO + 6H_2O$

Given : Ammonia used in the reaction = 100 cm^3

From the equation, 4 vol. of NH_3 requires 5 vol. of O_2 for its oxidation.

$$\therefore \quad \text{1 vol. will require} = \frac{5}{4}$$

$$\text{Thus 100 cm}^3 \text{ of ammonia will require} = \frac{5}{4} \times 100$$

$$= 125 \text{ cm}^3 \text{ of oxygen}$$

62. (i) An organic compound with vapour density = 94 contains. C = 12.67%, H = 2.13%, and Br = 85.11%. Find the molecular formula. [Atomic mass : C = 12, H = 1, Br = 80]

(ii) Calculate the mass of

(a) 10^{22} atoms of sulphur.

(b) 0.1 mole of carbon dioxide.

[Atomic mass : S = 32, C = 12 and O = 16 and Avogadro's Number = 6 × 10^{23}]

Sol. (i)

Elements	% Ratio	Atomic mass	Relative no. of atoms	Simplest ratio
C	12.67	12	12.67/12 = 1.055	1.055/1.055 = 1
H	2.13	1	2.13/1 = 2.13	2.13/1.055 = 2
Br	85.11	80	85.11/80 = 1.063	1.063/1.055 = 1

∴ Empirical formula of the compound is CH_2Br

$$\text{Molecular formula} = (\text{Empirical formula})_n$$

$$n = \frac{\text{M.W.}}{\text{Empirical formula weight}}$$

$$= \frac{2 \times \text{V.D.}}{\text{Empirical formula weight}}$$

$$= \frac{2 \times 94}{(12 + 2 + 80)} = \frac{2 \times 94}{94} = 2$$

$$\therefore \quad \text{Molecular formula} = (CH_2Br)_2 = C_2H_4Br_2$$

(ii) (a) 1 mole of sulphur = 6 × 10^{23} atoms = 32 g of sulphur

$$\therefore \quad 10^{22} \text{ atoms} = \frac{32 \times 10^{22}}{6 \times 10^{23}} = \frac{32}{60} = 0.533 \text{ g}$$

(b) 1 mole of carbon dioxide (CO_2) = 12 + (2 × 16) = 44 g

∴ 0.1 mole of carbon dioxide = 0.1 × 44 = 4.4 g

63. Calculate the percentage of :*

(i) Fluorine (ii) Sodium and (iii) Aluminium

in sodium aluminium fluoride [Na_3AlF_6], to the nearest whole number.

[Atomic Mass : Na = 23, Al = 27, F = 19]

Sol. Molecular weight of Na_3AlF_6

$$3 \times 23 + 27 + 6 \times 19 = 210$$

Now,

(i) Percentage of fluorine = $\frac{(6\times19)}{210}\times100 = 54.2\%$

(ii) Percentage of sodium = $\frac{(3\times23)}{210}\times100 = 32.8\ \%$

(iii) Percentage of aluminium = $\frac{27}{210}\times100 = 12.8\ \%$

64. (i) State the volume occupied by 40 gm of methane at STP, if its vapour density (V.D.) is 8.*

(ii) Calculate the number of moles present in 160 gm of NaOH.

[Atomic Mass : Na = 23, H = 1, O = 16]

Sol. (i) Molecular weight = 2 × Vapour density at STP

In case of methane, substituting the given information –

Molecular weight = 2 × 8 = 16

Molecular weight of methane = 16 which means 16 g of methane occupy 22.4 L volume at STP.

So, 40 g of methane will occupy = $\frac{22.4}{16} \times 40 = 56$ L

(ii) Number of moles = $\frac{\text{Given weight of substance}}{\text{Molecular weight of substance}}$

Molecular weight of NaOH = 23 + 16 + 1 = 40

No. of moles = $\frac{160}{40} = 4$

There are 4 moles of NaOH in 160 g of NaOH

65. Copper sulphate solution reacts with sodium hydroxide solution to form a precipitate of copper hydroxide according to the equation :*

$$2NaOH + CuSO_4 \rightarrow Na_2SO_4 + Cu(OH)_2 \downarrow$$

(i) What mass of copper hydroxide is precipitated by using 200 gm of sodium hydroxide ?

[H = 1, O = 16, Na = 23, S = 32, Cu = 64]

(ii) What is the colour of the precipitate formed ?

Sol. (i) The given equation is $2NaOH + CuSO_4 \rightarrow Na_2SO_4 + Cu(OH)_2 \downarrow$

Molecular weight of NaOH, Sodium hydroxide = 23 + 16 + 1 = 40

Molecular weight of $Cu(OH)_2$, Copper hydroxide = 64 + 16 + 1 + 16 + 1 = 98

2 × 40 = 80 g of NaOH is used to precipitate 98 g of $Cu(OH)_2$

Hence, 200 g of NaOH will be used to precipitate (98/80) 200 g of $Cu(OH)_2$ = 245 g of $Cu(OH)_2$

So, 245 g of copper hydroxide would be prepared using 200 g of sodium hydroxide.

(ii) The precipitate of copper hydroxide is bluish green solid or Pale blue solid.

66. Find the **empirical formula** and the **molecular formula** of an organic compound from the data given below :*

C = 75.92%, H = 6.32% and N = 17.76%

The vapour density of the compound is 39.5

[C = 12, H = 1, N = 14]

Sol. Given : C = 75.92%, H = 6.32% and N = 17.76%

Let us assume that the weight of compound is 100 g. So, in that 100 g C is 75.92 g, H is 6.32 g and N is 17.76 g.

* **Frequently asked previous years Board Exam Questions.**

Now, lets calculate the moles of each element present in the compound-

75.92 g C × (1 mol C/12 g C) = 6.32

6.32 g H × (1 mol H/1 g H) = 6.32

17.76 g N × (1 mol N/14 g N) = 1.26

Next, dividing all the mole numbers by the smallest among them, which is 1.26. This division yields – 5 mol C, 5 mol H and 1 mol N

So, the compound has the empirical formula –

C_5H_5N.

Now, we know that

Molecular mass of a gas = 2 × vapour density of the gas

= 2 × 39.5

= 79

So, Molecular mass of the given compound is 79.

Empirical formula mass = (12 × 5) + (1 × 5) + (14 × 1) = 79

As, Empirical formula mass = Molecular mass

So, in this case Empirical formula = Molecular formula = C_5H_5N.

67. Aluminium carbide reacts with water according to the following equation :*

$$Al_4C_3 + 12H_2O \rightarrow 4Al(OH)_3 + 3CH_4$$

(i) What mass of aluminium hydroxide is formed from 12 g of aluminium carbide ?

(ii) What volume of methane at S.T.P. is obtained from 12 g of aluminium carbide ?

[Relative molecular weight of Al_4C_3 = 144; $Al(OH)_3$ = 78]

Sol. (i) $Al_4C_3 + 12H_2O \longrightarrow 4Al(OH)_3 + 3CH_4$

One mole of Al_4C_3 gives 4 moles of $Al(OH)_3$.

i.e., 144 g of Al_4C_3 gives 4 × 78 g of $Al(OH)_3$

So, 12 g of Al_4C_3 gives $\frac{312 \times 12}{144}$ g of $Al(OH)_3$ = 26 g of $Al(OH)_3$.

(ii) One mole of Al_4C_3 gives 3 moles of methane

12 g of Al_4C_3 gives $\frac{48 \times 12}{144}$ g of CH_4 = 4 g

Now, 16 g of methane has volume 22.4 L (at STP, the volume of one mole of any gas is 22.4 L)

4 g of methane would occupy 5.6 L.

So, 5.6 L of methane would be obtained from 12 g of Al_4C_3.

67. Ethane burns in oxygen to form CO_2 and H_2O according to the equation :*

$$2C_2H_6 + 7O_2 \rightarrow 4CO_2 + 6H_2O$$

If 1250 cc of oxygen is burnt with 300 cc of ethane.

Calculate :

(i) the volume of CO_2 formed.

(ii) the volume of unused O_2.

Sol. The given equation is :

$$2C_2H_6 + 7O_2 \rightarrow 4CO_2 + 6H_2O$$

(i) So, according to above equation, 2 V (volumes) of ethane reacts to give 4 V of carbon dioxide. So, 300 cc of ethane would give 600 cc of carbon dioxide.

(ii) Also, 2 V (volumes) of ethane reacts with 7 V of oxygen.

300 cc of ethane is 2 V, so oxygen required for 300 cc of ethane is $\frac{300 \times 7}{2}$ = 1050 cc

The remaining oxygen would be :

1250 cc – 1050 cc = 200 cc

❑

*** Frequently asked previous years Board Exam Questions.**

Figure or Table Based Questions

| Set 15 |

Chapter 1. Periodic Properties and Variations of Properties

1. Consider the section of the periodic table given below :

Group Numbers	I A 1	II A 2	III A 13	IV A 14	V A 15	VI A 16	VII A 17	O 18
	Li		D			O	J	Ne
	A	Mg	E	Si		H	K	
	B	C		F	G			L

In this table B does not represent Boron
C does not represent Carbon
F does not represent Fluorine
H does not represent Hydrogen
K does not represent Potassium

You must see the position of the element in the periodic table. Some elements are given in their own symbol and position in the periodic table. While others are shown with a letter. With reference to the table :

(i) Which is the most electronegative?

(ii) How many valence electrons are present in G?

(iii) Write the formula of the compound between B and H.

Ans. (i) J (ii) 5 (iii) B_2H

2. Given below is the part of Periodic Table :

Li	Be	B	C	N	O	F
Na	Mg	Al	Si	P	S	Cl

(i) How does metallic character of an element change as one moves from
(a) Left to right in period? (b) Top to bottom in group?

(ii) How does the valency of elements change with respect to hydrogen as one moves from left to right in period?

(iii) (a) What is the valency of element silicon?
(b) Will it form a covalent or electrovalent bond with hydrogen?

(iv) Which are the most metallic and the most non-metallic elements in above table?

Ans. (i) (a) As one moves from left to right in period, the metallic character of an element decreases.
(b) As one moves from top to bottom in period, the metallic character of the elements increases.

(ii) The valency of elements with respect to hydrogen is stated as under :

I group + 1 II group + 2
III group + 3 IV group + 4
V group − 3 VI group − 2
VII group − 1

(iii) (a) The valency of element silicon is + 4.
(b) Silicon will form a covalent compound with hydrogen.

(iv) Sodium is the most metallic and chlorine is most non-metallic element.

3. The diagram given below is a part of Periodic Table. Study the table and answer the questions given below the table :

1																	2 He
3	4 Be											5	6	7	8	9	10
11	12											13	14 Si	15	16 S	17	18
19	20 Ca	21	22	23	24 Cr	25	26	27	28	29	30	31	32	33	34	35	36 Kr

(i) Name two elements in same group of Periodic Table.
(ii) Name the transition metal.
(iii) Name one elements, whose oxide has a very low melting point.
(iv) Name the element, whose oxide is acidic.
(v) Name an element, which reacts vigorously with water.
(vi) Name an element, which form basic oxide.
(vii) Which element shows variable oxidation state?
(viii) Which element is used in semiconductors?
(ix) Which elements posses complete valence shell?
(x) Which element forms very corrosive acid?

Ans. (i) Oxygen and sulphur (ii) Chromium
(iii) Calcium (iv) Sulphur
(v) Calcium (vi) Calcium
(vii) Chromium (viii) Silicon
(ix) Helium and Krypton (x) Chromic acid produced by chromium

4. In the portion of the Periodic Table given below, the letters A, B, ... represent the elements in periods 2 and 3 and groups 1, 2, 13, 14, 15, 16, 17 and 18 which are not the usual symbols of the elements.

	1	2	13	14	15	16	17	18
Period 2	A	B	C	D	E	F	G	H
Period 3	I	J	K	L	M	N	O	P
	Q	R	S	T	U			

Study the table and answer the following questions :
(i) Which is the most electropositive element ?
(ii) Which is the most electronegative element?
(iii) Which elements have properties that are similar to those of the element O?
(iv) Which elements are the noble gases?
(v) Which elements have a valency of 4?
(vi) Which is more metallic, Q or R?
(vii) Which is more non-metallic, E or M?
(viii) What is the collective name for the elements G and O?
(ix) Which elements are represented by the letters I, J, N and O?

Ans. (i) Q, (ii) G, (iii) G, (iv) H and P, (v) D, L and T, (vi) Q, (vii) E, (viii) The halogens, (ix) I : Sodium, J : Magnesium; N : Sulphur; O : Chlorine.

5. The electronegativities (according to pauling) of the elements in period 3 of the portion of Periodic Table are as follows when the elements arranged in alphabetical order :

Al	Cl	Mg	Na	P	S	Si
1.5	3.0	1.2	0.9	2.1	2.5	1.8

(i) Arrange the elements in the order in which they occur in the Periodic Table from left to right. (The group 1 element first, followed by the group 2 element and so on, up to group 7)
(ii) The element below sodium in the same group would be expected to have a .... (lower/higher) electronegativity than sodium and the element above chlorine would be expected to have a .... (lower/higher) ionization potential than chlorine.

Ans. (i) Arrangement of elements in the IIIrd group of Periodic Table is given below : Na, Mg, Al, Si, P, S, Cl.

(ii) Lower, Higher.

6. The questions refers to the elements of the Periodic Table with atomic numbers from 3 to 18. Some of the elements are shown by letters, but letters are not usual symbols of the elements.

Which of these :

3	4	5	6	7	8	9	10	11	12	13	14	15	16	17	18
A	B	C	D	E	F	G	H	I	J	K	L	M	N	O	P

(i) Is a noble gas?
(ii) Is a halogen?
(iii) Is an alkali metal?
(iv) Is an element with valency 4?

Ans. (i) P is a noble gas with atomic number 18, *i.e.*, 2, 8, 8.
H is a also a noble gas with atomic number 10, *i.e.*, 2, 8.

(ii) G is halogen with atomic number 9, *i.e.*, 2, 7.
O is also a halogen with atomic number 17, *i.e.*, 2, 8, 7.

(iii) A and I are alkali metals.

(iv) D and L are elements with valency 4.

Q. 7.

Group Number	IA 1	IIA 2	IIIA 13	IVA 14	VA 15	VIA 16	VIIA 17	0 18
2nd period	Li		D			O	J	Ne
	A	Mg	E	Si		H	M	
	R	T	I		Q	u		y

- In this table H does not represent hydrogen.
- Some elements are given in their own symbol and position in the periodic table.
- While others are shown with a letter.

With reference to the table answer the following questions :

(i) Identify the most electronegative element.

(ii) Identify the most reactive element of group 1.

(iii) Identify the element from period 3 with least atomic size.

(iv) How many valence electrons are present in Q?

(v) Which element from group 2 would have the least ionization energy?

(vi) Identify the noble gas of the fourth period.

(vii) In the compound between A and H what type of bond would be formed and give the molecular formula for the same?

Ans. (i) J (ii) R (iii) M
(iv) 5 (v) T (vi) *y*
(vii) Ionic bond. Molecular formula $\rightarrow A_2H$

Q. 8.

Elements	A	B	C
Mass numer	23	20	35
Number of neutrons	12	10	18

Study the table above and answer the following questions carefully :

(i) Write the atomic number and electronic configuration of elements A, B and C.

(ii) To which groups do A, B and C belong?

(iii) To which periods do A, B and C belong?

(iv) Which amongst A, B and C is (i) an alkali metal, (ii) noble gas, (iii) halogen?

Ans. (i) Atomic number of A = (Mass number – No. of neutrons)

$$= 23 - 12$$

$$= 11 \Rightarrow \begin{pmatrix} 2 & 8 & 1 \\ K, & L, & M \end{pmatrix}$$

Atomic number of B = (Mass number – No. of neutrons)

$$= 20 - 10$$

$$= 10 \Rightarrow \begin{pmatrix} 2 & 8 \\ K, & L \end{pmatrix}$$

Atomic number of C = (Mass number – No. of neutrons)

$$= 35 - 18$$

$$= 17 \Rightarrow \begin{pmatrix} 2 & 8 & 7 \\ K, & L, & M \end{pmatrix}$$

(ii) A belong to IA group — B belong to zero group
C belongs to VII A group — **Criterion :** No. of valence electrons

(iii) A belongs to 3rd period — B belongs to 2nd period
C belongs to 3rd period — **Criterion :** No. of shells

(iv) A is an alkali metal — B is a noble gas
C is a halogen

9. (i) Use the letters only written in the Periodic Table given below to answer the questions that follow :

GROUPS

PERIODS	I	II	(transition elements)	III	IV	V	VI	VII	0
1									L
2	Q			E	G	J	Z	M	
3	R								
4	T								
5									

(a) State the number of valence electrons in atom J.
(b) Which element shown forms ions with a single negative charge?
(c) Which metallic element is more reactive than R?
(d) Which element has its electrons arranged in four shells?

(ii) The following tables shows the electronic configuration of the elements W, X, Y, Z :

Elements	W	X	Y	Z
Electronic configurations	2, 8, 1	2, 8, 7	2, 5	1

Answer the following questions based on the table above :

(a)What type of bond is formed between :
(1) W and X (2) Y and Z

(b) What is the formula of the compound formed between :
(1) X and Z (2) W and X

Ans. (i) (a) Five valence electrons (b) M
(c) T (d) T

(ii) (a) (1) Ionic bond (2) Covalent bond
(b) (1) ZX (2) WX

10. The following table represent the elements and the atomic number :*

With reference to this, answer the following using only the alphabets given in the table.

Element	*Atomic number*
P	*13*
Q	*7*
R	*10*

(i) Which element combines with hydrogen to form a basic gas ?

(ii) Which element has an electron affinity zero ?

(iii) Name the element, which forms an ionic compound with chlorine.

Ans. (i) Q (QH_3 is the basic gas) with Z = 7

(ii) R with Z = 10

(iii) P with Z = 13

*** Frequently asked previous years Board Exam Questions.**

11. Study the extract of the Periodic Table given below and answer the questions that follow. Give the alphabet corresponding to the elements in question. DO NOT repeat an element.*

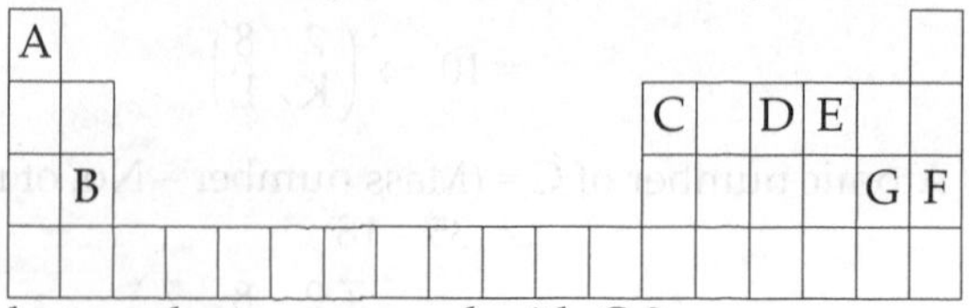

(i) Which element forms electrovalent compound with G ?
(ii) The ion of which element will migrate towards the cathode during electrolysis ?
(iii) Which non-metallic element has the valency of 2 ?
(iv) Which is an inert gas ?

Ans. (i) A would form electrovalent compound with G.
(ii) B ion would travel to cathode during electrolysis.
(iii) E has valency of 2.
(iv) F is an inert gas.

Chapter 2. Chemical Bonding

1. With the help of the figure, explain that methane molecule is a non-polar covalent compound.

Ans. A covalent bond formed between two atoms is non-polar if the electrons involved in bond formation are equally shared between two atoms having similar electronegativities. For example, in case of methane molecule, the four electron pairs shared between one carbon atom and four hydrogen atoms lie exactly in between one carbon atom and four hydrogen atoms and hence forms non-polar bonds. The three-dimensional structure of methane molecule has the carbon atom at the centre of the tetrahedron and four hydrogen atoms are located at the corners of the tetrahedron. In such a configuration, none of the participating atoms is more electrically charged. So, methane is a non-polar covalent compound.

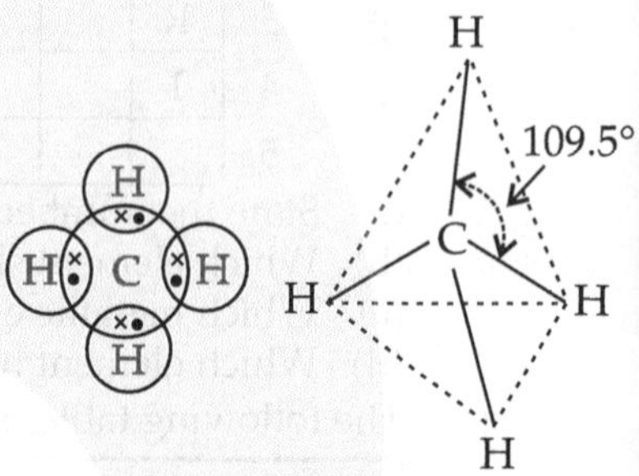

CH_4–molecule

Electronegativity of carbon = 2.5

Electronegativity of hydrogen = 2.1

2. With the help of the figure show HCl is a polar molecule.

Ans. In HCl molecule the strong nuclear charge of chlorine atom attracts the electron of hydrogen far away from its nucleus, with the result the hydrogen atom develops a slight positive charge (d^+) and chlorine atom develops a slight negative charge (δ^-)

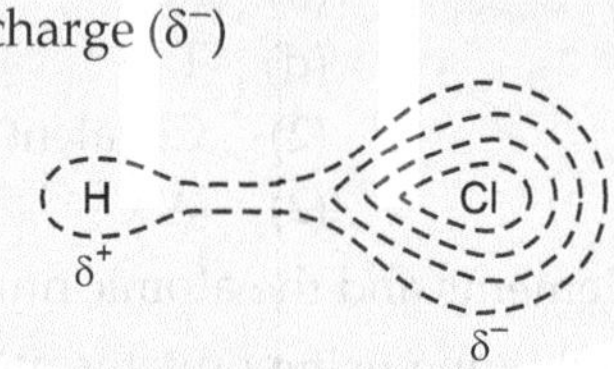

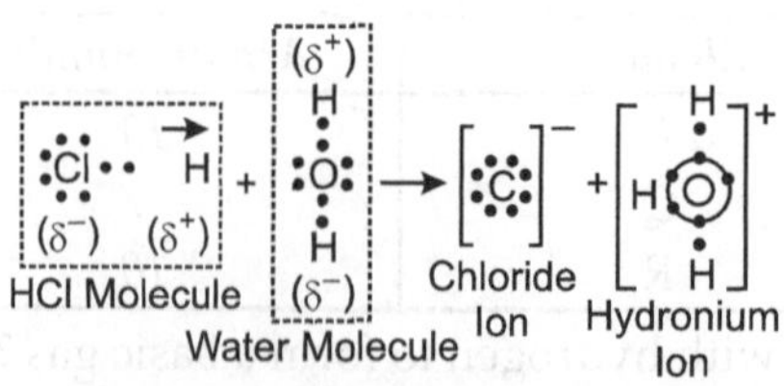

3. Explain how polar covalent compound HCl which is a bad conductor in pure and liquid state, ionises in water.

Ans. A molecule of HCl is a polar covalent, *i.e.*, the electron pair shared between hydrogen and chlorine is far away from hydrogen than chlorine, with the result that hydrogen atom has a slight positive charge on it and the atom of chlorine has slight negative charge on it as :

* **Frequently asked previous years Board Exam Questions.**

$$\underset{(\delta^-)}{:\ddot{\underset{..}{Cl}}} \;-\; \underset{(\delta^+)}{H}$$

Hydrogen chloride molecule

When hydrogen chloride is added to water, then water molecules bombard it from all directions. If the alignment of a water molecule is such that slightly negative oxygen atom faces the slightly positive hydrogen atom of HCl then, the slightly positive hydrogen atom is swallowed by water molecule in the form of single proton to form hydronium ion (H_3O^+). The residual chlorine atom takes away the electron of hydrogen atom along with it so as to form chloride ion as illustrated below :

$$\underset{\text{Hydrogen chloride molecule}}{H - \ddot{\underset{..}{Cl}}:} + \underset{\text{water molecule}}{H - \ddot{\underset{..}{O}} - H} \longrightarrow \underset{\text{hydronium ion}}{\left[H - \overset{..+}{\underset{|}{O}} - H\right]}\ \text{(with H below O)} + \underset{\text{chloride ion}}{:\ddot{\underset{..}{Cl}}:^-}$$

$$HCl + H_2O \rightleftharpoons H_3O^+ + Cl^-$$

4. How does common salt solution ionises in water which is a polar covalent compound?

Ans. When the sodium chloride crystals come in contact with water, they are bombarded by water molecules from all possible directions.

When the molecules of water bombard the sodium ion, such that the slightly negatively charged oxygen atom face sodium ion, then they exert an electrostatic pull on it. If sufficient number of water molecules strike a particular sodium ion in same fashion, then it can be pulled out of the crystal of sodium chloride to form free sodium ion.

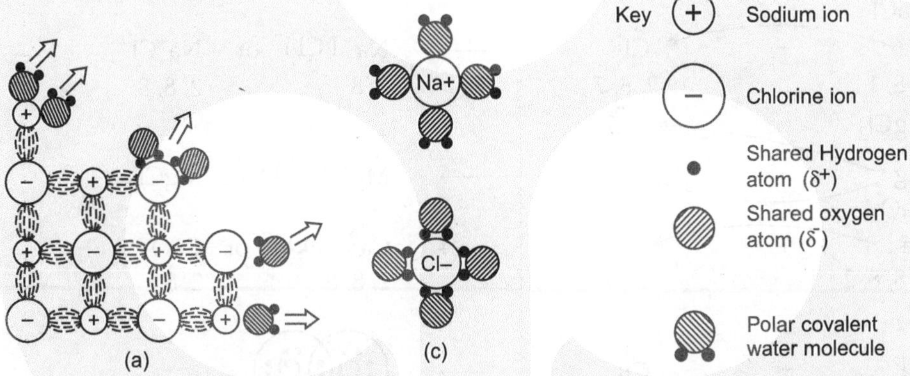

Similarly, if the molecules of water bombards the chloride ion, such that the slightly positively charged hydrogen atoms face chloride ion then they exert an electrostatic pull on it. If sufficient number of water molecules strike a particular chloride ion in same fashion, then it can be pulled out of the crystal of sodium chloride to form free chloride ion.

5. Draw dot diagrams to illustrate the structure of the molecules of :

(i) Ammonia (ii) Carbon dioxide

(iii) Methane (iv) Water

Ans. (i) Ammonia molecule

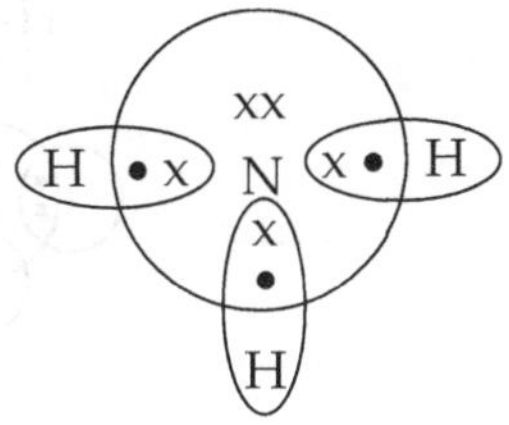

(ii) Carbon dioxide molecule

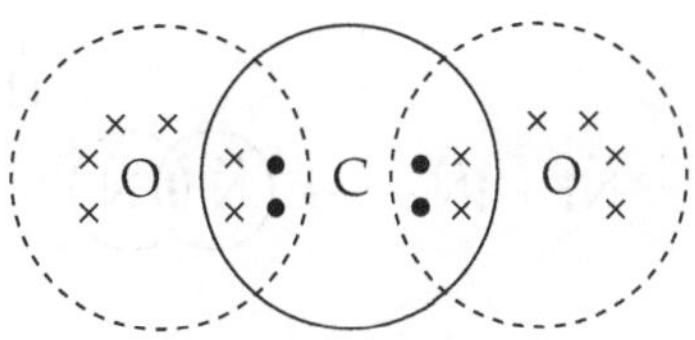

(iii) Methane molecule

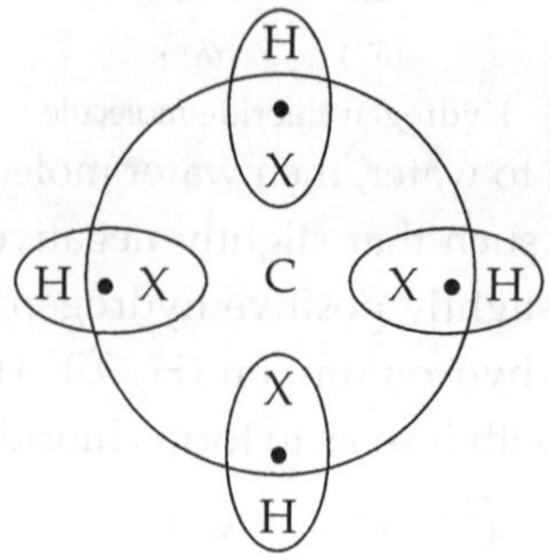

(iv) Water molecule

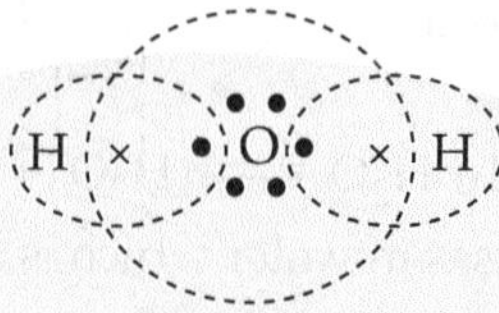

6. Give the electron dot structures of :

(i) NaCl (ii) $MgCl_2$

(iii) CaO (iv) Cl_2

(v) H_2O (vi) NH_3

(vii) N_2

Ans. (i) NaCl

Na· + ·C̈l: ⟶ Na^+ [:C̈l:]$^-$ or Na^+Cl^-

2, 8, 1 2, 8, 7 2, 8 2, 8, 8

(ii) $MgCl_2$

Mg : + ·C̈l: ·C̈l: ⟶ Mg^+ 2 [:C̈l:]$^-$ or $MgCl_2$

(iii) CaO

Ca : + Ö: ⟶ Ca^{2+} [:Ö:]$^-$ or CaO

2, 8, 8, 2 2, 6 2, 8, 8 2, 8

(iv) Cl_2

:C̈l· + ·C̈l: ⟶

2, 8, 7 2, 8, 7

or

Cl – Cl

(v) H_2O

H· + :Ö + ·H ⟶

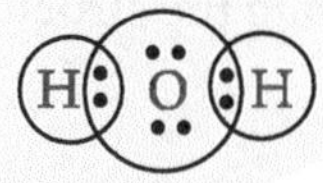

1 2, 6 1

or

H–O–H

(vi) NH_3

·N̈· + 3H ⟶

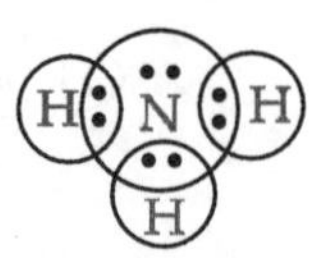

2, 5 1

or

H—N̈—H
|
H

(vii) N⋮ + ⋮N ⟶

7. Draw an electron dot diagram to show the structure of hydronium ion. State the type of bonding present in it.

Ans. Formation of hydronium ion :

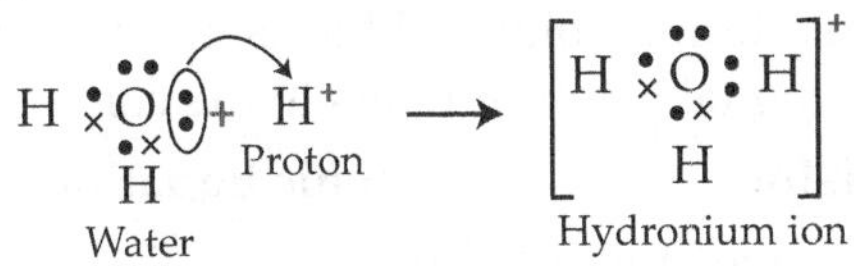

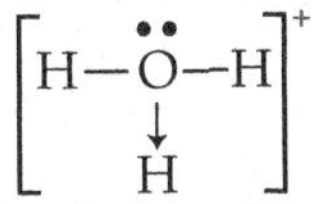

Hydronium ion

The type of bonding present in hydronium ion is coordinate bonding.

8. By drawing an electron dot diagram, show the lone pair effect leading to the formation of ammonium ion from ammonia gas and hydrogen ion.

Ans.

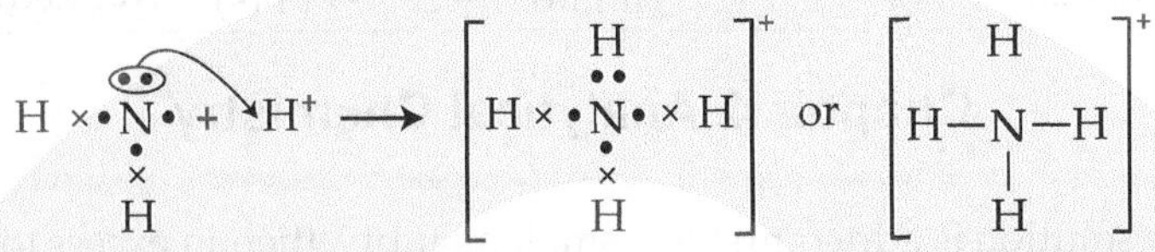

NH_3 has one lone pair of electrons which it donates to hydrogen atom forming a coordinate bond. The arrow represents a coordinate bond. The arrow points from donor to acceptor atom.

Chapter 3. Study of Acids, Bases and Salts

1. Draw the structure of the stable positive ion formed when an acid dissolves in water.

Ans. H—Ö:→H^+
 |
 H

(H_3O^+) Hydronium ion.

2. Some methods used for the laboratory preparation of salts are :

A : metal + acid B : carbonate + acid
C : precipitation (double decomposition) D : direct combination E : titration

Copy and complete the following table :

Salt	Method of Preparation
Ammonium sulphate Calcium carbonate Iron(III) chloride Lead nitrate Zinc sulphate	

Ans.

Salt	Method of Preparation
Ammonium sulphate	E : Titration
Calcium carbonate	C : Precipitation
Iron(III) chloride	D : Direct combination
Lead nitrate	B : Carbonate + acid
Zinc sulphate	A : Metal + acid

3. The pH values of three solution, A, B and C are given in the table. Answer the following questions :*

Solution	pH value
A	12
B	2
C	7

(i) Which solution will have no effect on litmus solution ?

(ii) Which solution will liberate CO_2 when reacted with sodium carbonate ?

(iii) Which solution will turn red litmus solution blue ?

* Frequently asked previous years Board Exam Questions.

Ans. (i) Solution C would have no effect on litmus solution as its pH is 7 and hence it is neutral.
(ii) Solution B would liberate CO_2 when reacted with sodium carbonate as it is acidic solution and ha pH 2.
(iii) Solution A would turn red litmus solution blue as it is basic in nature and has pH 12.

4. Copy and complete the following table which refers to the conversion of ions to neutral particles :*

Conversion	Ionic Equation	Oxidation/Reduction
Chloride ion to chlorine molecule	(i) ________	(ii) ________
Lead (II) ion to lead	(iii) ________	(iv) ________

Ans.

Conversion	Ionic Equation	Oxidation/Reduction
Chloride ion to chlorine molecule	(i) $Cl^- - e^- \rightarrow \frac{1}{2} Cl_2(g)$	(ii) Oxidation
Lead(II) ion to lead	(iii) $Pb^{2+} + 2e^- \rightarrow Pb(s)$	(iv) Reduction

Chapter 4. Analytical Chemistry

1. Sodium hydroxide solution is added first in a small quantity, then in excess to the aqueous salt solution of copper(II) sulphate, zinc nitrate, lead nitrate, calcium chloride and iron(III) sulphate. Copy the following table and write the colour of the precipitate in (i) to (v) and the nature of the precipitate (soluble or insoluble) in (vi) to (x).

Aqueous salt solution	Colour of ppt. when NaOH is added in a small quantity	Nature of ppt. (soluble or insoluble) when NaOH is added in excess
Copper(II) sulphate	(i)	(vi)
Zinc nitrate	(ii)	(vii)
Lead nitrate	(iii)	(viii)
Calcium chloride	(iv)	(ix)
Iron(III) sulphate	(v)	(x)

Ans.

Aqueous salt solution	Colour of ppt. when NaOH is added in a small quantity	Nature of ppt. (soluble or insoluble)when NaOH is added in excess
Copper(II) sulphate	Blue	Insoluble
Zinc nitrate	White	Soluble
Lead nitrate	White	Soluble
Calcium chloride	White	Insoluble
Iron(III) sulphate	Reddish Brown	Insoluble

2. Three test tubes contain calcium nitrate solution, zinc nitrate solution and lead nitrate solution respectively. Each solution is divided into two portions. Describe the effect of :
(i) Adding sodium hydroxide solution to each portion in turn till it is in excess.
(ii) Adding ammonium hydroxide to each portion till it is in excess.

Ans.

Solution	Effect of adding sodium hydroxide solution		Effect of adding ammonium hydroxide solution	
	Small amount	In excess	Small amount	In excess
Calcium nitrate	A white ppt.	No change.	No precipitate.	No change.
Zinc nitrate	A white ppt.	White ppt. dissolves, gives a colourless solution.	A white ppt.	A white ppt. dissolves to give a colourless solution.
Lead nitrate	A white ppt.	White ppt. dissolves to give a colourless solution.	A white ppt.	No change.

Figure or Table Based Questions

3. The following table shows the tests a student performed on four aqueous solutions A, B, C and D. Write down on your answer sheet the observations (i) to (iv) that were made.

Test	Observations	Conclusions
(i) To solution A, calcium nitrate solution and sodium hydroxide solution were added.	(i)	A contains Ca^{2+} ions
(ii) To solution B sodium hydroxide solution was added.	(ii)	B contains Fe^{3+} ions
(iii) To solution C ammonium hydroxide was added slowly till in excess.	(iii)	C contains Cu^{2+} ions
(iv) To solution D sodium hydroxide solution were added.	(iv)	D contains Pb^{2+} ions

Ans. Observations :

(i) White ppt. of calcium hydroxide is formed.

(ii) Reddish brown ppt. of $Fe(OH)_3$ is formed.

(iii) Pale blue ppt. is formed which gives deep blue solution with excess of ammonium hydroxide.

(iv) A white ppt. of lead hydroxide is formed.

4. Sodium hydroxide solution is added to the solutions containing the ions mentioned in list X. Y gives the details of the precipitate. Match the ions with their coloured precipitates.

	List X	List Y		List X	List Y
(i)	Pb^{2+}	A. Reddish brown	(iv)	Fe^{3+}	D. White soluble in excess
(ii)	Fe^{2+}	B. White insoluble in excess	(v)	Cu^{2+}	E. White soluble in excess
(iii)	Zn^{2+}	C. Dirty green	(iv)	Ca^{2+}	F. Blue

Ans. (i) $Pb^{2+} \rightarrow$ White soluble in excess
(ii) $Fe^{2+} \rightarrow$ Dirty green
(iii) $Zn^{2+} \rightarrow$ White soluble in excess
(iv) $Fe^{3+} \rightarrow$ Reddish brown
(v) $Cu^{2+} \rightarrow$ Blue
(vi) $Ca^{2+} \rightarrow$ White insoluble in excess.

5. Certain blank spaces are left in the following table and these are labelled as A, B, C, D and E. Identify each of them.

	Lab preparation of	Reactants used	Products formed	Drying agent	Method of collection
(i)	HCl gas	$NaCl + H_2SO_4$	A	conc. H_2SO_4	B
(ii)	NH_3 gas	C	$Mg(OH)_2$ NH_3	D	E

Ans. (i) (A) $NaHSO_4 + HCl$ (B) Upward displacement of air
(ii) (C) $Mg_3N_2 + H_2O$ (D) CaO
(E) Downward displacement of air

6. The following table shows the tests a student performed on four different aqueous solutions which are X, Y, Z and W. Based on the observations provided, identify the cation present.

Chemical test	Observation	Conclusion
To solution 'X' ammonium hydroxide is added in minimum quantity first and then in excess.	A dirty white ppt. is formed which dissolves in excess to form a clear solution.	(i)
To solution 'Y' ammonium hydroxide is added in minimum quantity first and then in excess.	A pale blue ppt. is formed which dissolves in excess to form a clear inky blue solution.	(ii)

To solution 'W' a small quantity of sodium hydroxide solution is added and then in excess.	A white ppt. is formed which remain insoluble.	(iii)
To a salt 'Z' calcium hydroxide solution is added and then heated.	A pungent smelling gas turning moist red litmus paper blue is obtained.	(iv)

Ans. (i) X – Zn^{2+}

(ii) Y – Cu^{2+}

(iii) W – Ca^{2+}

(iv) Z – NH_4^+

Chapter 6. Electrolysis

1.

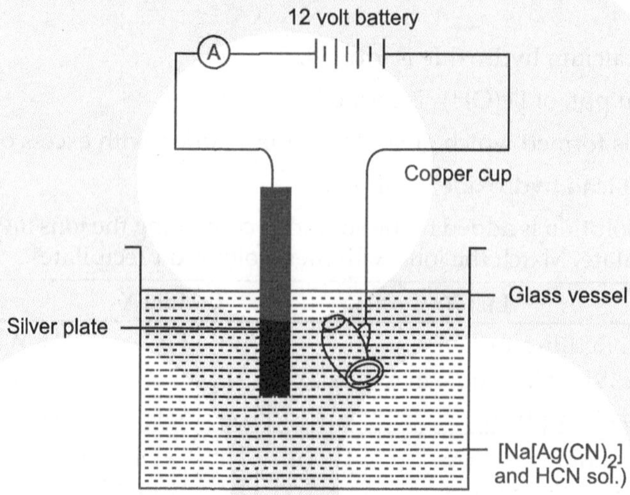

Fig. Silver electrolpating

(i) Name the cathode and anode used during electroplating of silver.

(ii) Name the electrolyte used in this processes.

(iii) Give the dissociation reactions taking place.

(iv) Give the reactions occuring at cathode and anode.

(v) The overall strength of silver ions remain constant in the reaction. Why?

Ans. (i) **Cathode :** Highly cleaned article such as copper cup.

Anode : A plate or rod of silver.

(ii) Sodium argento cyanide ($Na[Ag(CN)_2]$) and hydrocyanic acid (HCN).

(iii)

$$Na[Ag(CN)_2] \rightleftharpoons Na^+ + Ag^+ + 2CN^-$$

$$HCN \rightleftharpoons H^+ + CN^-$$

$$H_2O \rightleftharpoons H^+ + OH^-$$

(iv) **At cathode :**

$$Ag^+ + e^- \longrightarrow Ag(\downarrow)$$

At anode :

$$Ag - e^- \longrightarrow Ag^+$$

(v) Overall strength of silver ions does not change in the electrolyte as the number of Ag^+ ions entering the electrolyte is equal to the number of Ag^+ ions discharged at cathode.

2. Copper sulphate solution is electrolysed using copper electrodes. Study the diagram given below and answer the question that follows :

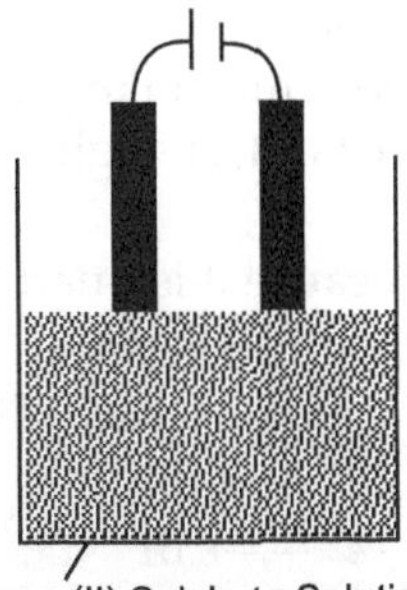

Copper(II) Sulphate Solution

(i) Which electrode to your left or right is known as the oxidising electrode and why?
(ii) Write the equation representing the reaction that occurs.
(iii) State two appropriate observations for the above electrolysis reaction.

Ans. (i) Electrode on the left side is the oxidising electrode because anode is always connected to positive terminal of the battery, where copper atoms lose electrons at this electrode.

(ii) At the Cathode: $Cu^{2+} + 2e^- \longrightarrow Cu$

At the Anode: $Cu - 2e^- \longrightarrow Cu^{2+}$

(iii) Reddish brown copper metal is deposited at cathode and blue colour of aqueous copper(II) sulphate solution remains unchanged.

3. (i) Study the diagram given below and answer the questions that follows :

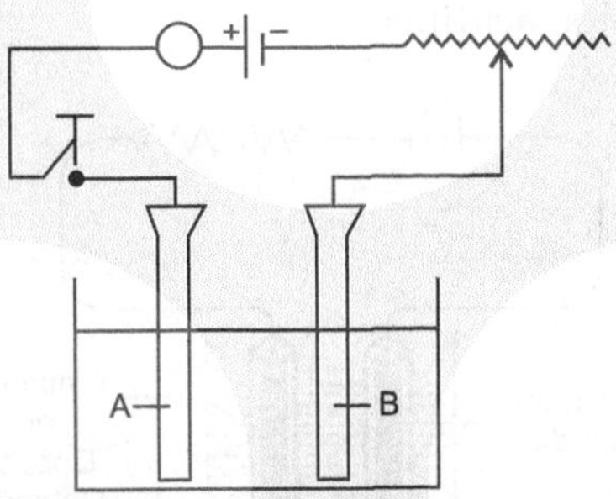

(a) Give the names of the electrodes A and B.
(b) Which electrode is the oxidizing electrode?

(ii) A strip of copper is placed in four different colourless salt solutions. They are KNO_3, $AgNO_3$, $Zn(NO_3)_2$, $Ca(NO_3)_2$. Which one of the solutions will finally turn blue?

(iii) Write the equations of the reactions which take place at the cathode and anode when acidified water is electrolysed.

Ans. (i) (a) A—Anode
B—Cathode
(b) A

(ii) $AgNO_3$ solution

(iii) **Cathode reaction**

$$2H^+ + 2e^- \longrightarrow 2H \longrightarrow H_2$$

Anode reaction

$$4OH^- - 4e^- \longrightarrow H_2O + O_2$$

4. Study the given figure and answer the question that follow :

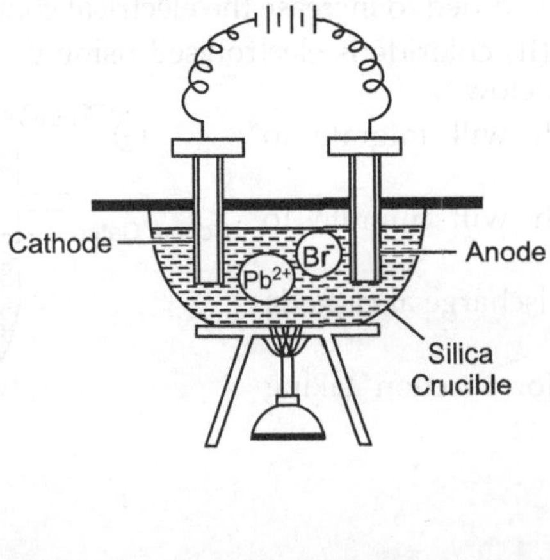

(i) Why silica crucible is used in this type of electrolysis?
(ii) Which anode is preferred and why?
(iii) Why crucible is heated from outside?
(iv) Write the equations of the reaction which take place at the cathode and anode.

Ans. (i) Silica is non-reactive. It can withstand at high temperature and is almost a non-conductor of electricity.
(ii) The graphite anode is preferred because it is unaffected by the reactive bromine vapours.
(iii) The crucible is heated from outside to keep lead bromide in the molten state. So that the ions become free.
(iv) At the cathode : $Pb^{2+} + 2e^- \longrightarrow Pb$
At the cathode : $Br^- - e^- \longrightarrow Br$
$Br + Br \longrightarrow Br_2$

5. An electrolytic cell is set up using two platinum electrodes and an aqueous solution of copper(II) sulphate.
(i) Draw a labelled diagram of the electrolytic cell.
(ii) Name the ions present in the cell.
(iii) Name the ions migrating towards the anode.
(iv) Name the ions migrating towards the cathode.
(v) Name the ions which will not be discharged at electrodes during electrolysis.
(vi) Write the reaction at the cathode.
(vii) Write the reaction at the anode.
(viii) Name the spectator ions in the solution.
(ix) Why the electrolytic solution acidified.

Ans. (i)

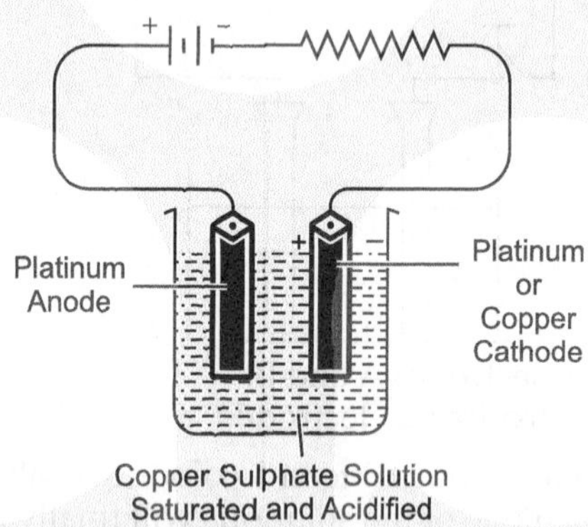

(ii) The ions present in the cell are Cu^{2+}, H^+, SO_4^{2-} and OH^-.
(iii) The ions migrating towards the anode are SO_4^{2-} and OH^+.
(iv) The ions migrating towards the cathode is $Cu^{2+}$and H^+.
(v) The ions that will not be discharged at electrode during electrolysis are H^+ and OH^-.
(vi) Reaction at cathode :

$$Cu^{2+} - 2e^- \longrightarrow Cu$$

(vii) Reaction at anode :

$$4OH^- - 4e^- \longrightarrow 4OH$$
$$2OH + 2OH \longrightarrow 2H_2O + O_2$$

(viii) The spectator ions present in the solution are SO_4^{2-} and OH^-.
(ix) Electrolytic solution is acidified to increase the electrical conductivity of electrolyte.

6. A saturated aqueous copper(II) chloride is electrolysed using graphite anode and copper cathode as illustrated in diagram given below :
(i) Name the ions which will migrate to cathode.
(ii) Name the ions which will migrate to anode.
(iii) Which ion is likely to discharge at cathode and why?
(iv) Write ionic equation for reaction taking place at cathode.

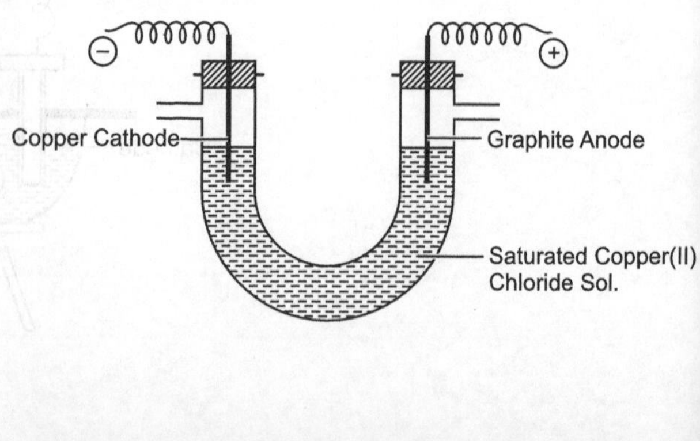

(v) Which ion is likely to discharge at anode and why?

(vi) Write ionic equation for reaction taking place at anode.

(vii) If electric current is passed for a very long time, state colour change taking place in electrolyte. State one reason for the change.

(viii) Give one reason for using graphite anode, rather than copper anode.

(ix) Name the gas liberated at cathode after the colour changes in electrolyte.

Ans. (i) Copper ions (Cu^{2+}) and hydrogen ions H^+ migrate to cathode.

(ii) Chloride ions (Cl^-) and hydroxyl ions OH^- migrate to anode.

(iii) Copper ions (Cu^{2+}) are likely to discharge at cathode, because their position is lower than hydrogen ions H^+ in electrochemical series.

(iv) $Cu^{2+} + 2e^- \longrightarrow Cu$

(v) Hydroxyl ions (OH^-) are likely to discharge at anode, because their position is lower than chloride ion (Cl^-) in electrochemical series.

(vi) $4OH^- - 4e^- \longrightarrow 2H_2O + O_2$ (g)

(vii) The electrolyte gets decolourised. It is because, the blue colour of electrolyte is due to the presence of Cu^{2+} ions. As Cu^{2+} ions discharge at cathode, therefore, their concentration in electrolyte decreases. Thus, gradually blue colour fades away.

(viii) In such a situation the copper atoms on copper anode ionise and enter into electrolyte. Thus, size of copper anode gradually decreases. This is not possible in case of graphite anode.

(ix) Hydrogen gas is liberated at cathode.

7. Compare the compounds carbon tetrachloride and sodium chloride with regard to solubility in water and electrical conductivity.

Ans.

Property	Carbon tetrachloride	Sodium chloride
Solubility in water	Insoluble in water	Soluble in water
Electrical conductivity	Non-conductor of electricity	Good conductor of electricity

Chapter 7. Metallurgy

1. The given sketch of an electrolytic cell used in the extraction of aluminium :

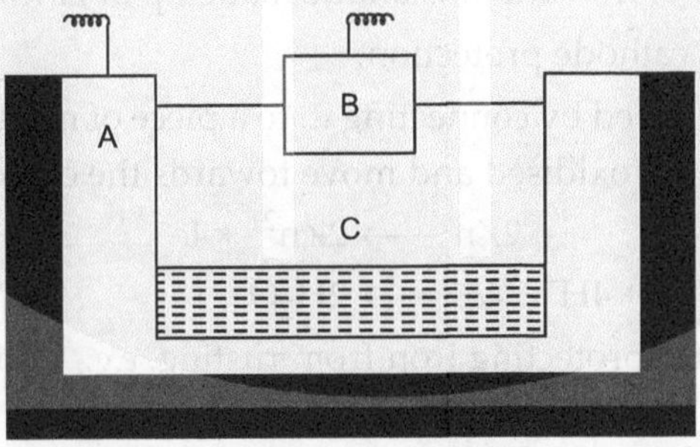

(i) What is the substance of which the electrode A and B are made?

(ii) At which electrode (A or B) is the aluminium formed?

(iii) What are the two aluminium compound in the electrolyte C?

(iv) Why is it necessary for electrode B to be continuously replaced?

Ans. (i) Carbon (Graphite)

(ii) A

(iii) Aluminium oxide, Alumina or cryolite (sodium aluminium fluoride).

(iv) It is necessary for electrode B to be continuously replaced because it burns away in the presence of oxygen produced or consumed.

2. The given figure illustrates the refining of aluminium by Hoope's process.

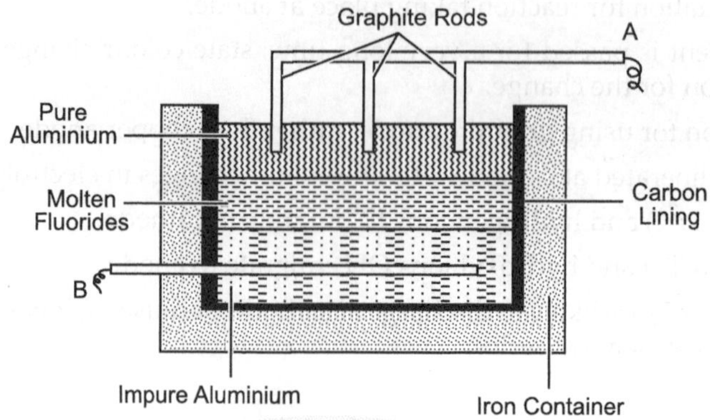

(i) Which of A and B is the cathode and which one is the anode?

(ii) What is the electrolyte in the tank?

(iii) What material is used for the cathode?

Ans. (i) A—Cathode, B—Anode.

(ii) Mixture of fluorides.

(iii) Graphite

3. Study the given figure and answer the following questions :

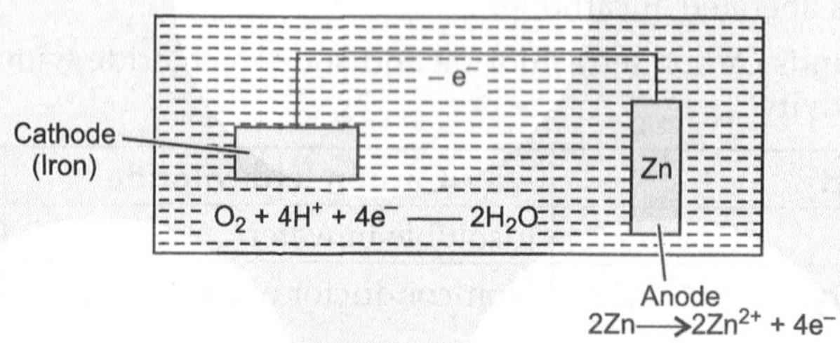

(i) What are the elements (any two) of which the electrodes are made?

(ii) Name the process shown.

(iii) Explain the process and write the reactions which takes place at the anode and cathode.

(iv) Where this method is used?

Ans. (i) A is cathode made up of iron and B is anode made up of zinc or magnesium.

(ii) The process shown is cathode protection.

(iii) The iron cathode is to saved by connecting it, to a piece of more electropositive metal (anode). The electrons from anode get oxidised and move towards the cathode, to cover it.

At anode : $2Zn \longrightarrow 2Zn^{2+} + 4e^-$

At cathode : $O_2 + 4H^+ + 4e^- \longrightarrow 2H_2O$

(iv) This method is used for protecting iron from rusting. *e.g.*, underground sewer pipes and storage tanks.

4. (i) Name the process and the element extracted by the above process as shown in the figure.

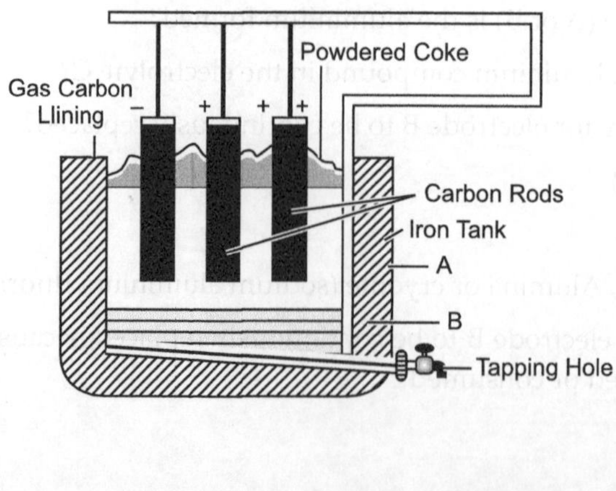

(ii) Give the function of three components of electrolyte.

(iii) Why is electrolyte covered with coke?

(iv) Write the electrolytic reaction taking place at cathode?

Ans. (i) Aluminium, Hall and Herault's process.

(ii) (a) Alumina (Al_2O_3) is the main aluminium yielding compound.

(b) Cryolite [Na_3AlF_6] acts as a solvent and lowers the fusion temperature from 2050°C to 950°C.

(c) Fluorspar (CaF_2) acts as a solvent and increases the conductivity of electrolytic mixture.

(iii) A layer of powdered coke is sprinkled over the surface of the electrolyte mixture because it reduces the heat loss by radiation and prevents carbon anode from burning in air.

(iv) At cathode $Al^{3+} + 3e^- \longrightarrow \underset{\text{Aluminium}}{Al}$

Aluminium formed sinks to the bottom of the tank and is periodically tapped off.

5. List 1 contains the metals or alloys, (i), (ii), (iii), (iv), (v) and list 2 contains their uses A, B, C, D, E.

List 1	List 2
(i) Aluminium	A. Steel making
(ii) Lead	B. Aeroplane wings
(iii) Brass	C. Galvanizing
(iv) Iron	D. Radiation shield
(v) Zinc	E. Electrical fittings

Copy and complete the following table writing down the letter for the correct use of each metal. An answer may be used only once. The first has been done for you.

Metal	(i)	(ii)	(iii)	(iv)	(v)
Use	B				

Ans.

Metal	(i)	(ii)	(iii)	(iv)	(v)
Use	B	D	E	A	C

6. The table below compare some properties of metals and non-metals. Write down the missing statements (i) to (iv).

Metal	Non-Metal
(i)	Poor conductor of heat
(ii) Malleable	....
(iii) Forms positive ions	....
(iv)	Form acidic oxide.

Ans. (i) Metals are good conductor of heat.

(ii) Non-metals are non-malleable [Brittle].

(iii) Non-metals form negative ions [Anions].

(iv) Metals form basic oxides.

7. An aqueous solution of nickel (II) sulphate was electrolyzed using nickel electrodes. Observe the diagram and answer the questions that follow :*

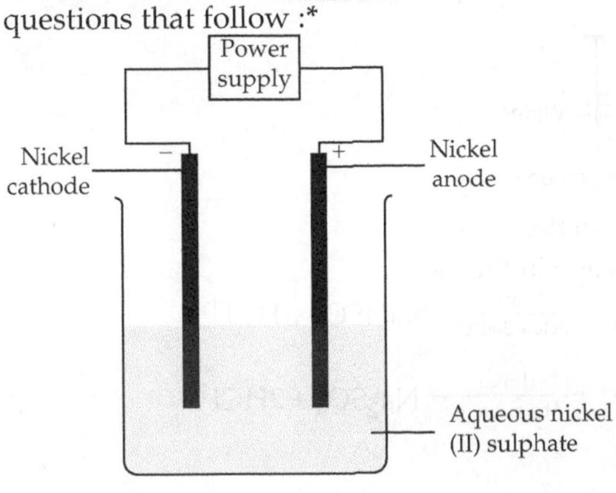

* **Frequently asked previous years Board Exam Questions.**

(i) What do you observe at the **cathode** and **anode** respectively ?
(ii) Name the cation that remains as a spectator ion in the solution.
(iii) Which equation for the reaction at the anode is correct ?
1. $Ni \rightarrow Ni^{2+} + 2e^-$ 2. $Ni + 2e^- \rightarrow Ni^{2+}$
3. $Ni^{2+} \rightarrow Ni + 2e^-$ 4. $Ni^{2+} + 2e^- \rightarrow Ni$

Ans. (i) At cathode reduction takes place and deposition of Ni takes place as Ni^{2+} ions from solution convert to Ni metal.

$$Ni^{2+} + 2e^- \rightarrow Ni$$

At anode oxidation takes place and Ni metal converts to Ni^{2+} ions.

$$Ni \rightarrow Ni^{2+} + 2e^-$$

(ii) H^+ is the spectator cation.
(iii) 1. $Ni \rightarrow Ni^{2+} + 2e^-$

Chapter 8. (a) Study of Compounds : Hydrogen Chloride

1. Refer to the flow chart diagram below and give balanced equations with conditions, if any, for the following conversions A to D.

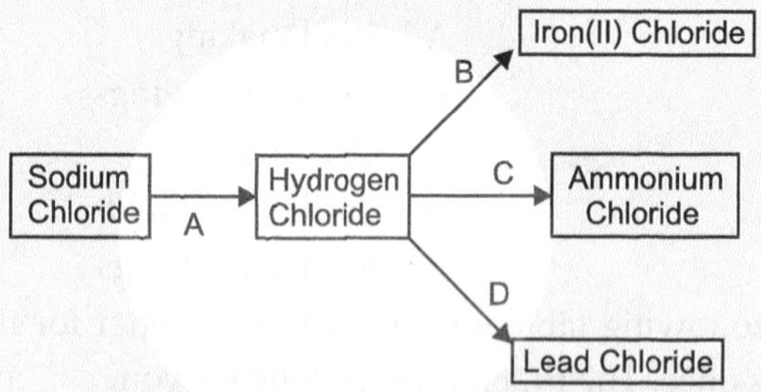

Ans. A. $NaCl + \underset{\text{Conc.}}{H_2SO_4} \xrightarrow{\Delta < 200°C} NaHSO_4 + \underset{\text{Hydrogen chloride}}{HCl\uparrow}$

B. $2HCl + Fe \longrightarrow \underset{\text{Iron(II) chloride}}{FeCl_2} + H_2$

C. $HCl_{(g)} + NH_{3(g)} \longrightarrow \underset{\text{Ammonium chloride}}{NH_4Cl}$

D. $\underset{\text{(dil.)}}{2HCl} + Pb(NO_3)_{2\,aq} \longrightarrow \underset{\text{Lead chloride}}{PbCl_2\downarrow} + 2HNO_3$

2. In the laboratory preparation of hydrochloric acid, HCl gas is dissolved in water.
(i) Draw a diagram to show the arrangement used for the absorption of HCl in water.
(ii) Why is such an arrangement necessary? Give two reasons.
(iii) Write the chemical equations for the laboratory preparation of HCl gas when the reactants are :
(A) below 200°C (B) above 200°C

Ans. (i)

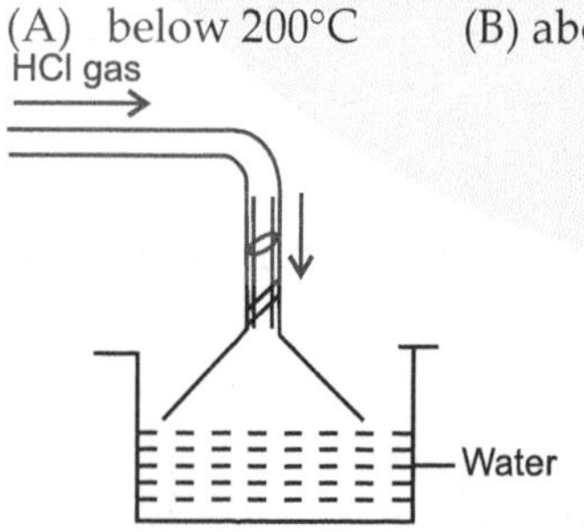

Inverted funnel arrangement

(ii) (a) To check back suction.
(b) To check its escape in the air.

(iii) (A) $NaCl + \underset{\text{conc.}}{H_2SO_4} \xrightarrow[\text{Below 200°C}]{\text{Heated}} NaHSO_4 + HCl\uparrow$

(B) $2NaCl + \underset{\text{conc.}}{H_2SO_4} \xrightarrow[\text{Above 200°C}]{\text{Heated}} Na_2SO_4 + 2HCl\uparrow$

3. The diagram shows an apparatus for the laboratory preparation of hydrogen chloride.

(i) Identify A and B.

(ii) Write the equation for the reaction.

(iii) How would you check whether or not the gas jar is filled with hydrogen chloride?

(iv) What does the method of collection tell you about the density of hydrogen chloride?

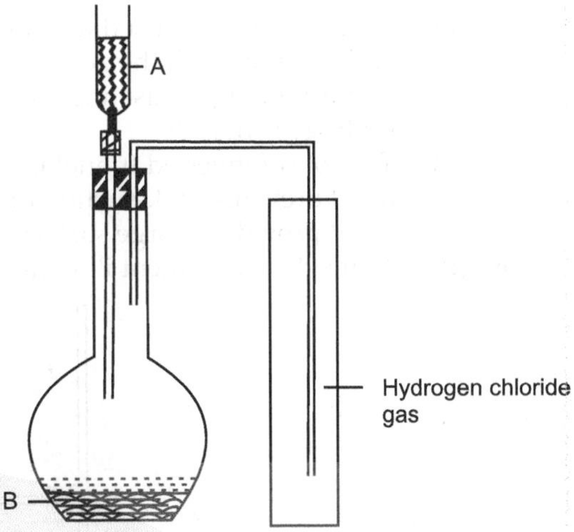

Ans. (i) A → Conc. H_2SO_4, B → NaCl

(ii) $NaCl + H_2SO_4 \rightarrow NaHSO_4 + HCl$ (NaHSO₄: sodium hydrogen sulphate)

(iii) If a moist blue litmus paper is brought near the mouth of gas jar it turns red, then the gas jar is filled with HCl.

(iv) Hydrogen chloride is denser than air.

4. Study the figure given below and answer the questions that follow :

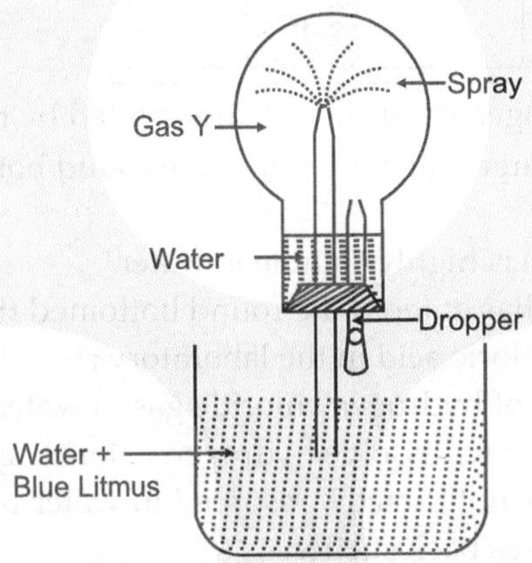

(i) Identify the gas Y.

(ii) What property of gas Y does this experiment demonstrate?

(iii) Name another gas which has the same property and can be demonstrated through this experiment.

Ans. (i) Y is Hydrochloride (HCl) gas.

(ii) Gas Y is highly soluble in water.

(iii) Ammonia gas.

5. The given figure shown is for the preparation of an acid.

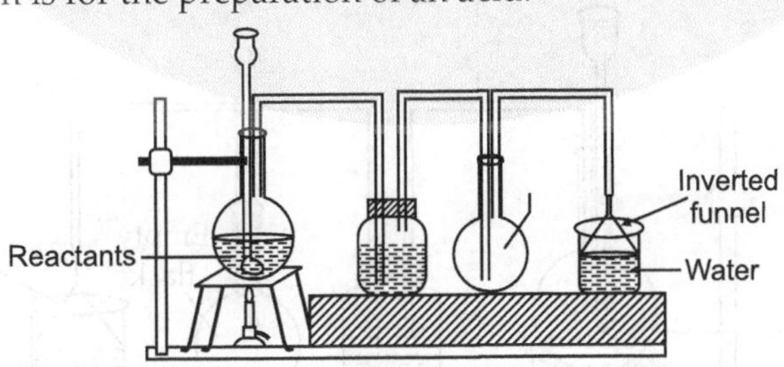

(i) Name the acid prepared by this method.

(ii) Name the reactants used.

(iii) Why an empty flask is used?

(iv) What is the drying agent used? Why is this drying agent chosen?

(v) What is the role of inverted funnel in the arrangement?

Ans. (i) Hydrochloric acid.

(ii) Sodium chloride and conc. sulphuric acid.

(iii) An empty flask is used to prevent back suction. If back suction occurs, the water will collect in it and will not reach the generating flask.

(iv) The drying agent used is concentrated sulphuric acid. It is used because it does not react with hydrogen chloride.

(v) The role of inverted funnel in the arrangement is :

(a) It prevents back-suction of water.

(b) It provides a large surface area for absorption of HCl gas.

6. (i) Name the experiment illustrated in the diagram.

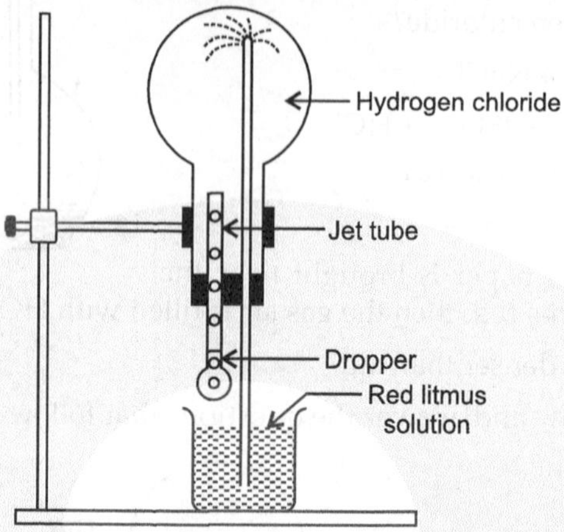

(ii) Which property of hydrogen chloride is demonstrated by this experiment.

(iii) State the colour of the water that has entered the round bottomed flask.

Ans. (i) Fountain experiment.

(ii) Hydrogen chloride (HCl) is highly soluble in water.

(iii) The colour of water that has entered the round bottomed flask is red.

6. For the preparation of hydrochloric acid in the laboratory :*

(i) Why is direct absorption of hydrogen chloride gas in water not feasible ?

(ii) What arrangement is done to dissolve hydrogen chloride gas in water ?

Ans. (i) Hydrogen chloride gas is not directly absorbed in water because hydrogen chloride gas is higly soluble in water and causes back suction.

(ii) Hydrogen chloride gas is produced by reacting sodium chloride and sulphuric acid in a reaction vessel, the outlet from the vessel containing hydrogen chloride gas is put into another vessel containing sulphuric acid which helps to obtain dry hydrogen chloride gas. The dry gas then reaches to the vessel containing water through an empty vessel (this empty vessel is kept for accommodation of any back suction of water during absorption of hydrogen chloride gas in water). After travelling the empty vessel, hydrogen chloride gas is introduced to the vessel containing water through a pipe fitted with a funnel at the end and over the water vessel, this ensures maximum surface area for hydrochloric acid gas absorption in water.

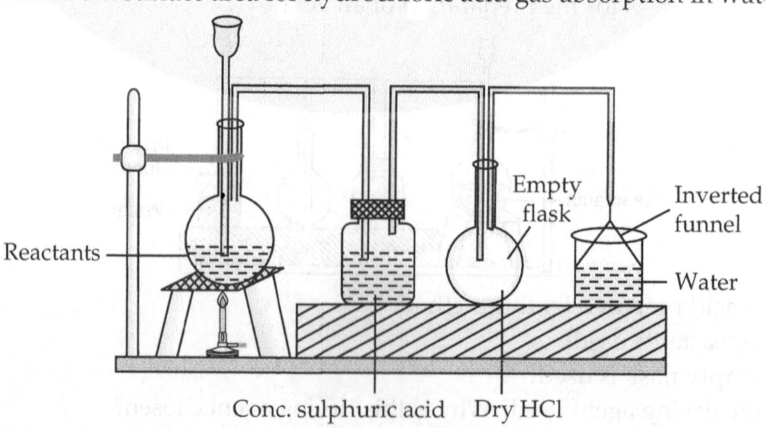

* Frequently asked previous years Board Exam Questions.

Chapter 8. (b) Study of Compounds : Ammonia and Nitric Acid

1.

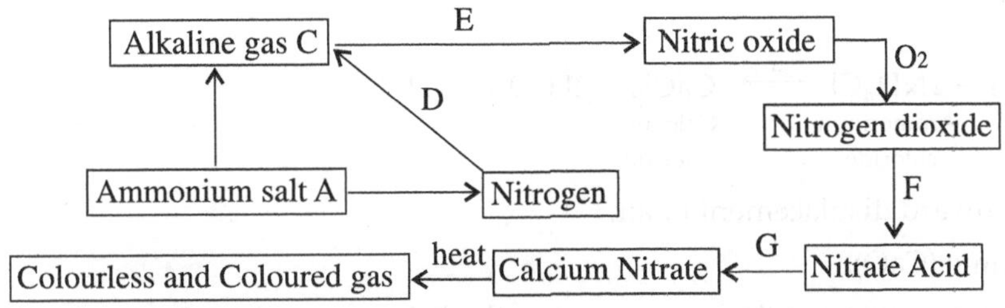

(i) Name (a) the ammonium salt A, (b) alkaline gas C.

(ii) How the conversion D is carried out? State all the conditions like temperature, pressure and catalyst.

(iii) (a) How is C converted to nitric oxide? Write the equation.

(b) What is the name of the process?

(c) How is temperature maintained in above process?

(d) Write the equation for conversion of F and G.

(iv) Name the coloured gas and the colourless gas.

Ans. (i) (a) Ammonium nitrate, NH_4NO_3

(b) Ammonia.

(ii) D is mixed with hydrogen in the ratio of 1 : 3, compressed to a pressure of 200 to 500 atmosphere and passes over a catalyst (iron) heated to 450 to 500°C.

(iii) (a) By passing the gas with excess of air over platinum gauze heated to 800°C.

$$4NH_3 + 5O_2 \longrightarrow 4NO + 6H_2O$$

(b) Ostwald process.

(c) The oxidation of ammonia to nitric oxide is exothermic reaction and once the reaction is started it maintains the temperature of the platinum gauze.

(d)
$$2N_2O + 2H_2O + O_2 \longrightarrow 4HNO_3$$
$$Ca(OH)_2 + 2HNO_3 \longrightarrow Ca(NO_3)_2 + 2H_2O$$

(iv) **Coloured gas :** Nitrogen dioxide and

Colourless gas : Oxygen

2. The diagram shows an experimental set up for the laboratory preparation of a pungent smelling gas. The gas is alkaline in nature.

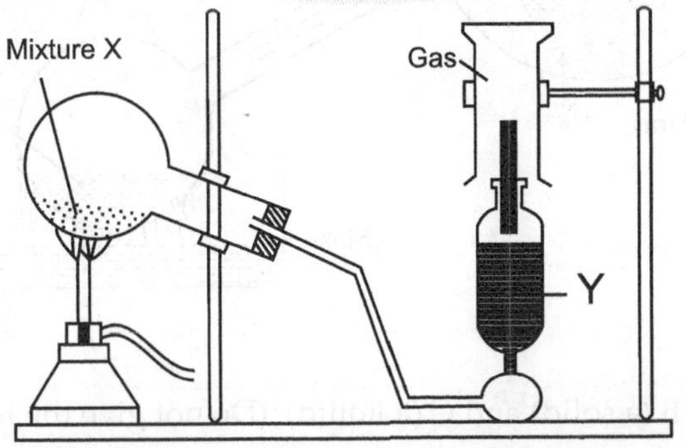

(i) Name the gas collected in the jar.

(ii) Write the balanced equation for the above preparation.

(iii) How is the gas being collected?
(iv) Name the drying agent used.
(v) How will you find that the jar is full of gas?

Ans. (i) Ammonia.

(ii) $Ca(OH)_2 + 2NH_4Cl \xrightarrow{\Delta} CaCl_2 + 2H_2O + 2NH_3\uparrow$
Calcium oxide, Ammonium chloride, Calcium chloride

(iii) By downward displacement of air.
(iv) Quick lime (CaO).
(v) By bringing a wet red litmus paper near the brim (mouth) of the gas jar. When jar is full of gas litmus will turn blue.

3. Dry ammonia gas is passed over black substance as shown in figure below :

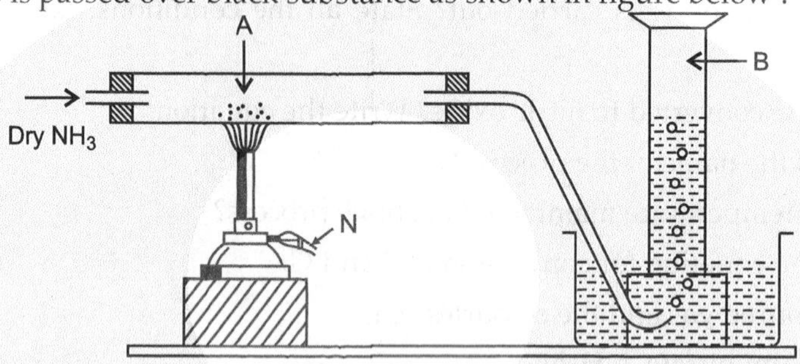

(i) Name the black substance A.
(ii) Name the gas evolved B.
(iii) Write a balanced equation for the reaction of ammonia with A.
(iv) What do you observe when ammonia is passed over copper oxide?
(v) State the property illustrated in the above reaction (iii).

Ans. (i) Copper(II) oxide
(ii) Nitrogen gas
(iii) $3CuO + 2NH_3 \xrightarrow{\Delta} 3Cu + 3H_2O + N_2$
(iv) The black mass changes to reddish powdery mass.
(v) Ammonia is a reducing agent.

4. The figure given below illustrates the apparatus used in the laboratory preparation of nitric acid.

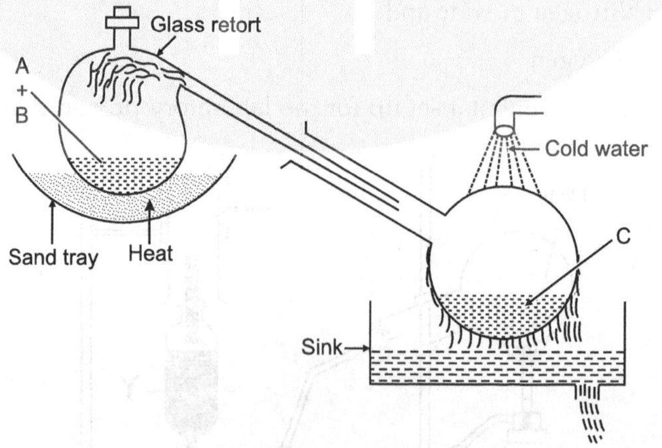

(i) Name A (a liquid), B (a solid) and C (a liquid). (Do not give the formulae)
(ii) Write an equation to show how nitric acid undergoes decomposition.
(iii) Write the equation for the reaction in which copper is oxidized by concentrated nitric acid.

Figure or Table Based Questions

Ans. (i) A—Conc. Sulphuric acid, B—Potassium nitrate or Sodium nitrate, C—Nitric acid

(ii) $4HNO_3 \longrightarrow 2H_2O + 2NO_2 + O_2$

(iii) $Cu + 4HNO_3 \longrightarrow Cu(NO_3)_2 + 2H_2O + 2NO_2$

5. (i) Draw a neat sketch for the manufacture of ammonia by Haber's process.

(ii) Discuss the principles involved in Haber's process.

Ans. (i)

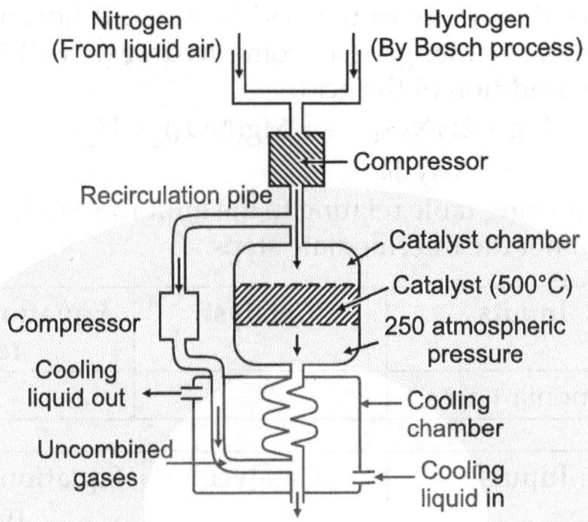

(ii) The principles involved in Haber's process are :

In this process, a mixture of pure, dry nitrogen and hydrogen in the ratio of 1 : 3 is compressed to a pressure of 200 to 900 atmospheres. The compressed gases are passed over heated catalyst (finally divided molybdenum, *i.e.*, Fe/Mo) at 450° to 500°C.

$$\underset{\text{1 vol.}}{N_2} + \underset{\text{3 vols.}}{3H_2} \rightleftharpoons \underset{\text{2 vols.}}{2NH_3} \; ; \Delta H = -92{\cdot}38 \text{ kJ}$$

The above reaction is exothermic and reversible. It is accompanied by decrease in volume. Ammonia formed is immediately removed from the catalytic chamber and allowed to condense to form liquid ammonia.

6. Study the scheme for dilute nitric acid.

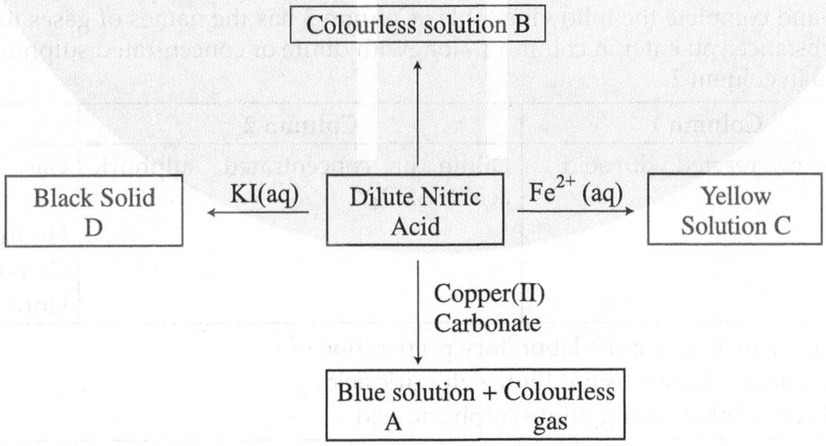

(i) Give the name or formula of

(a) Solution A (b) Solution B

(c) A cation in solution C (d) Solid D

(ii) Which property of nitric acid is shown by its reaction with iron(II) ions and aqueous potassium iodide solution?

(iii) Describe one other reaction of dilute nitric acid not shown in the reaction scheme, that is typical of a strong acid.

Ans. (i) (a) Copper(II) nitrate $[Cu(NO_3)_2]$ (c) Fe^{3+}
(b) Ammonium nitrate (d) Iodine

(ii) Oxidising property of nitric acid.

(iii) It is a characteristic of strong acids that when dilute, they react with the more electropositive metals, liberating hydrogen gas.

$$Zn + 2H^+ \longrightarrow Zn^{2+} + H_2$$

This reaction can not occur with nitric acid as it is a powerful oxidising agent. Any hydrogen initially produced is at once oxidised by more of the acid to water. If however very dilute nitric acid is used (about 1%) with magnesium or manganese, some hydrogen will be produced, escaping oxidation because of the very dilute condition of the acid.

$$\underset{\text{Very dil.}}{Mg + 2HNO_3} \longrightarrow Mg(NO_3)_2 + H_2$$

7. Copy and complete the following table relating to the important industrial process. Output refers to the product of the process and not the intermediate steps.

Name of process	Inputs	Catalyst	Equation for catalyst reaction	Output
	Ammonia + air			Nitric acid

Ans.

Name of process	Inputs	Catalyst	Equation for catalysed reaction	Output
Ostwald's process	Ammonia + air	Platinum	$4NH_3 + 5O_2 \xrightarrow[800°C]{Pt.} 4NO + 6H_2O + \text{heat}$	Nitric acid

8. Copy and complete the following table relating to important industrial process :

Name of process	Temperature	Catalyst	Equation for catalysed reaction
Haber's process			

Ans.

Name of process	Temperature	Catalyst	Equation for catalysed reaction
Haber's process	450° – 500 °C	Finely divided iron	$N_2 + 3H_2 \longrightarrow 2NH_3 + \text{Heat}$

Chapter 8. (c) Study of Compounds : Sulphuric Acid

1. (i) Copy and complete the following table : Column 3 has the names of gases to be prepared using the substance you enter in column 1 alongwith dilute or concentrated sulphuric acid as indicated by you in column 2.

Column 1	Column 2	Column 3
Substance reacted with acid	Dilute or concentrated sulphuric acid (H_2SO_4)	Gas
(a)		Hydrogen
(b)		Carbon dioxide
(c)		Only chlorine

(ii) Write the equations for the laboratory preparation of :

(a) Sodium sulphate using dilute sulphuric acid.

(b) Lead sulphate using dilute sulphuric acid.

Ans. (i)

Column 1	Column 2	Column 3
Substance reacted with acid	Dilute or concentrated sulphuric acid (H_2SO_4)	Gas
(a) Zinc	Dilute H_2SO_4	Hydrogen
(b) Copper carbonate	Dilute H_2SO_4	Carbon dioxide
(c) Sodium chloride + Manganese oxide	Concentrated H_2SO_4	Only chlorine

(ii) (a) $2NaOH + H_2SO_4 \longrightarrow Na_2SO_4 + 2H_2O$

(b) $Pb(OH)_2 + H_2SO_4 \longrightarrow PbSO_4 + 2H_2O$

2. Study the diagram given below, which illustrates the manufacture of sulphuric acid.

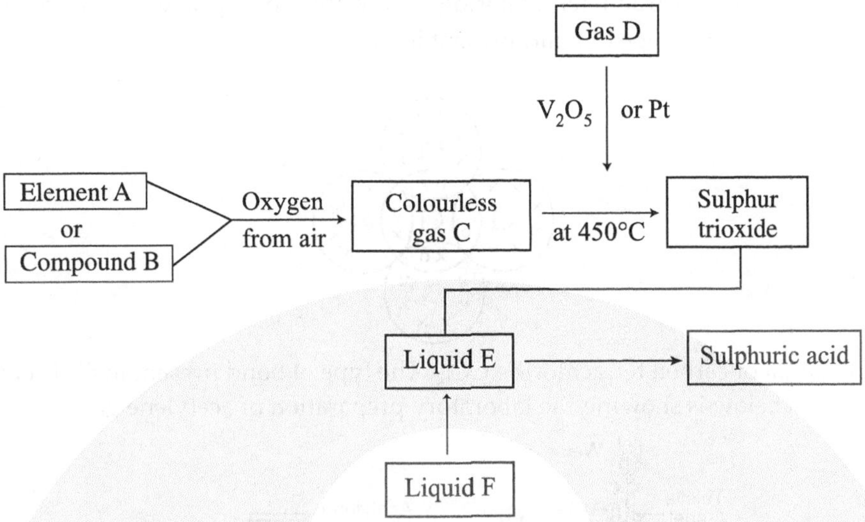

(i) Write the names of the substances A to F.

(ii) Describe how gas C could be identified.

(iii) Explain the purpose of V_2O_5 or Pt.

Ans. (i) A—Sulphur B—Iron pyrites

C—Sulphur D—Oxygen

E—Concentrated sulphuric acid F—Water

(ii) The gas C will turn acidified potassium dichromate paper green.

(iii) V_2O_5 or Pt acts as a catalyst and increases the rate of formation of sulphur trioxide from sulphur dioxide and oxygen.

3. (i) Name the catalyst which helps in the conversion of sulphur dioxide to sulphur trioxide in step C.

(ii) In the contact process for the manufacture of sulphuric acid, sulphur trioxide is not converted to sulphuric acid by reacting it with water. Instead a two step procedure is used. Write the equations for the two steps involved in D.

(iii) What type of substance will liberate sulphur dioxide from sodium sulphite in step E?

(iv) Write the equation for the reaction by which sulphur dioxide is converted to sodium sulphite in step F.

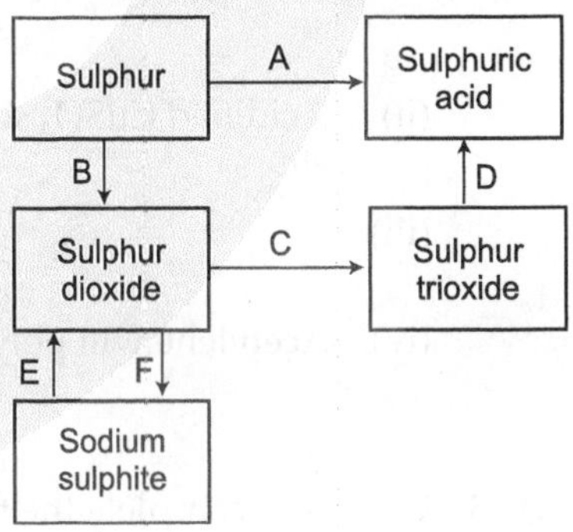

Ans. (i) Platinum (Pt)

(ii)
$$SO_3 + H_2SO_4 \longrightarrow \underset{\text{(Oleum)}}{H_2S_2O_7}$$

$$H_2S_2O_7 + H_2O \longrightarrow \underset{\text{(Sulphuric acid)}}{2H_2SO_4}$$

(iii) Any mineral acid HCl, H_2SO_4 or HNO_3

$$CaSO_3 + 2HCl \longrightarrow CaCl_2 + H_2O + SO_2$$

$$CaSO_3 + H2SO_4 \longrightarrow CaSO_4 + H_2O + SO_2$$

$$CaSO_3 + 2HNO_3 \longrightarrow Ca(NO_3)_2 + H_2O + SO_2$$

(iv)
$$2NaOH + SO_2 \longrightarrow Na_2SO_3 + H_2O$$

Chapter 9. Organic Chemistry

Q. 1. Methane is the first member of alkane, when it is treated with excess of chlorine in the presence of diffused sunlight forms carbon tetrachloride. Draw the appropriate structural formula of carbon tetrachloride and state the type of bond present in it.

Ans.

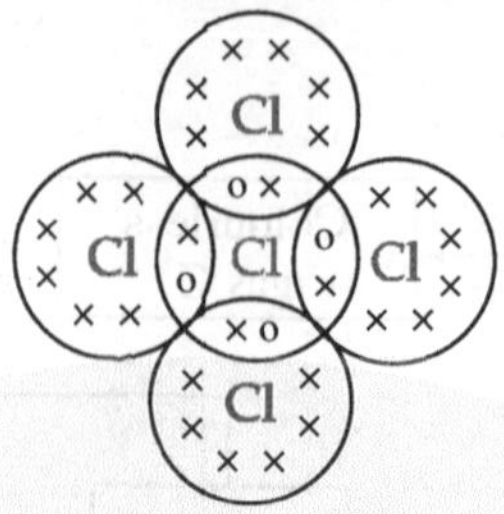

Structural formula of carbon tetrachloride CCl_4. The type of bond present in CCl_4 is covalent bond.

2. The figure given below is showing the laboratory preparation of acetylene gas.

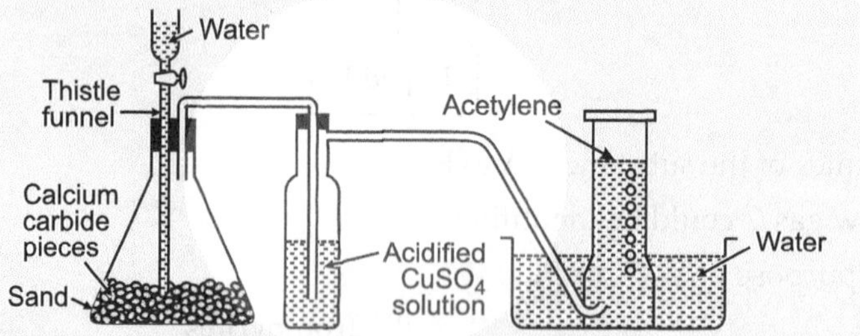

(i) How it is prepared in the laboratory?
(ii) What is the function of acidified copper sulphate solution?
(iii) Give a reaction in which acetylene gas is prepared by synthesis reaction.
(iv) What happens when acetylene is heated in a copper tube at 600°C?

Ans. (i) Laboratory preparation of acetylene gas (ethyne gas) : When calcium carbide is treated with water, it forms calcium hydroxide, with the liberation of acetylene.

$$\underset{\text{Calcium carbide}}{CaC_2} + 2H_2O \longrightarrow \underset{\text{Calcium hydroxide}}{Ca(OH)_2} + \underset{\text{Acetylene}}{C_2H_2\uparrow}$$

(ii) Acidified $CuSO_4$ solution is used to absorb impurities of phosphene hydrogen sulphide, ammonia etc.

(iii)
$$2C + H_2 \xrightarrow{\Delta} \underset{\text{Acetylene}}{CH \equiv CH}$$

(iv) Acetylene will polymerise in the copper tube to form benzene.

$$\underset{\text{Acetylene}}{3C_2H_2} \xrightarrow[600°C]{\text{Cu tube}} \underset{\text{Benzene}}{C_6H_6}$$

Q. 3. Copy and complete the following table which relates to three homologous series of Hydrocarbons :

General formula	C_nH_{2n-2}	C_nH_{2n}	C_nH_{2n+2}
IUPAC name of the homologous series			
Characteristic bond type			Single bonds
IUPAC name of the first member of the series			
Type of reaction with chlorine.		Addition	

Ans.

General formula	C_nH_{2n-2}	C_nH_{2n}	C_nH_{2n+2}
IUPAC name of the homologous series	Alkynes	Alkynes	Alkanes

Characteristic bond type	Triple bond	Double bond	Single bonds
IUPAC name of the first member of the series	Ethyne	Ethene	Methane
Type of reaction with chlorine.	Addition	Addition	Substitution

Q . 4. Match the options A to E with the statement (i) to (v).

A	Alkynes	(i) No. of molecules in 22.4 dm^3 of carbon dioxide at STP
B	Alkanes	(ii) An element with electronic configuration 2, 8, 8, 3
C	Iron	(iii) C_nH_{2n+2}
D	6.023×10^{23}	(iv) C_nH_{2n-2}
E	Metal	(v) The metal that forms two types of ion

Ans.

A	Alkynes	(iv) C_nH_{2n-2}
B	Alkanes	(iii) C_nH_{2n+2}
C	Iron	(v) The metal that forms two types of ion
D	6.023×10^{23}	(i) No. of molecules in 22.4 dm^3 of carbon dioxide at STP
E	Metal	(ii) An element with electronic configuration 2, 8, 8, 3

Chapter 10. Practical Chemistry

1. Salts A, B, C, D and E undergo reactions (i) to (v) respectively. Identify the anion present in these salts on the basis of these reactions. Tabulate your answers in the format given below :

(i) When silver nitrate solution is added to a solution of A, a white precipitate, insoluble in dilute nitric acid, is formed.

(ii) Addition of dilute hydrochloric acid to B produces a gas which turns lead acetate paper black.

(iii) When a freshly prepared solution of ferrous sulphate is added to a solution of C and concentrated sulphuric acid is gently poured from the side of the test-tube, a brown ring is formed.

(iv) When dilute sulphuric acid is added to D, a gas is produced which turns acidified potassium dichromate solution from orange to green.

(v) Addition of dilute hydrochloric acid to E produces an effervescence. The gas produced turns lime water milky but does not affect acidified potassium dichromate solution.

Salt	Anion
A	
B	
C	
D	
E	

Ans. (i) A—Chloride (Cl^-) (ii) B—Sulphide (S^{2-})

(iii) C—Nitrate (NO_3^-) (iv) D—Sulphite (SO_3^{2-}) (v) E—Carbonate (CO_3^{2-})

2. Sodium hydroxide solution is added first in a small quantity, then in excess to the aqueous salt solutions of copper(II) sulphate, zinc nitrate, lead nitrate, calcium chloride and iron(III) sulphate. Copy the following table and write the colour of the precipitate in (i) to (v) and the nature of the precipitate (soluble or insoluble) in (vi) to (x).

Aqueous salt solution	Colour of precipitate when NaOH is added in a small quantity	Nature of precipitate (soluble or insoluble) when NaOH is added in excess
Copper(II) sulphate	(i)	(vi)
Zinc nitrate	(ii)	(vii)
Lead nitrate	(iii)	(viii)
Calcium chloride	(iv)	(ix)
Iron(III) sulphate	(v)	(x)

Ans.

Aqueous salt solution	Colour of precipitate when NaOH is added in a small quantity	Nature of precipitate (soluble or insoluble) when NaOH is added in excess
Copper(II) sulphate Zinc nitrate Lead nitrate Calcium chloride Iron(III) sulphate	Blue White White White Reddish Brown	Insoluble Soluble Soluble Insoluble Insoluble

3. Match the gases given in column I to the identification of the gases mentioned in column II :*

Column I	Column II
(i) Hydrogen sulphide	A. Turns acidified potassium dichromate solution green.
(ii) Nitric oxide	B. Turns lime water milky.
(iii) Carbon dioxide	C. Turns reddish brown when it reacts with oxygen.
(iv) Sulphur dioxide	D. Turns moist lead acetate paper silvery black.

Ans.

Column I	Column II
(i) Hydrogen sulphide	D. Turns moist lead acetate paper silvery black.
(ii) Nitric oxide	C. Turns reddish brown when reacts with oxygen.
(iii) Carbon dioxide	B. Turns lime water milky.
(iv) Sulphur dioxide	A. Turns acidified potassium dichromate solution green.

❑

* **Frequently asked previous years Board Exam Questions.**

ICSE Solved Paper 2020

(Two Hours)

Answer to this paper must be written on the paper provided separately.

*You will **not** be allowed to write during the first 15 minutes.*

This time is to be spent in reading the Question paper.

The time given at the head of this Paper is the time allowed for writing the answers.

Section I** is compulsory. Attempt **any four** questions from **Section II.

The intended marks for questions or parts of questions are given in brackets [].

SECTION-I (40 Marks)

*Attempt **all** questions from this Section.*

Question 1.

(a) *Choose the correct answer from the options given below :* **[5]**

(i) *The element with highest ionization potential, is :*

(A) Hydrogen *(B) Caesium*

(C) Radon *(D) Helium*

(ii) *The inert electrode used in the electrolysis of acidified water, is :*

(A) Nickel *(B) Platinum*

(C) Copper *(D) Silver*

(iii) *A compound with low boiling point, is :*

(A) Sodium chloride

(B) Calcium chloride

(C) Potassium chloride

(D) Carbon tetrachloride

(iv) *The acid which can produce carbon from cane sugar, is :*

(A) Concentrated Hydrochloric acid

(B) Concentrated Nitric acid

(C) Concentrated Sulphuric acid

(D) Concentrated Acetic acid

(v) *The organic compound having a triple carbon-carbon covalent bond, is :*

(A) C_3H_4 *(B)* C_3H_6

(C) C_3H_8 *(D)* C_4H_{10}

(b) *State one relevant observation for each of the following reactions :* **[5]**

(i) *Action of concentrated nitric acid on copper.*

(ii) *Addition of excess ammonium hydroxide into copper sulphate solution.*

(iii) *A piece of sodium metal is put into ethanol at room temperature.*

(iv) *Zinc carbonate is heated strongly.*

(v) *Sulphide ore is added to a tank containing oil and water, and then stirred or agitated with air.*

(c) *Write a balanced chemical equation for each of the following reactions :* **[5]**

(i) *Reaction of carbon powder and concentrated nitric acid.*

(ii) *Reaction of excess ammonia with chlorine.*

(iii) *Reaction of lead nitrate solution with ammonium hydroxide.*

(iv) *Producing ethane from bromo ethane using Zn/Cu couple in alcohol*

(v) *Complete combustion of ethane.*

(d) **(i)** *Draw the structural formula for each of the following :* **[5]**

1. *2, 2 dimethyl pentane*

2. *Methanol*

3. *Iso propane*

(ii) *Write the IUPAC name for the following compounds :*

1. *Acetaldehyde*

2. *Acetylene*

(e) *State one relevant reason for each of the following :* **[5]**

(i) *Graphite anode is preferred to platinum in the electrolysis of molten lead bromide.*

(ii) *Soda lime is preferred to sodium hydroxide in the laboratory preparation of methane.*

(iii) *Hydrated copper sulphate crystals turn white on heating.*

(iv) *Concentrated nitric acid appears yellow, when it is left for a while in a glass bottle.*

(v) *Hydrogen chloride gas fumes in moist air.*

(f) *Calculate :* **[5]**

(i) *The amount of each reactant required to produce 750 ml of carbon dioxide, when two volumes of carbon monoxide combine with one volume of oxygen to produce two volumes of carbon dioxide :*

$$2CO + O_2 \rightarrow 2CO_2$$

(ii) *The volume occupied by 80 g of carbon dioxide at STP.*

(iii) Calculate the number of molecules in 4.4 gm of CO_2.

[Atomic mass of C = 12, O = 16]

(iv) State the law associated in question no. (f) (i) above.

(g) *Give one word or a phrase for the following statements :* **[5]**

(i) The chemical bond formed by a shared pair of electrons, each bonding atom contributing one electron to the pair.

(ii) Electrode used as cathode in electrorefining of impure copper.

(iii) The substance prepared by adding other metals to a base metal in appropriate proportions to obtain certain desirable properties.

(iv) The tendency of an atom to attract electrons to itself when combined in a compound.

(v) The reaction in which carboxylic acid reacts with alcohol in the presence of conc. H_2SO_4 to form a substance having a fruity smell.

(h) *Fill in the blanks from the choices given in brackets :* **[5]**

(i) The polar covalent compound in gaseous state that does not conduct electricity is ....

(carbon tetra chloride, ammonia, methane)

(ii) A salt prepared by displacement reaction is ....

(ferric chloride, ferrous chloride, silver chloride)

(iii) The number of moles in 11 gm of nitrogen gas is ....

(0.39, 0.49, 0.29) [atomic mass of N = 14]

(iv) An alkali which completely dissociates into ions is .... *(ammonium hydroxide, calcium hydroxide, lithium hydroxide)*

(v) An alloy used to make statues is ....

(bronze, brass, fuse metal)

Answers 1.

(a) **(i)** (D) Helium

(ii) (B) Platinum

(iii) (D) Carbon tetrachloride

(iv) (C) Concentrated Sulphuric acid

(v) (A) C_3H_4

(b) **(i)** Action of concentrated nitric acid on copper : Reddish brown fumes of NO_2 are produced when copper reacts with conc. nitric acid :

$$Cu(s) + 4HNO_3 \rightarrow Cu(NO_3)_2 + 2H_2O + 2NO_2$$

(ii) Addition of excess ammonium hydroxide into copper sulphate solution leads to formation of a deep blue coloured solution.

When ammonium hydroxide is added in the solution of copper sulphate drop-wise, a pale blue precipitate of copper hydroxide is obtained. The equation for this follows :

$$CuSO_4 + 2NH_4OH \rightarrow Cu(OH)_2 + (NH_4)_2SO_4 + 4H_2O$$

When ammonium hydroxide is added in excess, the precipitate dissolves and gives a deep blue solution of tetraammine copper (II) sulphate. The equation for this follows :

$$Cu(OH)_2 + (NH_4)_2SO_4 + 2NH_4OH \rightarrow [Cu(NH_3)_4]SO_4 + H_2O$$

Hence, the product formed is a complex named as tetraammine copper (II) sulphate.

(iii) When a piece of sodium metal is put into ethanol at room temperature hydrogen gas is produced which can be identified by a pop sound and it extinguishes a burning splinter.

$$\underset{\text{Sodium}}{2Na(s)} + \underset{\text{Ethanol}}{2C_2H_5OH(l)} \rightarrow \underset{\text{Sodium ethoxide}}{2C_2H_5ONa(l)} + \underset{\text{Hydrogen}}{H_2(g)}$$

(iv) White coloured zinc carbonate is heated strongly to give pale yellow zinc oxide and carbon dioxide gas which extinguishes wooden splinter.

$$ZnCO_3 \rightarrow ZnO + CO_2 \uparrow$$

(v) Sulphide ore is added to a tank containing oil and water and then stirred or agitated with air to generate a froth.

This is known as froth floatation where the sulphide ore particles are preferentially wetted by oil while gangue particles are preferentially wetted by water. A mixture of water and pine oil is taken in the tank. The powdered Sulphide ore is dropped in. Compressed air is blown in through the agitator. The agitator is rotated several times. Froth containing ore starts rising up.

(c) **(i)** Carbon powder reacts with concentrated nitric acid to give carbon dioxide, nitrogen dioxide and water.

$$C(s) + 4HNO_3(l) \rightarrow CO_2(g) + 4NO_2(g) + 2H_2O(l)$$

(ii) Excess of ammonia reacts with chlorine to give nitrogen and ammonium chloride.

$$8NH_3(g) + 3Cl_2(g) \rightarrow N_2(g) + \underset{\text{Solid white fog}}{6NH_4Cl}$$

(iii) Lead nitrate solution reacts with ammonium hydroxide to give lead hydroxide and ammonium nitrate.

$$Pb(NO_3)_2 + 2NH_4OH \rightarrow Pb(OH)_2 + 2NH_4NO_3$$

(iv) Production of ethane from bromoethane using Zn/Cu couple in ethanol gives H_2 gas for the reaction.

$$\underset{\text{Bromoethane}}{C_2H_5Br} + H_2 \xrightarrow{Zn/Cu,\ C_2H_5OH} \underset{\text{Ethane}}{C_2H_6} + HBr$$

(v) Complete combustion of ethane gives carbon dioxide and water.

$$2C_2H_6 + 7O_2 \longrightarrow 4CO_2 + 6H_2O$$

Ethane Oxygen Carbon dioxide Water

(d) **(i) 1.** 2, 2-Dimethyl pentane

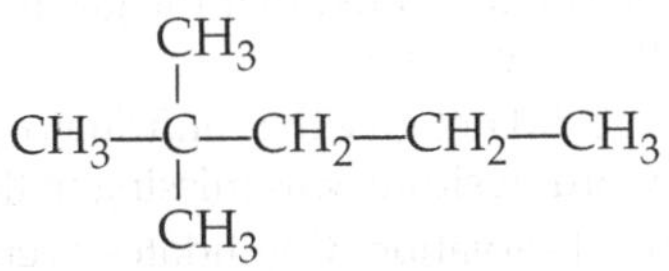

2. Methanol

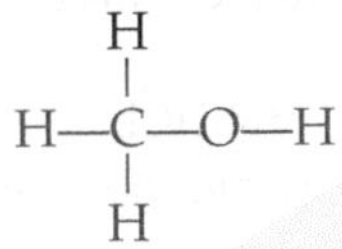

3. Isopropane

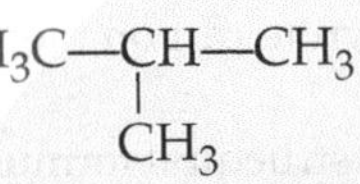

(ii) 1. IUPAC name of acetaldehyde (CH_3CHO) is Ethanal.

2. IUPAC name of acetylene ($HC \equiv CH$) is Ethyne.

(e) **(i)** Graphite anode is preferred in the electrolysis of molten lead bromide, because graphite remains unaffected by the reactive bromine vapours which are released at the anode.

(ii) Soda lime is preferred to sodium hydroxide in the laboratory preparation for methane because Sodium hydroxide is deliquescent and absorbs water from atmosphere.

$$CH_3COONa \xrightarrow{NaOH/CaO} CH_4 + Na_2CO_3$$

(iii) Hydrated copper sulphate crystals turn white on heating due to the loss of water molecules upon heating.

$$CuSO_4 . 5H_2O \underset{\text{add } H_2O}{\overset{\Delta}{\rightleftharpoons}} CuSO_4 + 5H_2O$$

(iv) Concentrated nitric acid appears yellow, when it is left for a while in a glass bottle because Nitric acid has tendency to decompose slowly in presence of sunlight and produce nitrogen dioxide gas, which is reddish brown in colour. This liberated NO_2 gas dissolves in nitric acid and gives it a yellowish colour.

$$4HNO_3 \rightarrow 4NO_2 + 2H_2O + O_2$$

(v) Hydrogen chloride gas fumes in moist air because it is highly soluble in water and when it comes in contact with air containing water droplets it dissolves in water and forms mist, which appears as white fumes.

(f) **(i)** $2CO + O_2 \rightarrow 2CO_2$

Here, 2 moles of CO react with 1 mole of O_2

At STP, 1 mole of any ideal gas takes up 22.4 L.

So, 44.8 L of CO reacts with 22.4 L of O_2 to give 44.8 L of CO_2.

So, 750 ml of CO_2 production will need 750 ml of CO and 375 ml of O_2

(ii) Molecular weight of CO_2 = 44 g

So, weight of one mole CO_2 gas is 44 g

Or, 44 g of CO_2 occupies 22.4 L at STP

$\therefore$ 80 g will occupy $\frac{22.4}{44} \times 80$

$= 40.72$ L

or $= 40720$ mL

(iii) Weight of one mole CO_2 gas is 44 g

Or, 44 g CO_2 contains 6.02×10^{23} CO_2 molecules.

So, 4.4 g of CO_2 will contain 6.023×10^{22} CO_2 molecules.

(iv) Gay Lussac's law of combining volumes of gases is associated here and it states that : "When gases react, they do so in volumes which bears a simple whole number ratio to one another and to the volumes of the products, if gaseous, provided the temperature and pressure of the reacting gases and their products remain constant".

(g) **(i)** Covalent bond

(ii) Pure (thin block) of copper

(iii) Alloys

(iv) Electronegativity

(v) Esterification

(h) **(i)** ammonia

(ii) ferric chloride

(iii) 0.39

(iv) calcium hydroxide

(v) bronze

Comments of Examiners

1. (a) (i) Most of the candidates answered it correctly. Some got confused with Hydrogen.

(ii) This was answered correctly by most of the candidates. A few candidates chose Nickel as the option.

(iii) Some candidates chose the incorrect option NaCl as they had no idea that CCl_4 is a covalent compound.

(iv) Most of them the candidates answered correctly. A few selected concentrated Nitric acid.

(v) This part was well attempted by most of the candidates, but a few chose C_3H_6.

(b) A common error in most of the observations was that candidates either wrote the equation or named the product formed instead of stating the observations. In some cases, incomplete observations were written.

(i) The word 'gas' was missing in many answers.

(ii) Some candidates did not write the solubility of the blue precipitate in excess of ammonium hydroxide solution; instead of writing deep or inky blue solution some wrote "deep blue precipitate is formed".

(iii) A large number of candidates identified it as H2. Some candidates stated that the gas burns with a blue flame instead of stating that a burning splint is put off with a pop sound.

(iv) Some candidates gave only the name of the product and some did not give the confirmatory test for carbon dioxide. A few candidates wrote that it turns yellow. The word 'residue' was missing in the observations.

(v) Many explained the Froth Flotation method instead of the observation. Candidates even wrote that the ore mixes with oil and settles down.

(c) (i) Some candidates did not balance the equations correctly. Some wrote CO instead of CO_2, or NO instead of NO_2.

(ii) Some candidates wrote the products with excess Chlorine as NCl_3 and HCl.

(iii) Most candidates wrote this chemical equation correctly. A few wrote the formula of lead hydroxide incorrectly.

(iv) Many candidates were confused with the condition given - Zn/Cu couple in alcohol.

(v) A few candidates were confused regarding the products and wrote carbon monoxide instead of carbon dioxide.

(d) (i) 1. Some candidates wrote the condensed formula instead of the structural formula. Many candidates did not understand the difference between straight and branched chain. Some showed carbon skeleton but missed showing the complete structural formula.

2. Most candidates drew the structural formula correctly. A few candidates drew the condensed formula instead.

3. This part of the question was well attempted by most candidates.

(ii) 1. Some candidates wrote the IUPAC name as methanal instead of ethanal.

2. Most candidates attempted well but some were confused between 'ene' and 'yne'.

(e) (i) Key words were missing in answers.

(ii) Most candidates mentioned that graphite is cheaper and easily available or inert or that platinum is expensive, instead of stating that graphite remains unaffected by bromine vapour.

Most candidates mentioned that soda lime acts as a catalyst. Very few candidates gave correct reasoning.

(iii) The words water of crystallization was missing in most of the answers. Instead candidates wrote moisture is removed. A few even used an incorrect term 'Amorphous'

(iv) Many candidates did not mention the word 'dissolves' or dissolution of the gas. Instead, they simply wrote, due to NO_2.

(v) Most candidates gave a relevant reason. Some wrote, "Because it comes in contact with ammonia gas". A few even wrote that it is due to the reaction between HCl and NH_3. Several candidates got confused and wrote "dense white fumes are formed".

(f) (i) A number of candidates solved this question correctly. Some calculated only the volume of oxygen and missed calculating the volume of carbon monoxide. Some of the candidates instead of calculating volume, calculated mass. Others ignored the ratio.

(ii) Most candidates calculated the volume occupied by 80 g of carbon dioxide at STP correctly, but a few made errors in the calculation of M.W of CO_2.

(iii) Some candidates made errors is calculating the number of molecules in 4.4 g of Carbon dioxide. Many calculated the number of moles instead of molecules. A few did not use Avogadro's number.

(iv) Most of the candidates did not state the law. They only named it.

(g) (i) Several candidates answered this question correctly, however a few got confused by coordinate bonding.

(ii) A few candidates missed associating the word 'pure'.

(iii) Only a few candidates answered incorrectly.

(iv) Some candidates wrote electron affinity instead of electronegativity.

(v) Majority of the candidates gave correct answers. A few mentioned it as 'ester' or neutralization reaction

(h) (i) Most of the candidates answered correctly. Some chose methane instead of ammonia.

(ii) Some candidates were not able to distinguish between soluble and insoluble salts. A few even chose ferric chloride instead of ferrous chloride.

(iii) Most of the candidates answered this question correctly.

(iv) Many candidates were able to select the correct answer in this part possibly due to the offered options. Very few wrote ammonium hydroxide.

(v) This question was attempted correctly by most candidates. A few however chose Brass.

Answering Tips

- Learn the trends in the periodic properties of elements in the Periodic Table across a period as well as down a group thoroughly with a help of chart. Emphasize on exceptions.
- Stress upon the differences between active and inert electrode in the process of electrolysis, with examples of inert and active electrodes.
- Prepare a table of difference between the covalent and electrovalent compounds on the basis of their properties. Comparative study of properties of acids is essential. .
- Practice for observation-based and reason-based questions. Stress upon related observations like colour changes, formation of precipitates, gases temperature, etc., while reading the respective chapters.
- Perform experiments related to the solubility of precipitates and tests of gases, difference between residue, solution, precipitate, etc. for better understanding. This would help in easy memorising the properties of compounds and their reactions.
- Memorise the reactivity of metals with concentrated Nitric acid, with relevant equations.
- Remember that burning of hydrocarbons under complete oxidation, produces carbon dioxide and water.
- Understand the logic of balancing chemical equations. Writing of balanced equations should be practiced.
- With the help of examples, understand the difference between structural formula and condensed formula.
- Familiarize with the basic rules of the IUPAC nomenclature, the selection of longest chain and the method of numbering. Clarify the differences in the general formula of alkanes, alkenes and alkynes
- Understand the differences in the functional groups especially –OH and –CHO group. Ensure that the students grasp the fact that alcohol ends in 'ol' while aldehydes in 'al'.
- Remember that in the structure of organic compounds, the valency of all the 'C' atoms is satisfied.
- Emphasise on the advantages and disadvantages of the electrode used in the electrolysis process.
- Lay stress on the fact that graphite is unaffected by bromine vapours.
- Remember that Water of Crystallization is responsible for the colour and crystalline structure of a hydrated salt. Stress upon the appropriate key words like 'deliquescent', 'anhydrous', 'solubility', etc.
- Learn how, by changing the conditions of the reaction, the products change. Stress upon mentioning the concentration of reactants, temperature, pressure, catalyst, etc. if any, in each reaction.
- Learn how to identify the laws to be applied, if the reactants involved are gases.
- Acquaint with the following terms: number of moles, Avogadro's number of atoms / molecules, Relative atomic mass (RAM), Relative molecular mass (RMM), molar mass, molar volume, etc.
- Remember that for electro-refining any metal the cathode should always be the pure metal.
- Stress upon the differences between the terms electronegativity and electron affinity with suitable examples. Similarly, the difference between the terms 'ester' and 'esterification'.
- Learn the differences between polar and non-polar covalent compounds with relevant examples.
- Understand the ionization of alkali clearly and point out the difference between strong & weak alkali and solubility of alkali metals in water.
- Familiarise with the composition of alloys and their properties

SECTION-II (40 Marks)

*Attempt any **four** questions from this Section*

Question 2.

(a) *The following table represent the elements and the atomic number :* **[3]**

With reference to this, answer the following using only the alphabets given in the table.

Element	*Atomic number*
P	13
Q	7
R	10

(i) *Which element combines with hydrogen to form a basic gas ?*

(ii) *Which element has an electron affinity zero ?*

(iii) *Name the element, which forms an ionic compound with chlorine.*

(b) *Draw the electron dot diagram for the compounds given below. Represent the electrons by (.) and (x) in the diagram :* **[3]**

[*Atomic No. Ca* = 20, *O* = 8, *Cl* = 17, *H* = 1]

(i) *Calcium oxide*

(ii) *Chlorine molecule*

(iii) *Water molecule*

(c) *Choose the correct word which refers to the process of electrolysis from A to E, to match the description (i) to (iv) :* **[4]**

A : Oxidation B : Cathode C : Anode

D : An electrolyte E : Reduction

(i) Conducts electricity in aqueous or in molten state.

(ii) Loss of electron takes place at anode.

(iii) A reducing electrode.

(iv) Electrode connected to the positive end or terminal of the battery.

Answers 2.

(a) **(i)** Q (QH_3 is the basic gas) with Z = 7

(ii) R with Z = 10

(iii) P with Z = 18

(b) **(i)** Electron dot diagram for Calcium oxide

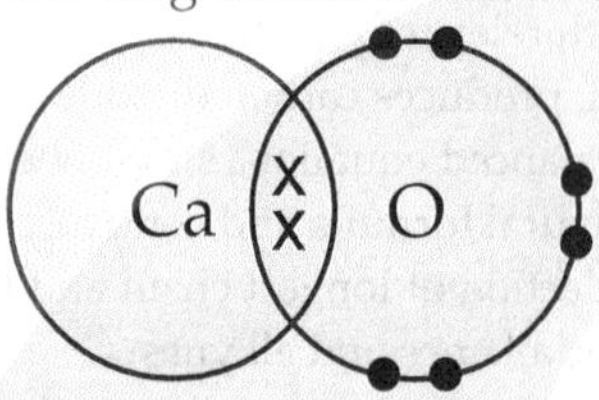

(ii) Electron dot diagram for chlorine molecule

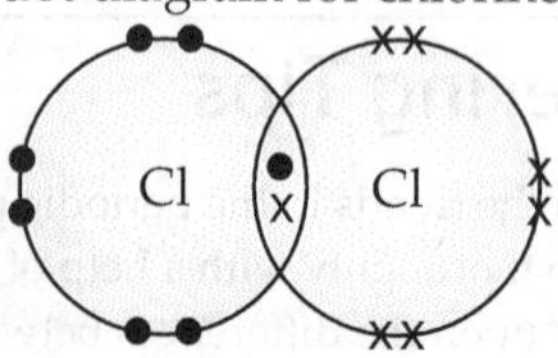

(iii) Electron dot diagram for water molecule

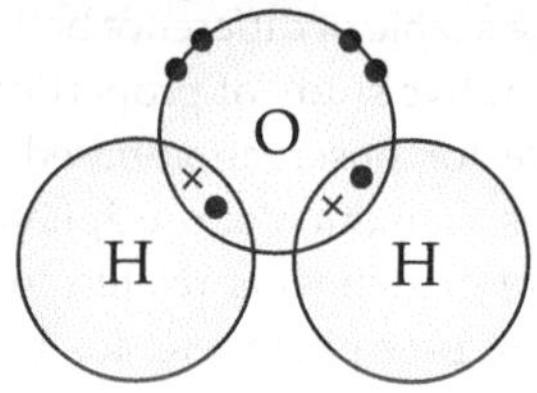

(c) **(i)** Conducts electricity in aqueous or in molten state – D: An electrolyte

(ii) Loss of electron takes place at anode – A: Oxidation

(iii) A reducing electrode – B: Cathode

(iv) Electrode connected to the positive end or terminal of the battery – C: Anode

Comments of Examiners

2. (a) Most of the candidates attempted this part correctly. Some candidates identified the elements instead of choosing the given alphabets.

(b) (i) A number of candidates did not draw arrows to show transfer of electrons. Some drew only one arrow instead of two to show transfer of two valence electrons of calcium. In several cases, electrons of calcium were not represented differently from oxygen, or the charge on calcium ion and oxide ion was not written. Some drew it as a covalent molecule.

(ii) Many candidates drew orbit structure or showed the electron pair along with the bond. Several candidates were confused regarding the number of electrons that were shared.

(iii) Some candidates forgot to draw the lone pair of electrons on water.

(c) Most candidates were able to answer this part correctly. However, some were not able to distinguish between oxidation and reduction reactions.

Answering Tips

- Follow the instructions given in the question paper carefully.
- Remember to use the letters given in the question.
- While explaining, focus on the following points:
  - Relation between the number of shells and period number.
  - Relation between the number of valence electrons and the group to which the element belongs.
  - Periodic properties.
- Understand the importance of arrows and the charge on ion in the electron dot diagrams of electrovalent bonds and practice them regularly. Understand the difference between dot diagram and orbital structure. Represent the electrons by (.) and (x) in the diagram.
- Understand the concept of oxidation / reduction on the basis of loss/gain of electrons.
- Learn all the key terms in electrolysis.

Question 3.

(a) *Baeyer's process is used to concentrate bauxite ore to alumina. :* **[3]**

Give balanced chemical equations for the reaction taking place for its conversion from bauxite to alumina.

(b) *Complete the following by selecting the correct option from the choices given :* **[3]**

(i) *pH of acetic acid is greater than dilute sulphuric acid. So acetic acid contains .... concentration of H^+ ions. (greater, same, low)*

(ii) *The indicator which does not change colour on passage of HCl gas is .... (methyl orange, moist blue litmus, phenolphthalein)*

(iii) *The acid which cannot act as an oxidizing agent is ....(conc. H_2SO_4, conc. HNO_3, conc. HCl)*

(c) *Match the gases given in column I to the identification of the gases mentioned in column II :* **[4]**

Column I	Column II
(i) Hydrogen sulphide	*A. Turns acidified potassium dichromate solution green.*
(ii) Nitric oxide	*B. Turns lime water milky.*
(iii) Carbon dioxide	*C. Turns reddish brown when it reacts with oxygen.*
(iv) Sulphur dioxide	*D. Turns moist lead acetate paper silvery black.*

Answers 3.

(a) Chemical reactions related to Baeyer's process for the conversion from bauxite to alumina:

1. $Al_2O_3.2H_2O(s) + 2NaOH \xrightarrow{150° - 200° C} 2NaAlO_2 + H_2O$

2. $NaAlO_2 + 2H_2O \xrightarrow{50° - 60° C} Al(OH)_3 + NaOH$

3. $2Al(OH)_3 \xrightarrow[\text{Alumina}]{\text{Heat}} Al_2O_3 + 3H_2O$

(b) **(i)** low **(ii)** phenolphthalein

(iii) conc. HCl

(c)

Column I	Column II
(i) Hydrogen sulphide	D. Turns moist lead acetate paper silvery black.
(ii) Nitric oxide	C. Turns reddish brown when reacts with oxygen.
(iii) Carbon dioxide	B. Turns lime water milky.
(iv) Sulphur dioxide	A. Turns acidified potassium dichromate solution green.

Comments of Examiners

3. (a) Most candidates did not write all three equations correctly. A few did not balance the equations. Many candidates wrote the wrong formula for $NaAlO_2$. The third reaction was written correctly by most candidates.

(b) (i) Most candidates attempted it correctly.

(ii) This part was attempted correctly by most candidates.

(iii) In this part many candidates overlooked the word 'not' and instead wrote the oxidizing agent. Majority of the candidates selected the incorrect option conc. nitric acid instead of conc. hydrochloric acid.

(c) Most candidates answered correctly but some were not sure of the tests for gases. Some wrote that Nitric oxide turns lead acetate paper silvery black. Others got mixed up with the options of hydrogen sulphide and sulphur dioxide.

Answering Tips

- Learn the process of extraction thoroughly.
- Learn the chemical reactions with necessary conditions and their balanced chemical equations.
- Understand the concept between strong and weak acid on the basis of H+ ion concentration.
- Learn about the types of acids and bases on the basis of dissociation /ionization whether complete / partial or no dissociation / ionization.
- Point out the properties of nitric acid, sulphuric acid and hydrochloric acid for better understanding the differences between them.

Question 4.

(a) *Differentiate between the following pairs based on the information given in the brackets :* **[3]**

(i) *Conductor and electrolyte (conducting particles)*

(ii) *Cations and anions (formation from an atom)*

(iii) *Acid and Alkali (formation of type of ions)*

(b) *Draw the structures of isomers of pentane.* **[3]**

(c) *Hydrogen chloride gas is prepared in the laboratory using concentrated sulphuric acid and sodium chloride. Answer the questions that follow based on this reaction :* **[4]**

(i) *Give the balanced chemical equation for the reaction with suitable conditions(s) if any.*

(ii) *Why is concentrated sulphuric acid used instead of concentrated nitric acid ?*

(iii) *How is the gas collected ?*

(iv) *Name the drying agent not used for drying the gas.*

Answers 4.

(a) **(i)** Conductor has electrons as the conduction particles whereas ions are the conducting species in electrolyte.

(ii) Cations are formed when a neutral atom loses electrons whereas anions are formed when a neutral atom gains electron.

(iii) An acid dissociates to furnish H^+ ions and the conjugate base which is negatively charged and a base dissociates to give OH^- ions and a conjugate acid which is positively charged.

(b) Structures of isomers of pentane are as follows :

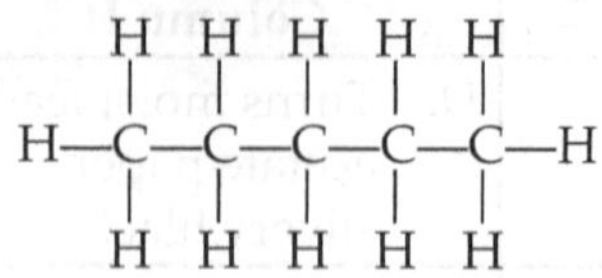

Pentane

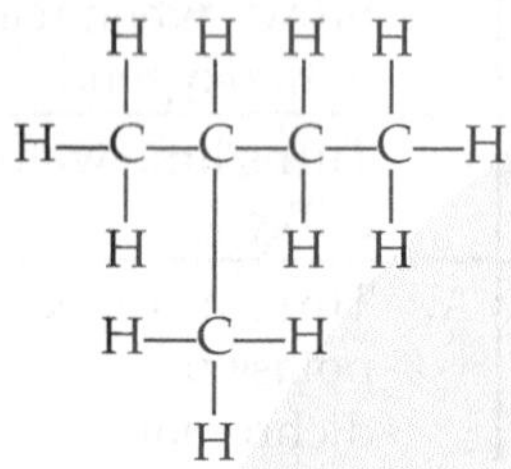

2-Methylbutane

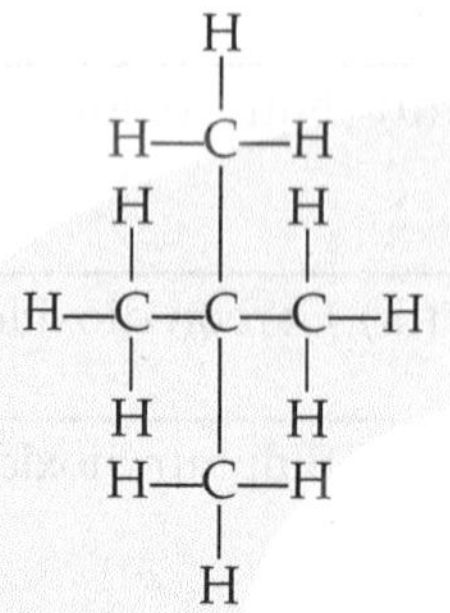

2, 2-Dimethylpropane

Structural isomers of C_5H_{12}

(c) (i) 1. $\underset{\text{conc.}}{NaCl + H_2SO_4} \xrightarrow[\text{below } 200°C]{\text{Heated}} NaHSO_4 + HCl\uparrow$

2. $\underset{\text{conc.}}{2NaCl + H_2SO_4} \xrightarrow[\text{above } 200°C]{\text{Heated}} Na_2SO_4 + 2HCl\uparrow$

(ii) Concentrated nitric acid is not used for preparation of HCl gas as it is very strong oxidising agent and will end up oxidising HCl gas.

(iii) HCl is collected by the upward displacement of air.

(iv) Calcium oxide.

Comments of Examiners

4. (a) (i) Many candidates could not answer this part correctly. The differences were not based on the specified parameters. Some wrote atoms in place of ions.

(ii) Most candidates wrote cations migrate towards cathode and anions towards anode or anions are negatively charged and cations are positively charged. Some were confused and wrote that cations gain electrons while anions lose electrons.

(iii) Answered correctly by majority of the candidates.

(b) Most candidates drew the first isomer correctly but could not draw the other two isomers. Some candidates drew only two isomers. Branched structures were not drawn correctly. Some missed out H atoms or the count of carbon atoms went incorrect. Many candidates did not understand the difference between straight and branched chain.

(c) (i) Several candidates did not mention the condition or wrote incorrect condition. Some used greater than sign instead of less than sign.

(ii) Almost all attempted correctly.

(iii) Some candidates wrote that the gas was collected by downward displacement of air.

(iv) Some candidates missed out the word 'not used', hence, identified the drying agent as Conc. H_2SO_4 instead of CaO and P_2O_5.

Answering Tips

- Point out the similarities and differences between electrolyte and non - electrolyte, cation and anion, conductors, and non-conductors. Identify the particles present in electrolyte, non-electrolyte and conductors, respectively.
- Learn isomerism thoroughly and practice to draw the related structures. Learn the isomers with both IUPAC names and common names.
- Point out the difference between condensed formula and structural formula with the help of examples. Videos showing isomerism help in understanding the arrangement of carbon atoms, resulting in formations of isomers.
- Learn about the physical and chemical properties of HCl gas and the method of collection.
- Learn the methods of preparation of compounds with relevant diagrams, for better understanding of the aspects /procedure involved in the preparation.

Question 5.

(a) *Distinguish between the following pairs of compounds using a reagent as a chemical test :* **[3]**

(i) *Calcium nitrate and Zinc nitrate solution.*

(ii) *Ammonium sulphate crystals and Sodium sulphate crystals.*

(iii) *Magnesium chloride and Magnesium nitrate solution.*

(b) *Calculate the percentage of :* **[3]**

(i) *Fluorine*

(ii) *Sodium and*

(iii) *Aluminium*

in sodium aluminium fluoride [Na_3AlF_6], to the nearest whole number. [Atomic Mass : Na = 23, Al = 27, F = 19]

(c) **(i)** *State the volume occupied by 40 gm of methane at STP, if its vapour density (V.D.) is 8.* **[4]**

(i) *Calculate the number of moles present in 160 gm of NaOH.*

[Atomic Mass : Na = 23, H = 1, O = 16]

Answers 5.

(a) **(i)** Calcium nitrate and Zinc nitrate solutions can be distinguish by reacting with ammonium hydroxide solution :

1. On adding ammonium hydroxide gelatinous white precipitates of zinc hydroxide are formed.

$Zn(NO_3)_2 + 2NH_4OH \longrightarrow Zn(OH)_2 + 2NH_4NO_3$

2. On adding excess of ammonium hydroxide, the precipitates dissolve forming a soluble complex.

$Zn(NO)_2 + 2NH_4NO_3 \longrightarrow 2NH_4OH$
$[Zn(NH_3)_4](NO_3)_2 + 4H_2O$

No visible reaction occurs when we add calcium nitrate to ammonium hydroxide.

$CaNO_3 + NH_4OH \longrightarrow$ No reaction

(ii) Ammonium sulphate crystals give pungent colourless gas Ammonia (NH_3) when heated. When NH_3 gas comes in contact with a glass rod dipped in HCl white fumes of NH_4Cl are produced. Sodium sulphate crystals do no undergo the above reaction sequence, hence can be differentiated from ammonium sulphate crystals.

(iii) Magnesium chloride reacts with silver nitrate solution to give precipitate of silver chloride, whereas magnesium nitrate does not react with silver nitrate solution to give a precipitate.

$$MgCl_2\,(aq) + 2AgNO_3\,(aq) \longrightarrow Mg(NO_3)_2\,(aq) + 2AgCl\,(s)\downarrow$$

$MgNO_3 + AgNO_3 \longrightarrow$ No reaction

(b) Molecular weight of Na_3AlF_6

$3 \times 23 + 27 + 6 \times 19 = 210$

Now,

(i) Percentage of fluorine $= \frac{(6\times19)}{210}\times100$
$= 54.2\%$

(ii) Percentage of sodium $= \frac{(3\times23)}{210}\times100$
$= 32.8\ \%$

(iii) Percentage of aluminium $= \frac{27}{210}\times100$
$= 12.8\ \%$

(c) **(i)** Molecular weight = 2 × Vapour density at STP

In case of methane, substituting the given information –

Molecular weight = 2 × 8
= 16

Molecular weight of methane = 16 which means 16 g of methane occupy 22.4 L volume at STP.

So, 40 g of methane will occupy $= \frac{22.4}{16} \times 40$
= 56 L

(ii) Number of moles

$$= \frac{\text{Given weight of substance}}{\text{Molecular weight of substance}}$$

Molecular weight of NaOH = 23 + 16 + 1
= 40

No. of moles $= \frac{160}{40} = 4$

There are 4 moles of NaOH in 160 g of NaOH

Comments of Examiners

5. (a) A number of candidates chose the incorrect reagent; others could not write the correct observations. Some candidates differentiated on the basis of thermal decomposition. Several wrote the reactions or named the product instead of writing the observations. Majority distinguished between each pair, using different chemical reagents; at times, the observations were not correct. Several candidates wrote the correct observation without mentioning the reagent.

(b) (i) Almost all the candidates answered this part correctly. However, a few calculated the molecular weight of Na_3AlF_6 incorrectly while others did not substitute the values properly.

(ii) Most of the candidates could do this part correctly. Some candidates, instead of multiplying 3 × 23 for sodium, calculated the percentage with only 23.

(iii) Almost all candidates attempted this subpart correctly.

(c) (i) Several candidates did the calculation taking the value of vapour density as '8', instead of calculating RMM= 2 × VD i.e., '16'

(ii) Some candidates calculated molecular weight of NaOH incorrectly. A large number of candidates made incorrect substitution and/or calculated mass instead of moles. Some even calculated number of molecules in place of moles.

Answering Tips

- Learn about different steps and techniques to differentiate salts.
- Clarify the concepts about the cation and anion test using sodium hydroxide and ammonium hydroxide.
- Remember to state the result of the test with both substances, besides mentioning the choice of reagent. Learn the various reagents and reactions involved in the anions' tests.
- Remember that conc. Sulphuric acid cannot be used if the salt provided is in solution form.
- Learn Analytical chemistry thoroughly. Insist upon the correlation between practical and theoretical knowledge.
- Practice on problems based on percentage composition as well as calculation of molecular weight.
- Practice numerical on concept of mole, mass-mass and mass-volume relationship.

Question 6.

(a) *Identify the salts* ***P, Q, R*** *from the following observations* **[3]**

(i) *Salt* ***P*** *has light bluish green colour. On heating, it produces a black coloured residue. Salt* ***P*** *produces brisk effervescence with dil. HCl and the gas evolved turns lime water milky, but no action with acidified potassium dichromate solution.*

(ii) *Salt* ***Q*** *is white in colour. On strong heating, it produces buff yellow residue and liberates reddish brown gas. Solution of salt Q produces chalky white insoluble precipitate with excess of ammonium hydroxide.*

(iii) *Salt* ***R*** *is black in colour. On reacting with concentrated HCl, it liberates a pungent greenish yellow gas which turns moist starch iodide paper blue black.*

(b) *Identify the substance underlined in each of the following :* **[3]**

(i) *The electrode that increases in mass during the electro-refining of silver.*

(ii) *The acid that is a dehydrating as well as a drying agent.*

(iii) *The catalyst used to oxidize ammonia into nitric oxide.*

(c) *Copy and complete the following paragraph using the options given in brackets :* **[4]**

Alkenes are a homologous series of (i) .... (saturated / unsaturated) hydrocarbons characterised by the general formula (ii) .... (C_nH_{2n+2}/ C_nH_{2n}). Alkenes undergo (iii) .... (addition/ substitution) reactions and also undergo (iv) .... (hydrogenation / dehydrogenation) to form alkanes.

Answers 6.

(a) (i) P is Copper carbonate.

$CuCO_3(s) \rightarrow CuO(s) + CO_2(g)$

Copper II carbonate (Bluish green) → Copper (II) oxide (Black) + Carbon dioxide(g)

$CuCO_3 + HCl \rightarrow CuCl_2 + CO_2 + H_2O$

(P) Effervescence

$CuCO_3 + K_2Cr_2O_7 \rightarrow$ No reaction

(ii) The salt Q is Lead nitrate

$$2Pb(NO_3)_2(s) \xrightarrow{heat} 2PbO(s) + 4NO_{2(g)} + O_2(g)$$

Lead nitrate, Colourless (Q) → Lead monoxide, Yellow residue + Nitrogen dioxide, Reddish brown + Oxygen

When reacted with ammonium hydroxide solution it gives a chalky white precipitate of lead hydroxide :

$Pb(NO_3)_2 + 2NH_4OH \rightarrow 2NH_4NO_3 + Pb(OH)_2\downarrow$

(Q) Chalky white ppt.

(iii) The salt R is MnO_2, which is black in colour and reacts with HCl to give Cl_2 gas

$MnO_2(s) + 4HCl(aq) \rightarrow MnCl_2(aq) + 2H_2O(l) + Cl_2(g)$

Chlorine gas is pungent and greenish yellow in colour. The chlorine gas oxidises some of the iodide ions in the starch iodide paper to create iodine diatomic molecules. These molecules react with the iodide ions and the starch to form a charge-transfer complex of blue colour.

(b) (i) Cathode

(ii) Conc. H_2SO_4

(iii) Platinum Rhodium catalyst in Ostwald's process.

(c) (i) unsaturated

(ii) C_nH_{2n}

(iii) addition

(iv) hydrogenation

Comments of Examiners

6. (a) Candidates who had knowledge of practical chemistry found this question easy. Some common mistakes made by candidates were as follows:
   - (i) Few candidates made errors in identifying the salt. Some identified it as calcium carbonate instead of copper carbonate.
   - (ii) Some candidates were confused and identified the salt as zinc nitrate.
   - (iii) Several candidates mistook the black compound to be Copper oxide.

   (b) (i) Many candidates gave incorrect answers and mentioned anode instead of cathode.
   - (ii) The word concentrated was missing in some cases. Candidates only mentioned the substance as sulphuric acid.
   - (iii) The part was well attempted by most candidates. Some candidates wrote Fe instead of platinum. A few mentioned it as V_2O_5.

   (c) Most of the candidates answered this question correctly. A few got confused in the following fill in the blanks and wrote:
   - [ii] C_nH_{2n+2}
   - [iii] substitution
   - [iv] Dehydrogenation.

Answering Tips

- ➤ Perform practical on the reaction of conc. HCl on the black compounds namely, CuO and MnO_2 to see the colour in each test and understand the difference between the compounds.
- ➤ Practice on identify salt from the observations given.
- ➤ Remember to write dilute or concentrated whenever reference is made to any acid.
- ➤ Tabulate the various catalysed reactions with conditions wherever mentioned and highlight the catalysts in each case, for easy recapitulation.
- ➤ Tabulate the properties of Hydrocarbons to enable students to compare and contrast between alkanes, alkenes and alkynes. Identify the feasibility of reactions of saturated and unsaturated hydrocarbons.
- ➤ Memorise the terms dehydrogenation and hydrogenation giving relevant examples.

Question 7.

(a) *Write balanced chemical equations, for the preparation of the given salts :* **[3]**

(i) to (iii) by using the methods A to C respectively :

A : Neutralization B : Precipitation C : Titration

(i) *Copper sulphate*

(ii) *Zinc carbonate*

(iii) *Ammonium sulphate*

(b) *Name the following elements :* **[3]**

(i) *An alkaline earth metal present in group 2 and period 3.*

(ii) *A trivalent metal used to make light tools.*

(iii) *A monovalent non-metal present in fluorspar.*

(c) *An aqueous solution of nickel (II) sulphate was electrolyzed using nickel electrodes. Observe the diagram and answer the questions that follow :* **[4]**

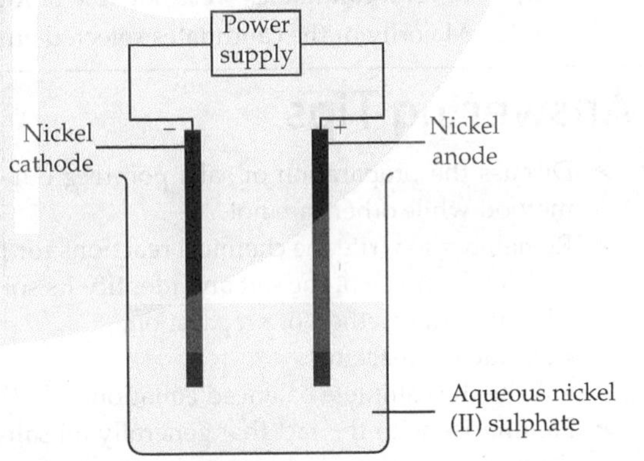

(i) *What do you observe at the* ***cathode*** *and* ***anode*** *respectively ?*

(ii) *Name the cation that remains as a spectator ion in the solution.*

(iii) *Which equation for the reaction at the anode is correct ?*

1. $Ni \rightarrow Ni^{2+} + 2e^-$

2. $Ni + 2e^- \rightarrow Ni^{2+}$

3. $Ni^{2+} \rightarrow Ni + 2e^-$

4. $Ni^{2+} + 2e^- \rightarrow Ni$

Answers 7.

(a) (i) Copper sulphate by neutralization

$$\underset{\text{Copper hydroxide}}{Cu(OH)_2} + \underset{\text{Sulphuric acid}}{H_2SO_4} \rightarrow \underset{\text{Copper sulphate}}{CuSO_4\downarrow} + 2H_2O$$

(ii) Zinc carbonate by precipitation

$$\underset{\text{Zinc nitrate}}{Zn(NO_3)_2\,(aq)} + \underset{\text{(aq) Sodium carbonate}}{2NaCO_3} \rightarrow \underset{\text{Zinc carbonate}}{Zn(CO_3)_2} + 2NaNO_3\,(aq)$$

(iii) Ammonium sulphate can be prepared by titration of NH_3 solution with dil. H_2SO_4 solution.

$$2NH_4OH\,(aq) + H_2SO_4\,(aq) \rightarrow \underset{\text{White powder}}{(NH_4)_2SO_4} + 2H_2O\,(l)$$

(b) (i) Magnesium or Mg

(ii) Aluminium or Al

(iii) Fluorine or F

(c) (i) At cathode reduction takes place and deposition of Ni takes place as Ni^{2+} ions from solution convert to Ni metal.

$$Ni^{2+} + 2e^- \rightarrow Ni$$

At anode oxidation takes place and Ni metal converts to Ni^{2+} ions.

$$Ni \rightarrow Ni^{2+} + 2e^-$$

(ii) H^+ is the spectator cation.

(iii) 1. $Ni \rightarrow Ni^{2+} + 2e^-$

Comments of Examiners

7. (a) Most candidates matched the salt with the method of preparation instead of writing balanced chemical equations. Common errors made in the subparts were as follows:
Reactants selected were incorrect.
   - (i) Some candidates used copper sulphate as the reactant or used sodium sulphate instead of sulphuric acid.
   - (ii) Some candidates took carbonic acid as one of the reactants.
   - (iii) Either the reactants selected were incorrect, or formula of ammonium sulphate written was incorrect in some cases.

(b) (i) Most candidates attempted this question correctly, but some wrote one element of period 2 and another element from group 3.
   - (ii) Some candidates could not relate the term trivalent metal and made errors.
   - (iii) Some candidates named calcium as they might have missed out on the word non - metal.

(c) Many candidates could not give the correct observation at the cathode/anode. Some candidates did not associate the correct electrode with the observation given.
   - (i) Some wrote the cathode and anode reactions, instead of the observations, while some wrote observations in the reverse way.
   - (ii) Several candidates were not able to identify the spectator ions - all possible ions were mentioned.
   - (iii) Majority of the candidates selected an incorrect option.

Answering Tips

- Discuss the preparation of salts pointing out the reasons why certain salts are prepared by using a certain method while others are not.
- Remember to write the chemical reactions for preparation of salts stepwise, that is:
  - Write the name of the salt and identify its solubility;
  - Identify the method of preparation;
  - Decide the reactants;
  - Write the complete balanced equation.
- Familiarise with the fact that generally all salts of sodium, potassium and ammonium are soluble and hence titration is employed as a procedure to carry out neutralization reactions for soluble base.
- Learn the meaning of the terms monovalent, divalent and trivalent.
- Write electrode equation correctly. Tabulate the various electrode reactions and highlight the observations at the cathode and anode in each case for easy recapitulations.
- Understand the process of electroplating with the help of a chart or a video, to observe the changes taking place at the respective electrodes.
- Familiarise with the reaction taking place at the cathode and anode respectively and the reactions in terms of electron transfer and how to represent them using equations.
- Revise frequently the terms Cations, Anions, Spectator ions, the concept of loss and gain of electrons at the Anode and Cathode, Oxidising and Reducing electrode, etc.